Structural Analysis

Structural Analysis

Structural Analysis
In Theory and Practice

Alan Williams, **Ph.D., S.E., C.ENG.**

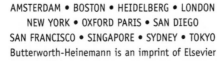

AMSTERDAM • BOSTON • HEIDELBERG • LONDON
NEW YORK • OXFORD PARIS • SAN DIEGO
SAN FRANCISCO • SINGAPORE • SYDNEY • TOKYO
Butterworth-Heinemann is an imprint of Elsevier

Butterworth-Heinemann is an imprint of Elsevier
30 Corporate Drive, Suite 400, Burlington, MA 01803, USA
Linacre House, Jordan Hill, Oxford OX2 8DP, UK

Library of Congress Cataloging-in-Publication Data
Application submitted

British Library Cataloguing-in-Publication Data
A catalogue record for this book is available from the British Library.

ISBN: 978-1-85617-550-0

For information on all Butterworth–Heinemann publications
visit our Web site at www.elsevierdirect.com

Printed and bound in the United Kingdom
Transferred to Digital Printing, 2011

Working together to grow
libraries in developing countries
www.elsevier.com | www.bookaid.org | www.sabre.org

ELSEVIER   BOOK AID
            International   Sabre Foundation

Contents

Foreword

It is with great pleasure that I write this foreword to *Structural Analysis: In Theory and Practice*, by Alan Williams. Like many other engineers, I have utilized Dr. Williams' numerous publications through the years and have found them to be extremely useful. This publication is no exception, given the extensive experience and expertise of Dr. Williams in this area, the credibility of Elsevier with expertise in technical publications internationally, and the International Code Council (ICC) with expertise in structural engineering and building code publications.

Engineers at all levels of their careers will find the determinate and indeterminate analysis methods in the book presented in a clear, concise, and practical manner. I am a strong advocate of all of these attributes, and I am certain that the book will be successful because of them. Coverage of many other important areas of structural analysis, such as Plastic Design, Matrix and Computer Methods, Elastic-Plastic Analysis, and the numerous worked-out sample problems and the answers to the supplementary problems greatly enhance and reinforce the overall learning experience.

One may ask why, in this age of high-powered computer programs, a comprehensive book on structural analysis is needed. The software does all of the work for us, so isn't it sufficient to read the user's guide to the software or to have a cursory understanding of structural analysis?

While there is no question that computer programs are invaluable tools that help us solve complicated problems more efficiently, it is also true that the software is only as good as the user's level of experience and his/her knowledge of the software. A small error in the input or a misunderstanding of the limitations of the software can result in completely meaningless output, which can lead to an unsafe design with potentially unacceptable consequences.

That is why this book is so valuable. It teaches the *fundamentals* of structural analysis, which I believe are becoming lost in structural engineering. Having a solid foundation in the fundamentals of analysis enables engineers to understand the behavior of structures and to recognize when output from a computer program does not make sense.

Simply put, students will become better students and engineers will become better engineers as a result of this book. It will not only give you a better understanding of structural analysis; it will make you more proficient and efficient in your day-to-day work.

David A. Fanella, Ph.D., S.E., P.E.
Chicago, IL
March 2008

Part One

Analysis of Determinate Structures

1 Principles of statics

Notation

F	force
f_i	angle in a triangle opposite side F_i
H	horizontal force
l	length of member
M	bending moment
P	axial force in a member
R	support reaction
V	vertical force
W_{LL}	concentrated live load
w_{DL}	distributed dead load
θ	angle of inclination

1.1   Introduction

Statics consists of the study of structures that are at rest under equilibrium conditions. To ensure equilibrium, the forces acting on a structure must balance, and there must be no net torque acting on the structure. The principles of statics provide the means to analyze and determine the internal and external forces acting on a structure.

For planar structures, three equations of equilibrium are available for the determination of external and internal forces. A statically determinate structure is one in which all the unknown member forces and external reactions may be determined by applying the equations of equilibrium.

An indeterminate or redundant structure is one that possesses more unknown member forces or reactions than the available equations of equilibrium. These additional forces or reactions are termed redundants. To determine the redundants, additional equations must be obtained from conditions of geometrical compatibility. The redundants may be removed from the structure, and a stable, determinate structure remains, which is known as the cut-back structure. External redundants are redundants that exist among the external reactions. Internal redundants are redundants that exist among the member forces.

1.2   Representation of forces

A force is an action that tends to maintain or change the position of a structure. The forces acting on a structure are the applied loads, consisting of both dead and imposed loads, and support reactions. As shown in Figure 1.1, the simply supported beam is loaded with an imposed load W_{LL} located at point 3 and with its own weight w_{DL}, which is uniformly distributed over the length of the beam. The support reactions consist of the two vertical forces located at the ends of the beam. The lines of action of all forces on the beam are parallel.

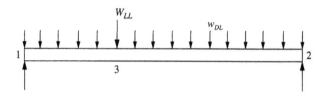

Figure 1.1

In general, a force may be represented by a vector quantity having a magnitude, location, sense, and direction corresponding to the force. A vector represents a force to scale, such as a line segment with the same line of action as the force and with an arrowhead to indicate direction.

The point of application of a force along its line of action does not affect the equilibrium of a structure. However, as shown in the three-hinged portal frame in Figure 1.2, changing the point of application may alter the internal forces in the individual members of the structure.

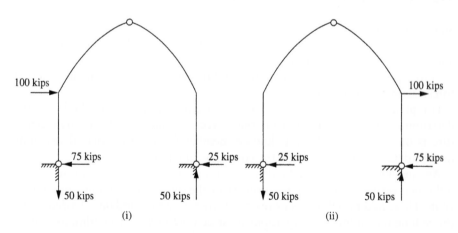

Figure 1.2

Collinear forces are forces acting along the same line of action. The two horizontal forces acting on the portal frame shown in Figure 1.3 (i) are collinear and may be added to give the single resultant force shown in (ii).

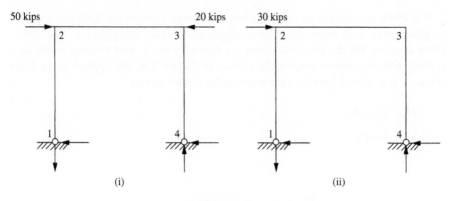

(i)   (ii)

Figure 1.3

Forces acting in one plane are coplanar forces. Space structures are three-dimensional structures and, as shown in Figure 1.4, may be acted on by non-coplanar forces.

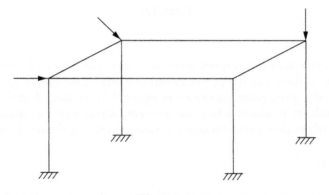

Figure 1.4

In a concurrent force system, the line of action of all forces has a common point of intersection. As shown in Figure 1.5 for equilibrium of the two-hinged arch, the two reactions and the applied load are concurrent.

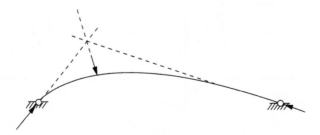

Figure 1.5

It is often convenient to resolve a force into two concurrent components. The original force then represents the resultant of the two components. The directions adopted for the resolved forces are typically the x- and y-components in a rectangular coordinate system. As shown in Figure 1.6, the applied force F on the arch is resolved into the two rectangular components:

$$H = F\cos\theta$$

$$V = F\sin\theta$$

Figure 1.6

The moment acting at a given point in a structure is given by the product of the applied force and the perpendicular distance of the line of action of the force from the given point. As shown in Figure 1.7, the force F at the free end of the cantilever produces a bending moment, which increases linearly along the length of the cantilever, reaching a maximum value at the fixed end of:

$$M = Fl$$

The force system shown at (i) may also be replaced by either of the force systems shown at (ii) and (iii). The support reactions are omitted from the figures for clarity.

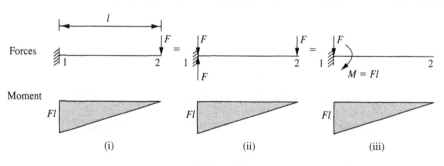

Figure 1.7

1.3 Conditions of equilibrium

In order to apply the principles of statics to a structural system, the structure must be at rest. This is achieved when the sum of the applied loads and support reactions is zero and there is no resultant couple at any point in the structure. For this situation, all component parts of the structural system are also in equilibrium.

A structure is in equilibrium with a system of applied loads when the resultant force in any direction and the resultant moment about any point are zero. For a system of coplanar forces this may be expressed by the three equations of static equilibrium:

$$\sum H = 0$$
$$\sum V = 0$$
$$\sum M = 0$$

where H and V are the resolved components in the horizontal and vertical directions of a force and M is the moment of a force about any point.

1.4 Sign convention

For a planar, two-dimensional structure subjected to forces acting in the xy plane, the sign convention adopted is shown in Figure 1.8. Using the right-hand system as indicated, horizontal forces acting to the right are positive and vertical forces acting upward are positive. The z-axis points out of the plane of the paper, and the positive direction of a couple is that of a right-hand screw progressing in the direction of the z-axis. Hence, counterclockwise moments are positive.

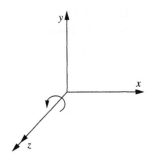

Figure 1.8

Example 1.1

Determine the support reactions of the pin-jointed truss shown in Figure 1.9. End 1 of the truss has a hinged support, and end 2 has a roller support.

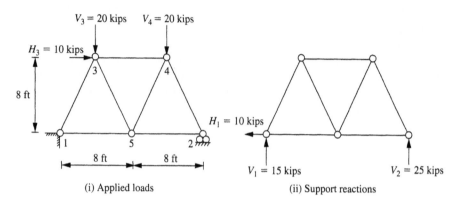

(i) Applied loads   (ii) Support reactions

Figure 1.9

Solution

To ensure equilibrium, support 1 provides a horizontal and a vertical reaction, and support 2 provides a vertical reaction. Adopting the convention that horizontal forces acting to the right are positive, vertical forces acting upward are positive, and counterclockwise moments are positive, applying the equilibrium equations gives, resolving horizontally:

$$H_1 + H_3 = 0$$
$$H_1 = -H_3$$
$$= -10 \text{ kips ... acting to the left}$$

Taking moments about support 1 and assuming that V_2 is upward:

$$16\,V_2 - 8\,H_3 - 4\,V_3 - 12\,V_4 = 0$$
$$V_2 = (8 \times 10 + 4 \times 20 + 12 \times 20)/16$$
$$= 25 \text{ kips ... acting upward as assumed}$$

Resolving vertically:

$$V_1 + V_2 + V_3 + V_4 = 0$$
$$V_1 = -25 + 20 + 20$$
$$= 15 \text{ kips ... acting upward}$$

The support reactions are shown at (ii).

1.5   Triangle of forces

When a structure is in equilibrium under the action of three concurrent forces, the forces form a triangle of forces. As indicated in Figure 1.10 (i), the three forces F_1, F_2, and F_3 are concurrent. As shown in Figure 1.10 (ii), if the initial point of force vector F_2 is placed at the terminal point of force vector F_1, then the force vector F_3 drawn from the terminal point of force vector F_2 to the initial point of force vector F_1 is the equilibrant of F_1 and F_2. Similarly, as shown in Figure 1.10 (iii), if the force vector F_3 is drawn from the initial point of force vector F_1 to the terminal point of force vector F_2, this is the resultant of F_1 and F_2. The magnitude of the resultant is given algebraically by:

$$(F_3)^2 = (F_1)^2 + (F_2)^2 - 2F_1F_2 \cos f_3$$

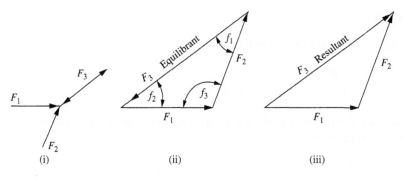

Figure 1.10

and:

$$F_3 = F_1 \sin f_3 \csc f_1$$

or:

$$F_1/\sin f_1 = F_2/\sin f_2$$
$$= F_3/\sin f_3$$

Example 1.2

Determine the angle of inclination and magnitude of the support reaction at end 1 of the pin-jointed truss shown in Figure 1.11. End 1 of the truss has a hinged support, and end 2 has a roller support.

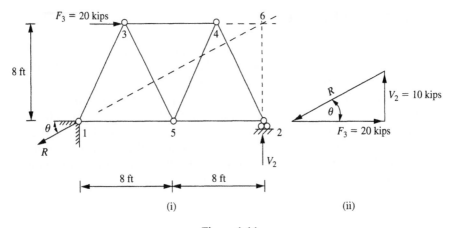

Figure 1.11

Solution

Taking moments about support 1 gives:

$$V_2 = 8 \times 20/16$$
$$= 10 \text{ kips } \dots \text{ acting vertically upward}$$

The triangle of forces is shown at (ii), and the magnitude of the reaction at support 1 is given by:

$$R^2 = (V_2)^2 + (F_3)^2 - 2V_2 F_3 \cos f_r$$
$$= 10^2 + 20^2 - 2 \times 10 \times 20 \cos 90°$$
$$R = (100 + 400)^{0.5}$$
$$= 22.36 \text{ kips}$$

The angle of inclination of R is:

$$\theta = \text{atan} (10/20)$$
$$= 26.57°$$

Alternatively, since the three forces are concurrent, their point of concurrency is at point 6 in Figure 1.11 (i), and:

$$\theta = \text{atan} (8/16)$$
$$= 26.57°$$

and

$$R = 20 \sin 90°/\sin 63.43°$$
$$= 22.36 \text{ kips}$$

The reaction R may also be resolved into its horizontal and vertical components:

$$H_1 = R\cos\theta$$
$$= 20 \text{ kips}$$
$$V_1 = R\sin\theta$$
$$= 10 \text{ kips}$$

1.6 Free body diagram

For a system in equilibrium, all component parts of the system must also be in equilibrium. This provides a convenient means for determining the internal forces in a structure using the concept of a free body diagram. Figure 1.12 (i) shows the applied loads and support reactions acting on the pin-jointed truss that was analyzed in Example 1.1. The structure is cut at section A-A, and the two parts of the truss are separated as shown at (ii) and (iii) to form two free body diagrams. The left-hand portion of the truss is in equilibrium under the actions of the support reactions of the complete structure at 1, the applied

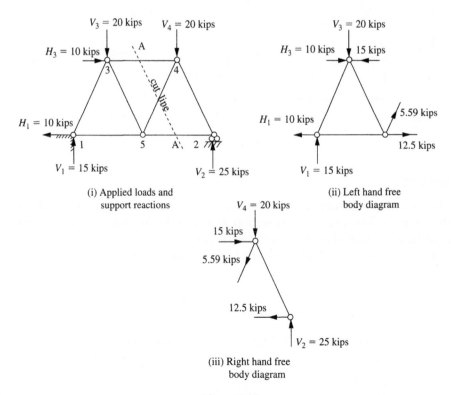

Figure 1.12

loads at joint 3, and the internal forces acting on it from the right-hand portion of the structure. Similarly, the right-hand portion of the truss is in equilibrium under the actions of the support reactions of the complete structure at 2, the applied load at joint 4, and the internal forces acting on it from the left-hand portion of the structure. The internal forces in the members consist of a compressive force in member 34 and a tensile force in members 45 and 25. By using the three equations of equilibrium on either of the free body diagrams, the internal forces in the members at the cut line may be obtained. The values of the member forces are indicated at (ii) and (iii).

Example 1.3

The pin-jointed truss shown in Figure 1.13 has a hinged support at support 1 and a roller support at support 2. Determine the forces in members 15, 35, and 34 caused by the horizontal applied load of 20 kips at joint 3.

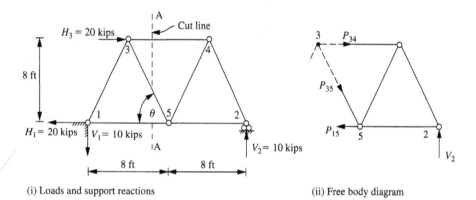

(i) Loads and support reactions   (ii) Free body diagram

Figure 1.13

Solution

The values of the support reactions were obtained in Example 1.2 and are shown at (i). The truss is cut at section A-A, and the free body diagram of the right-hand portion of the truss is shown at (ii).

Resolving forces vertically gives the force in member 35 as:

$$P_{35} = V_2/\sin\theta$$
$$= 10/\sin 63.43°$$
$$= 11.18 \text{ kips } ... \text{ compression}$$

Taking moments about node 3 gives the force in member 15 as:

$$P_{15} = 12V_2/8$$
$$= 12 \times 10/8$$
$$= 15 \text{ kips } ... \text{ tension}$$

Taking moments about node 5 gives the force in member 34 as:

$$P_{34} = 8V_2/8$$
$$= 8 \times 10/8$$
$$= 10 \text{ kips } ... \text{ compression}$$

1.7 Principle of superposition

The principle of superposition may be defined as follows: the total displacements and internal stresses in a linear structure corresponding to a system of applied forces is the sum of the displacements and stresses corresponding to each force applied separately. The principle applies to all linear-elastic structures in which displacements are proportional to applied loads and which are constructed from materials with a linear stress-strain relationship. This enables loading on a structure to be broken down into simpler components to facilitate analysis.

As shown in Figure 1.14, a pin-jointed truss is subjected to two vertical loads at (i) and a horizontal load at (ii). The support reactions for each loading

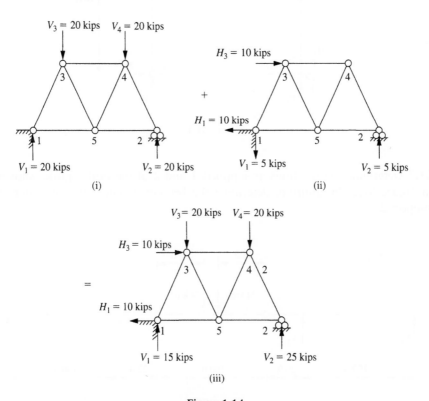

Figure 1.14

case are shown. As shown at (iii), the principle of superposition and the two loading cases may be applied simultaneously to the truss, producing the combined support reactions indicated.

Supplementary problems

S1.1 Determine the reactions at the supports of the frame shown in Figure S1.1.

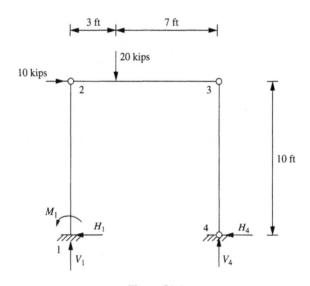

Figure S1.1

S1.2 Determine the reactions at supports 1 and 2 of the bridge girder shown in Figure S1.2. In addition, determine the bending moment in the girder at support 2.

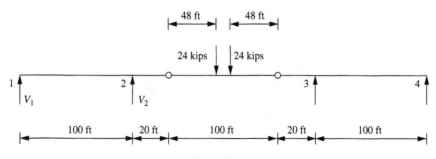

Figure S1.2

S1.3 Determine the reactions at the supports of the frame shown in Figure S1.3. In addition, determine the bending moment in member 32.

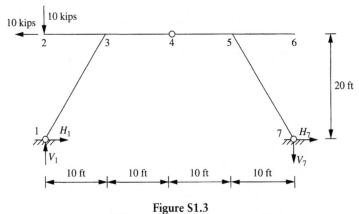

Figure S1.3

S1.4 Determine the reactions at the supports of the derrick crane shown in Figure S1.4. In addition, determine the forces produced in the members of the crane.

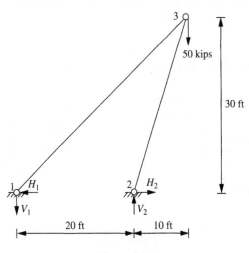

Figure S1.4

S1.5 Determine the reactions at the supports of the pin-jointed frame shown in Figure S1.5. In addition, determine the force produced in member 13.

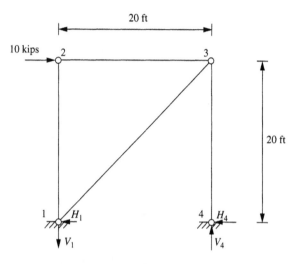

Figure S1.5

S1.6 Determine the reactions at the supports of the bent shown in Figure S1.6. In addition, determine the bending moment produced in the bent at node 3.

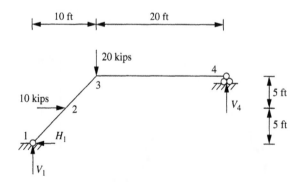

Figure S1.6

S1.7 Determine the reactions at the supports of the pin-jointed truss shown in Figure S1.7.

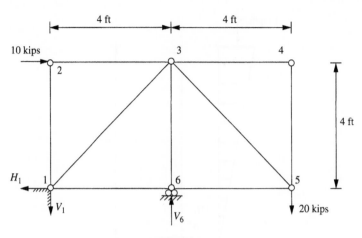

Figure S1.7

S1.8 Determine the reactions at the supports of the bent shown in Figure S1.8. The applied loading consists of the uniformly distributed load indicated.

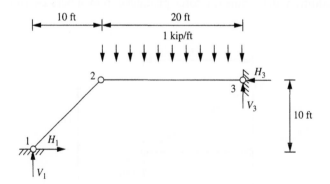

Figure S1.8

S1.9 Determine the reactions at the support of the cantilever shown in Figure S1.9. The applied loading consists of the distributed triangular force shown.

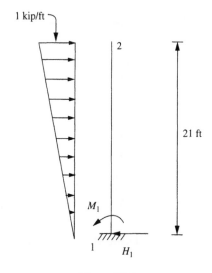

Figure S1.9

S1.10 Determine the reactions at the support of the jib crane shown in Figure S1.10. In addition, determine the force produced in members 24 and 34.

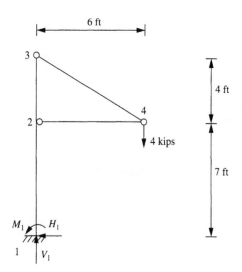

Figure S1.10

2 Statically determinate pin-jointed frames

Notation

a	panel width
F	force
H	horizontal force
h	truss height
l	length of member
M	bending moment
P	axial force in a member due to applied loads
P'	axial force in a member of the modified truss due to applied loads
R	support reaction
u	axial force in a member due a unit virtual load applied to the modified truss
V	vertical force
W_{LL}	concentrated live load
θ	angle of inclination

2.1 Introduction

A simple truss consists of a triangulated planar framework of straight members. Typical examples of simple trusses are shown in Figure 2.1 and are customarily used in bridge and roof construction. The basic unit of a truss is a triangle formed from three members. A simple truss is formed by adding members, two at a time, to form additional triangular units. The top and bottom members of a truss are referred to as chords, and the sloping and vertical members are referred to as web members.

Simple trusses may also be combined, as shown in Figure 2.2, to form a compound truss. To provide stability, the two simple trusses are connected at the apex node and also by means of an additional member at the base.

Simple trusses are analyzed using the equations of static equilibrium with the following assumptions:

- all members are connected at their nodes with frictionless hinges
- the centroidal axes of all members at a node intersect at one point so as to avoid eccentricities

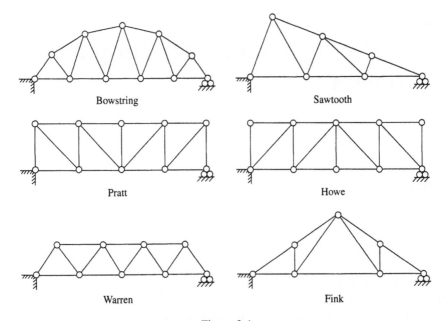

Bowstring   Sawtooth

Pratt   Howe

Warren   Fink

Figure 2.1

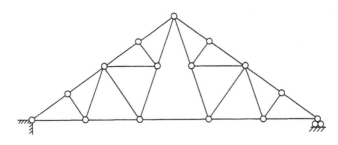

Figure 2.2

- all loads, including member weight, are applied at the nodes
- members are subjected to axial forces only
- secondary stresses caused by axial deformations are ignored

2.2 Statical determinacy

A statically determinate truss is one in which all member forces and external reactions may be determined by applying the equations of equilibrium. In a simple truss, external reactions are provided by either hinge supports or roller supports, as shown in Figure 2.3 (i) and (ii). The roller support provides only one degree of restraint, in the vertical direction, and both horizontal and

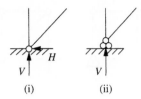

Figure 2.3

rotational displacements can occur. The hinge support provides two degrees of restraint, in the vertical and horizontal directions, and only rotational displacement can occur. The magnitudes of the external restraints may be obtained from the three equations of equilibrium. Thus, a truss is externally indeterminate when it possesses more than three external restraints and is unstable when it possesses less than three.

In a simple truss with j nodes, including the supports, $2j$ equations of equilibrium may be obtained, since at each node:

$$\Sigma H = 0$$
and $$\Sigma V = 0$$

Each member of the truss is subjected to an unknown axial force; if the truss has n members and r external restraints, the number of unknowns is $(n + r)$. Thus, a simple truss is determinate when the number of unknowns equals the number of equilibrium equations or:

$$n + r = 2j$$

A truss is statically indeterminate, as shown in Figure 2.4, when:

$$n + r > 2j$$

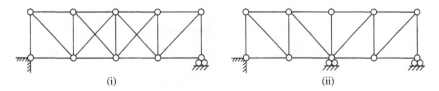

Figure 2.4

The truss at (i) is internally redundant, and the truss at (ii) is externally redundant.

A truss is unstable, as shown in Figure 2.5, when:

$$n + r < 2j$$

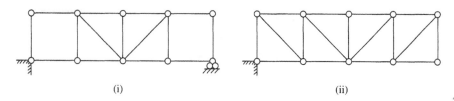

(i)   (ii)

Figure 2.5

The truss at (i) is internally deficient, and the truss at (ii) is externally deficient.

However, a situation can occur in which a truss is deficient even when the expression $n + r = 2j$ is satisfied. As shown in Figure 2.6, the left-hand side of the truss has a redundant member, while the right-hand side is unstable.

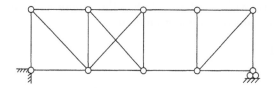

Figure 2.6

2.3 Sign convention

As indicated in Figure 2.7, a tensile force in a member is considered positive, and a compressive force is considered negative. Forces are depicted as acting from the member on the node; the direction of the member force represents the force the member exerts on the node.

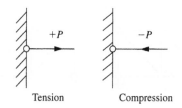

Tension   Compression

Figure 2.7

2.4 Methods of analysis

Several methods of analysis are available, each with a specific usefulness and applicability.

(a) Method of resolution at the nodes

At each node in a simple truss, the forces acting are the applied loads or support reactions and the forces in the members connected to the node. These forces constitute a concurrent, coplanar system of forces in equilibrium, and, by applying the equilibrium equations $\Sigma H = 0$ and $\Sigma V = 0$, the unknown forces in a maximum of two members may be determined. The method consists of first determining the support reactions acting on the truss. Then, each node, at which not more than two unknown member forces are present, is systematically selected in turn and the equilibrium equations applied to solve for the unknown forces.

It is not essential to resolve horizontally and vertically at all nodes; any convenient rectangular coordinate system may be adopted. Hence, it follows that for an unloaded node, when two of the three members at the node are collinear, the force in the third member is zero.

For the truss shown in Figure 2.8, the support reactions V_1 and V_9, caused by the applied load W_{LL}, are first determined. Selecting node 1 as the starting point and applying the equilibrium equations, by inspection it is apparent that:

$$P_{13} = 0$$

and $P_{12} = V_1 \ldots$ compression

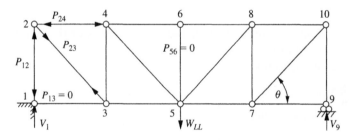

Figure 2.8

Node 2 now has only two members with unknown forces, which are given by:

$$P_{23} = P_{12}/\sin\theta \ldots \text{tension}$$

and $P_{24} = P_{23}\cos\theta \ldots$ compression

Nodes 3 and 4 are now selected in sequence, and the remaining member forces are determined. Since members 46 and 68 are collinear, it is clear that:

$$P_{56} = 0$$

This technique may be applied to any truss configuration and is suitable when the forces in all the members of the truss are required.

Example 2.1

Determine the forces produced by the applied loads in the members of the saw-tooth truss shown in Figure 2.9.

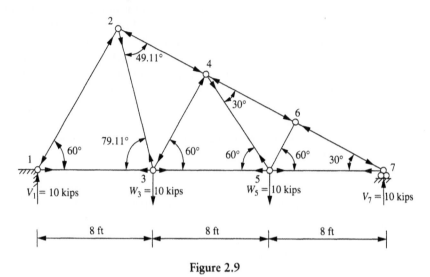

Figure 2.9

Solution

The support reactions are calculated, and the directions of the member forces are assumed as indicated.

Resolving forces at node 7:

$$P_{67} = V_7/\sin 30°$$
$$= 20 \text{ kips ... compression}$$
$$P_{57} = V_7/\tan 30°$$
$$= 17.32 \text{ kips ... tension}$$

Resolving forces at node 6:

$$P_{46} = P_{67}$$
$$= 20 \text{ kips ... compression}$$
$$P_{56} = 0$$

Resolving forces at node 5:

$$P_{45} = W_5/\sin 60°$$
$$= 11.55 \text{ kips ... tension}$$
$$P_{35} = P_{57} - W_5/\tan 60°$$
$$= 11.55 \text{ kips ... tension}$$

Resolving forces at node 4:

$$P_{34} = P_{45}/\sin 30°$$
$$= 5.77 \text{ kips} \ldots \text{compression}$$
$$P_{24} = P_{46} - P_{45} \sin 60°$$
$$= 10 \text{ kips} \ldots \text{compression}$$

Resolving forces at node 2:

$$P_{23} = P_{24}/\cos 49.11°$$
$$= 15.28 \text{ kips} \ldots \text{tension}$$

Resolving forces at node 1:

$$P_{12} = V_1/\sin 60°$$
$$= 11.55 \text{ kips} \ldots \text{compression}$$
$$P_{13} = V_1/\tan 60°$$
$$= 5.77 \text{ kips} \ldots \text{tension}$$

(b) Method of sections

This method uses the concept of a free body diagram to determine the member forces in the members of a specific panel of a truss. The method consists of first determining the support reactions acting on the truss. Then, a free body diagram is selected so as to cut through the panel; the forces acting on the free body consist of the applied loads, support reactions, and forces in the cut members. These forces constitute a coplanar system of forces in equilibrium; by applying the equilibrium equations $\Sigma M = 0$, $\Sigma H = 0$ and $\Sigma V = 0$, the unknown forces in a maximum of three members may be determined.

For the truss shown in Figure 2.10, the support reactions V_1 and V_9, caused by the applied load W_{LL}, are first determined. To determine the forces in

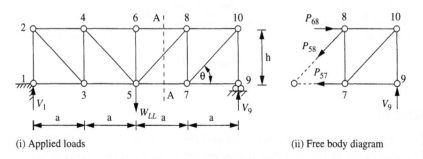

(i) Applied loads   (ii) Free body diagram

Figure 2.10

members 68, 57, and 58 of the truss, the structure is cut at section A-A, and
the right-hand portion separated as shown at (ii).

Resolving forces vertically gives the force in member 58 as:

$$P_{58} = V_9/\sin\theta \dots \text{tension}$$

Taking moments about node 8 gives the force in member 57 as:

$$P_{57} = aV_9/h \dots \text{tension}$$

Taking moments about node 5 gives the force in member 68 as:

$$P_{68} = 2aV_9/h \dots \text{compression}$$

This technique may be applied to any truss configuration and is suitable when
the forces in selected members of the truss are required. For the case of a truss
with non-parallel chords, by taking moments about the point of intersection of
two of the cut members the force in the third cut member may be obtained.

Example 2.2

Determine the forces produced by the applied loads in members 46, 56 and 57
of the sawtooth truss shown in Figure 2.11.

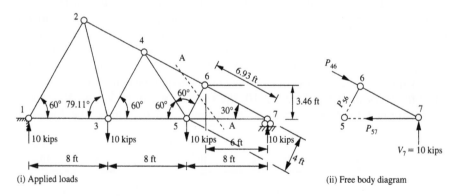

(i) Applied loads   (ii) Free body diagram

Figure 2.11

Solution

The support reactions are calculated and the truss cut at section A-A as shown.
The directions of the member forces on the right-hand free body diagram are
assumed as indicated in (ii).

Taking moments about node 7 gives the force in member 56 as:

$$P_{56} = 0/6.93$$
$$= 0$$

Taking moments about node 5 gives the force in member 46 as:

$$P_{46} = 8V_7/4$$
$$= 8 \times 10/4$$
$$= 20 \text{ kips} \dots \text{compression}$$

Taking moments about node 6 gives the force in member 57 as:

$$P_{57} = 6V_7/3.46$$
$$= 6 \times 10/3.46$$
$$= 17.32 \text{ kips} \dots \text{tension}$$

(c) Method of force coefficients

This is an adaptation of the method of sections, which simplifies the solution of trusses with parallel chords and with web members inclined at a constant angle. The forces in chord members are determined from the magnitude of the moment of external forces about the nodes of a truss. The forces in web members are obtained from knowledge of the shear force acting on a truss. The method constitutes a routine procedure for applying the method of sections.

The shear force diagram for the truss shown in Figure 2.12 is obtained by plotting at any section the cumulative vertical force produced by the applied loads on one side of the section. The force in a vertical web member is given directly, by the method of sections, as the magnitude of the shear force at the location of the member. Thus, the force in member 12 is equal to the shear force at node 1 and:

$$P_{12} = V_1$$
$$= 3W \dots \text{compression}$$

The force in member 34 is equal to the shear force at node 3 and:

$$P_{34} = V_1 - 2W$$
$$= W \dots \text{compression}$$

The force in member 56, by inspection of node 6, is:

$$P_{56} = 0$$

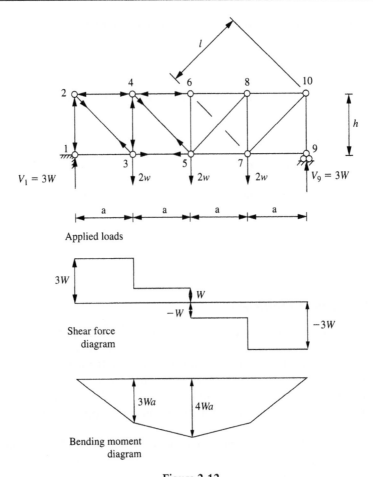

Figure 2.12

The force in a diagonal web member is given directly, by the method of sections, as the magnitude of the shear force in the corresponding panel multiplied by the coefficient l/h. Thus, the force in member 23 is equal to the shear force in the first panel multiplied by l/h and:

$$P_{23} = V_1 \times l/h$$
$$= 3W \times l/h \ldots \text{tension}$$

The force in member 45 is equal to the shear force in the second panel multiplied by l/h and:

$$P_{45} = (V_1 - 2W) \times l/h$$
$$= W \times l/h \ldots \text{tension}$$

The bending moment diagram for the truss shown in Figure 2.12 is obtained by plotting at each node the cumulative moment produced by the applied loads on one side of the node. The force in the bottom chord of a particular panel is given directly, by the method of sections, as the magnitude of the moment at the point of intersection of the top chord and the diagonal member in the panel multiplied by the coefficient $1/h$. Thus, the force in member 13 is equal to the bending moment at node 2 multiplied by $1/h$ and:

$$P_{13} = M_2 \times 1/h$$
$$= 0$$

The force in member 35 is equal to the bending moment at node 4 multiplied by $1/h$ and:

$$P_{35} = M_4 \times 1/h$$
$$= V_1 \times a \times 1/h$$
$$= 3Wa/h \dots \text{tension}$$

The force in the top chord of a particular panel is given directly, by the method of sections, as the magnitude of the moment at the point of intersection of the bottom chord and the diagonal member in the panel multiplied by the coefficient $1/h$. Thus, the force in member 24 is equal to the bending moment at node 3 multiplied by $1/h$ and:

$$P_{24} = M_3 \times 1/h$$
$$= V_1 \times a \times 1/h$$
$$= 3Wa/h \dots \text{compression}$$

The force in member 46 is equal to the bending moment at node 5 multiplied by $1/h$ and:

$$P_{46} = M_5 \times 1/h$$
$$= (V_1 \times 2a - 2W \times a) \times 1/h$$
$$= 4Wa/h \dots \text{compression}$$

This technique may be applied to any truss with parallel chords and with web members at a constant angle of inclination. It may be used when the forces in all members or in selected members of the truss are required.

Example 2.3

Determine the forces produced by the applied loads in the members of the truss shown in Figure 2.13.

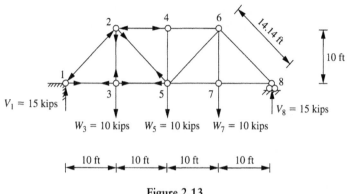

Figure 2.13

Solution

The support reactions are calculated as shown. By inspection, the force in the vertical web members is obtained directly. Thus, the force in member 23 is given as:

$$P_{23} = W_3$$
$$= 10 \text{ kips} \dots \text{tension}$$
$$P_{45} = 0$$

The force in the diagonal web member 12 is given by the magnitude of the shear force in the first panel multiplied by the coefficient l/h. Thus, the force in member 12 is:

$$P_{12} = V_1 \times l/h$$
$$= 15 \times 14.14/10$$
$$= 21.21 \text{ kips} \dots \text{compression}$$

The force in the diagonal web member 25 is given by the magnitude of the shear force in the second panel multiplied by the coefficient l/h. Thus, the force in member 25 is:

$$P_{25} = (V_1 - W_3) \times l/h$$
$$= 5 \times 14.14/10$$
$$= 7.07 \text{ kips} \dots \text{tension}$$

The force in the bottom chord member 13 is given by the magnitude of the moment at node 2 multiplied by the coefficient $1/h$. Thus, the force in member 13 is:

$$P_{13} = M_2 \times 1/h$$
$$= V_1 \times a \times 1/h$$
$$= 150 \times 1/10$$
$$= 15 \text{ kips} \ldots \text{tension}$$

Similarly, the force in the bottom chord member 35 is given by the magnitude of the moment at node 2 multiplied by the coefficient $1/h$. Thus, the force in member 35 is:

$$P_{35} = M_2 \times 1/h$$
$$= V_1 \times a \times 1/h$$
$$= 150 \times 1/10$$
$$= 15 \text{ kips} \ldots \text{tension}$$

The force in the top chord member 24 is given as the magnitude of the moment at node 5 multiplied by the coefficient $1/h$. Thus, the force in member 24 is:

$$P_{24} = M_5 \times 1/h$$
$$= (V_1 \times 2a - W_3 \times a) \times 1/h$$
$$= (300 - 100) \times 1/10$$
$$= 20 \text{ kips} \ldots \text{compression}$$

(d) Method of substitution of members

A complex truss, as shown in Figure 2.14 (i), has three or more connecting members at a node, all with unknown member forces. This precludes the use of the method of sections or the method of resolution at the nodes as a means of determining the forces in the truss. The technique consists of removing one of the existing members at a node so that only two members with unknown forces remain and substituting another member so as to maintain the truss in stable equilibrium.

The forces in members 45 and 59 are obtained by resolution of forces at node 5. However, at nodes 4 and 9, three unknown member forces remain, and these cannot be determined by resolution or by the method of sections. As shown at (ii), member 39 is removed, leaving only two unknown forces at node 9, which may be determined. To maintain stable equilibrium, a substitute member 38 is added to create a modified truss, and the original applied loads are applied to the modified truss. The forces P' in all the

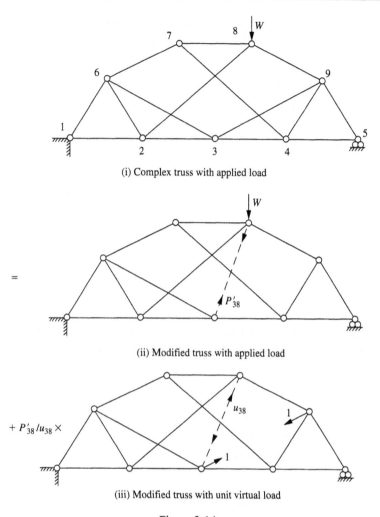

(i) Complex truss with applied load

(ii) Modified truss with applied load

(iii) Modified truss with unit virtual load

Figure 2.14

remaining members of the modified truss may now be determined. The force in member 38 is P'_{38}.

The applied loads are now removed, and unit virtual loads are applied to the modified truss along the line of action of the original member 39, as shown at (iii). The forces u in the modified truss are determined; the force in member 38 is $-u_{38}$. Multiplying the forces in system (iii) by P'_{38}/u_{38} and adding them to the forces in system (ii) gives the force in member 38 as:

$$P_{38} = P'_{38} + (-u_{38})P'_{38}/u_{38}$$
$$= 0$$

In effect, the substitute member 38 has been eliminated from the truss.

Hence, by applying the principle of superposition, the final forces in the original truss are obtained from the expression:

$$P = P' + uP'_{38}/u_{38}$$

where tensile forces are positive and compressive forces are negative. The final force in member 39 is:

$$P_{39} = 1 \times P'_{38}/u_{38}$$
$$= P'_{38}/u_{38}$$

Example 2.4

Determine the forces produced by the applied loads in members 49, 39, and 89 of the complex truss shown in Figure 2.15.

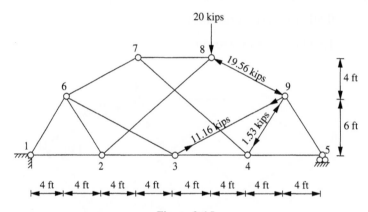

Figure 2.15

Solution

The modified truss shown in Figure 2.16 is created by removing member 39, adding the substitute member 38, and applying the 20 kips load. The member forces in the modified truss may now be determined; the values obtained are:

$$P'_{49} = 7.51 \text{ kips} \dots \text{tension}$$

$$P'_{89} = -13.98 \text{ kips} \dots \text{compression}$$

$$P'_{38} = 21.54 \text{ kips} \dots \text{tension}$$

The 20 kips load is removed from the modified truss, and unit virtual loads are applied at nodes 3 and 9 in the direction of the line of action of the force in

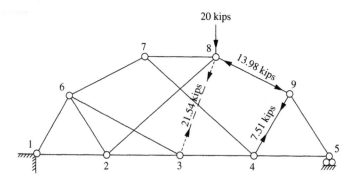

Figure 2.16

member 39. The member forces for this loading condition may now be determined; the values obtained are:

$$u_{49} = -0.81 \text{ kips} \dots \text{compression}$$

$$u_{89} = -0.50 \text{ kips} \dots \text{compression}$$

$$u_{38} = -1.93 \text{ kips} \dots \text{compression}$$

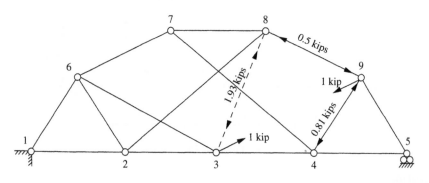

Figure 2.17

The multiplying ratio is given by:

$$P'_{38}/u_{38} = 21.54/1.93$$
$$= 11.16$$

The final member forces in the original truss are:

$$P_{49} = 7.51 + 11.16(-0.81)$$
$$= -1.53 \text{ kips} \dots \text{compression}$$

$$P_{89} = -13.98 + 11.16(-0.50)$$
$$= -19.56 \text{ kips} \dots \text{compression}$$
$$P_{39} = P'_{38}/u_{38}$$
$$= 11.16 \text{ kips} \dots \text{tension}$$

Supplementary problems

S2.1 Determine in Figure S2.1 the reactions at the supports of the roof truss shown and the forces in members 34, 38, and 78 caused by the applied loads.

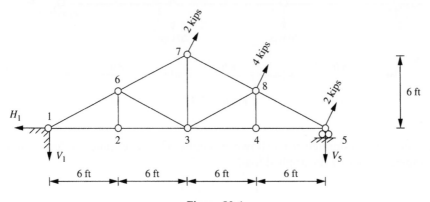

Figure S2.1

S2.2 For the pin-jointed truss shown in Figure S2.2 determine the forces in members 45, 411, and 1011 caused by the applied loads.

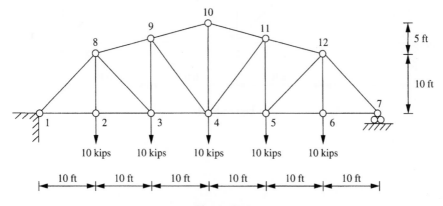

Figure S2.2

S2.3 For the roof truss shown in Figure S2.3 determine the forces in members 23, 27 and 67 caused by the applied loads.

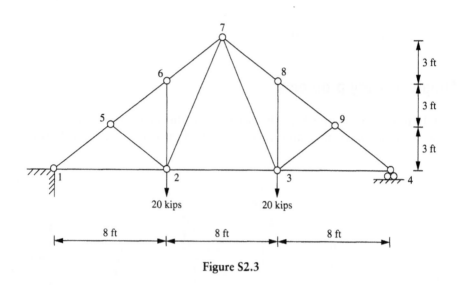

Figure S2.3

S2.4 For the pin-jointed truss shown in Figure S2.4 determine the forces in all members caused by the applied loads.

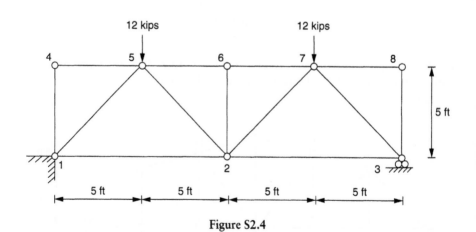

Figure S2.4

S2.5 For the roof truss shown in Figure S2.5 determine the forces in all members due to the applied loads.

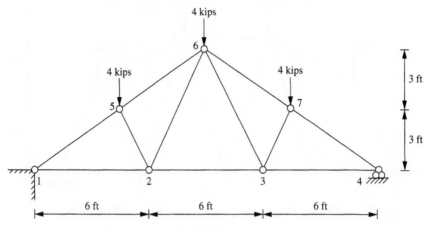

Figure S2.5

S2.6 For the truss shown in Figure S2.6 determine the forces in members 12, 114, 110, 23, 310, 315, 1014, and 1415 caused by the applied loads.

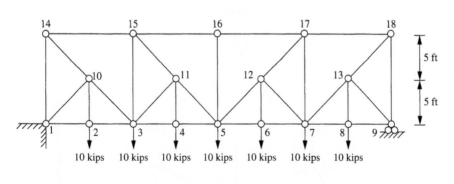

Figure S2.6

S2.7 For the roof truss shown in Figure S2.7 determine the forces in members 23, 27, 37, 78, and 67 caused by the applied loads.

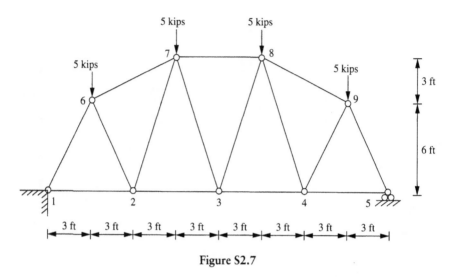

Figure S2.7

S2.8 For the roof truss shown in Figure S2.8 determine the forces in all members due to the applied loads.

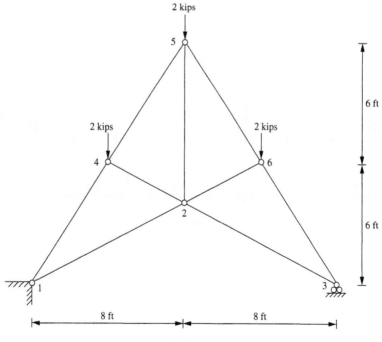

Figure S2.8

S2.9 For the roof truss shown in Figure S2.9 determine the forces in members 23, 26, 27, 67, and 37 caused by the applied loads.

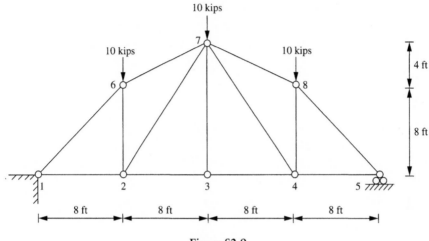

Figure S2.9

S2.10 For the roof truss shown in Figure S2.10 determine the force in members 49, 59, 89, and 78 due to the applied loads.

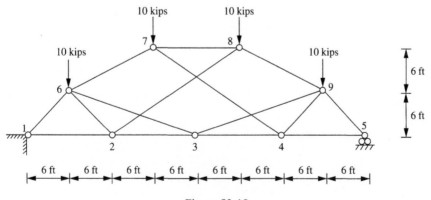

Figure S2.10

Figure S2.9

Figure S2.10

3 Elements in flexure

Notation

c rise of an arch
F force
H horizontal force, support reaction
h height of a rigid frame
l length of span
M bending moment
P axial force in a member
Q shear force in a member
R support reaction
V vertical force, support reaction
W concentrated load
w distributed load
δ displacement
θ rotation

3.1 Load intensity, shear force, and bending moment diagrams

A uniformly distributed load of magnitude $-w$ is applied to a simply supported beam as shown in Figure 3.1 (i). The sign convention adopted is that forces acting upward are defined as positive. The support reaction at end 1 of the beam is obtained by considering moment equilibrium about end 2. Hence:

$$M_2 = 0$$
$$= lV_1 + (-w)l^2/2$$

and:

$$V_1 = wl/2$$

Similarly:

$$V_2 = wl/2$$

The load intensity diagram is shown at (ii) and consists of a horizontal line of magnitude $-w$.

The shear force acting on any section A-A at a distance x from end 1 is defined as the cumulative sum of the vertical forces acting on one side of the

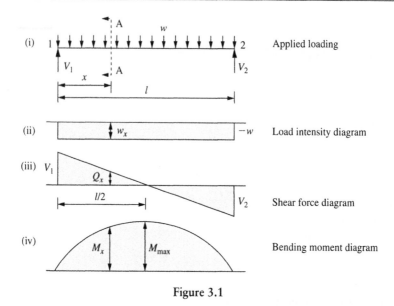

Figure 3.1

section. The vertical forces consist of the applied loads and support reactions. Considering the segment of the beam on the left of section A-A; the shear force is given by:

$$Q_x = V_1 + (-w)x$$
$$= V_1 - wx$$

where $(-wx)$ represents the area of the load intensity diagram on the left of section A-A from $x = 0$ to $x = x$. In general, the change in shear force between two sections of a beam equals the area of the load intensity diagram between the same two sections.

The variation of shear force along the length of the beam may be illustrated by plotting a shear force diagram as shown at (iii). The sign convention adopted is that resultant shear force upward on the left of a section is positive. The maximum shear force occurs at the location of zero load intensity, which is at the ends of the beam.

The bending moment acting on any section A-A at a distance x from end 1 is defined as the cumulative sum of the moments acting on one side of the section. Considering the segment of the beam on the left of section A-A, the bending moment is given by:

$$M_x = V_1 x + (-w)x^2/2$$
$$= V_1 x + (Q_x - V_1)x/2$$
$$= (V_1 + Q_x)x/2$$

where $(V_1 + Q_x)x/2$ represents the area of the shear force diagram on the left of section A-A from $x = 0$ to $x = x$. In general, the change in bending moment between two sections of a beam equals the area of the shear force diagram between the same two sections.

The variation of bending moment along the length of the beam may be illustrated by plotting a bending moment diagram as shown at (iv). The sign convention adopted is that a bending moment producing tension in the bottom fibers of the beam is positive. The maximum bending moment occurs at the location of zero shear force in the beam.

Example 3.1

For the simply supported beam shown in Figure 3.2 (i) draw the load intensity, shear force, and bending moment diagrams.

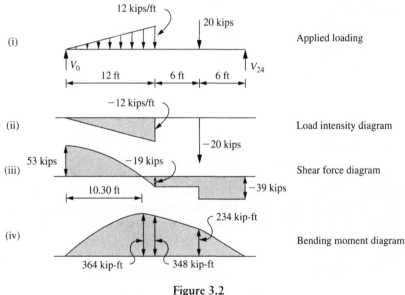

Figure 3.2

Solution

The intensity of loading diagram is shown at (ii) and consists of triangularly shaped sections varying from zero to $-12\,\text{kips/ft}$ over the left half of the beam and a concentrated force of $-20\,\text{kips}$ acting at $x = 18\,\text{ft}$.

The support reaction at the left end of the beam is derived by taking moments about the right end of the beam to give:

$$M_{24} = 0$$
$$= 24V_0 + (12/3 + 12)(-12 \times 12/2) + (-20 \times 6)$$

and:

$$V_0 = 53 \text{ kips}$$
$$V_{24} = 72 + 20 - V_0$$
$$= 39 \text{ kips}$$

The ordinate of the shear force diagram at $x = 0$ equals the support reaction at the left end of the beam. The change in shear force between $x = 0$ and $x = 12\,\text{ft}$ equals the area of the load intensity diagram over the initial 12 ft of the beam. Hence, the ordinate of the shear force diagram at $x = 12\,\text{ft}$ is given by:

$$Q_{12} = V_0 - 12 \times 12/2$$
$$= 53 - 72$$
$$= -19 \text{ kips}$$

The shear force diagram crosses the base line at a distance x from the end of the beam given by:

$$Q_x = 0$$
$$= V_0 - w_x x/2$$
$$= 53 - x^2/2$$

and:

$$x^2 = 106$$
$$x = 10.30\,\text{ft}$$

Between $x = 12\,\text{ft}$ and $x = 18\,\text{ft}$ the ordinate of the shear force diagram remains constant at a value of -19 kips since the load intensity is zero over this segment of the beam. At $x = 18\,\text{ft}$, the location of the concentrated load of 20 kips, the ordinate of the shear force diagram reduces to the value:

$$Q_{18} = -19 - 20$$
$$= -39 \text{ kips}$$

Between $x = 18\,\text{ft}$ and $x = 24\,\text{ft}$ the ordinate of the shear force diagram remains constant at a value of -39 kips. At $x = 24\,\text{ft}$ the ordinate of the shear force diagram changes to the value:

$$Q_{24} = -39 + 39$$
$$= 0 \text{ kips}$$

The bending moment ordinates at each end of the beam are zero. Between $x = 0$ and $x = 12\,\text{ft}$ the ordinate of the bending moment diagram is given by:

$$M_x = V_0 x - (w_x x/2)(x/3)$$
$$= 53x - x^3/6$$

The maximum bending moment over this length occurs when:

$$\mathrm{d}M_x/\mathrm{d}x = 0$$
$$= 53 - x^2/2$$

and:

$$x = 10.30\,\mathrm{ft}$$

This is the same location at which the shear force diagram crosses the base line, and the maximum bending moment is:

$$M_{max} = 53 \times 10.30 - (10.30)^3/6$$
$$= 546 - 182$$
$$= 364 \text{ kip-ft}$$

At $x = 12\,\mathrm{ft}$ the bending moment is:

$$M_{12} = 53 \times 12 - 12^3/6$$
$$= 636 - 288$$
$$= 348 \text{ kip-ft}$$

At $x = 18\,\mathrm{ft}$ the bending moment is:

$$M_{18} = M_{12} + (-19) \times 6$$
$$= 348 - 114$$
$$= 234 \text{ kip-ft}$$

and the bending moment decreases linearly between $x = 12\,\mathrm{ft}$ and $x = 18\,\mathrm{ft}$. At $x = 24\,\mathrm{ft}$ the bending moment is:

$$M_{24} = M_{18} + (-39) \times 6$$
$$= 234 - 234$$
$$= 0 \text{ kip-ft}$$

and the bending moment decreases linearly between $x = 18\,\mathrm{ft}$ and $x = 24\,\mathrm{ft}$.

3.2 Relationships among loading, shear force, and bending moment

A small element of the beam shown in Figure 3.1 is taken at a distance x from end 1. The forces acting on the element are shown in Figure 3.3. Resolving forces vertically gives the expression:

$$Q = (Q + \delta Q) + w\,\delta x$$

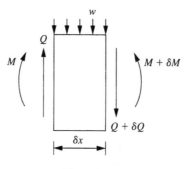

Figure 3.3

and:

$$\delta Q / \delta x = -w$$

The limiting condition is:

$$dQ/dx = -w$$

which indicates that the slope of the shear force diagram, at any section, equals the intensity of loading at that section. This also indicates that the shear force reaches a maximum or a minimum value where the load intensity diagram crosses the base line. Alternatively, since:

$$dQ = -w\ dx$$
$$\int dQ = \int -w\ dx$$
$$Q_2 - Q_1 = \int_{x_1}^{x_2} -w\ dx$$

where Q_1 = shear force in the beam at $x = x_1$, Q_2 = shear force in the beam at $x = x_2$, and the change in shear force between the two sections equals the area of the load intensity diagram between the two sections.

Taking moments about the lower right corner of the element gives the expression:

$$M = (M + \delta M) - Q\ \delta x + w\ (\delta x)^2/2$$

Neglecting the small value $(\delta x)^2$ gives:

$$\delta M / \delta x = Q$$

The limiting condition is:

$$dM/dx = Q$$

which indicates that the slope of the bending moment diagram at any section equals the force at that section. In addition, it indicates that the bending moment reaches a maximum or a minimum value where the shear force diagram crosses the base line. Alternatively, since:

$$dM = Q\,dx$$
$$\int dM = \int Q\,dx$$
$$M_2 - M_1 = \int_{x_1}^{x_2} Q\,dx$$

where M_1 = bending moment in the beam at $x = x_1$, M_2 = bending moment in the beam at $x = x_2$, and the change in bending moment between the two sections equals the area of the shear force diagram between the two sections.

Example 3.2

For the simply supported beam shown in Figure 3.2 use the load-shear-moment relationships to draw the shear force and bending moment diagrams.

Solution

The support reactions were determined in Example 3.1:

$$V_0 = 53 \text{ kips}$$
$$V_{24} = 39 \text{ kips}$$

The change in shear force between $x = 0$ and $x = 12\,\text{ft}$ is:

$$
\begin{aligned}
Q_{12} - V_0 &= \int_0^{12} -w\,dx \\
&= \int_0^{12} -12x/12\,dx \\
&= \int_0^{12} -x\,dx \\
&= [-x^2/2]_0^{12} \\
&= -72 \text{ kips}
\end{aligned}
$$

and:

$$
\begin{aligned}
Q_{12} &= 53 - 72 \\
&= -19 \text{ kips}
\end{aligned}
$$

At intermediate sections between $x = 0$ and $x = 12\,\text{ft}$, the ordinates of the shear force diagram are given by:

$$
\begin{aligned}
Q_x &= V_0 - wx/2 \\
&= 53 - x^2/2
\end{aligned}
$$

which is the equation of a parabola.

The shear force diagram crosses the base line at a distance x from the end of the beam, given by:

$$Q_x = 0$$
$$= V_0 - wx/2$$
$$= 53 - x^2/2$$

and:

$$x^2 = 106$$
$$x = 10.30 \text{ ft}$$

Between $x = 12$ ft and $x = 18$ ft the load intensity is zero, and the shear force diagram is a horizontal line. At $x = 18$ ft, a concentrated load of 20 kips is applied to the beam; this constitutes an infinite load intensity and produces a vertical step of -20 kips in the ordinate of the shear force diagram. The ordinate of the shear force diagram is given by:

$$Q_{18} = V_{12} - 20$$
$$= -19 - 20$$
$$= -39 \text{ kips}$$

The ordinate of the shear force diagram remains constant at a value of -39 kips to the end of the beam.

The change in bending moment between $x = 0$ and $x = 12$ ft is:

$$M_{12} - M_0 = \int_0^{12} Q \, dx$$
$$= \int_0^{12} (V_0 - wx/2) \, dx$$
$$= \int_0^{12} (53 - x^2/2) \, dx$$
$$= [53x - x^3/6]_0^{12}$$
$$= 636 - 288$$
$$= 348 \text{ kip-ft}$$

At intermediate sections between $x = 0$ and $x = 12$ ft, the ordinates of the bending moment diagram are given by:

$$M_x = V_0 x - (wx/2)(x/3)$$
$$= 53x - x^3/6$$

which is the equation of a cubic parabola.

The maximum bending moment occurs when:

$$dM_x/dx = 0$$
$$= 53 - x^2/2$$

and:

$x = 10.30$ ft

The change in bending moment between $x = 12$ ft and $x = 18$ ft is:

$$
\begin{aligned}
M_{18} - M_{12} &= \int_{12}^{18} Q \, dx \\
&= \int_{12}^{18} -19 \, dx \\
&= [-19x]_{12}^{18} \\
&= -19 \times 6 \\
&= -114 \, \text{kip-ft}
\end{aligned}
$$

Hence:

$$
\begin{aligned}
M_{18} &= 348 - 114 \\
&= 234 \, \text{kip-ft}
\end{aligned}
$$

The bending moment decreases linearly between $x = 12$ ft and $x = 18$ ft.

At $x = 18$ ft a concentrated load of 20 kips is applied, and the change in bending moment between $x = 18$ ft and $x = 24$ ft is:

$$
\begin{aligned}
M_{24} - M_{18} &= \int_{18}^{24} Q \, dx \\
&= \int_{18}^{24} -39 \, dx \\
&= [-39x]_{18}^{24} \\
&= -39 \times 6 \\
&= -234 \, \text{kip-ft}
\end{aligned}
$$

Hence:

$$
\begin{aligned}
M_{24} &= 234 - 234 \\
&= 0 \, \text{kip-ft}
\end{aligned}
$$

3.3 Statical determinacy

A statically determinate beam or rigid frame is one in which all member forces and external reactions may be determined by applying the equations of equilibrium. In a beam or rigid frame external reactions are provided by either hinge or roller supports or by a fixed end, as shown in Figure 3.4. The roller support

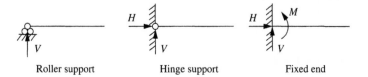

Roller support   Hinge support   Fixed end

Figure 3.4

provides only one degree of restraint, in the vertical direction, and both horizontal and rotational displacements can occur. The hinge support provides two degrees of restraint, in the vertical and horizontal directions, and only rotational displacement can occur. The fixed end provides three degrees of restraint, vertical, horizontal, and rotational. Identifying whether a structure is determinate depends on the configuration of the structure.

(a) Beam or rigid frame with no internal hinges

In a rigid frame with j nodes, including the supports, $3j$ equations of equilibrium may be obtained since, at each node:

$$\Sigma H = 0$$
$$\Sigma V = 0$$

and:

$$\Sigma M = 0$$

Each member of the rigid frame is subjected to an unknown axial and shear force and bending moment. If the rigid frame has n members and r external restraints, the number of unknowns is $(3n + r)$. Thus, a beam or frame is determinate when the number of unknowns equals the number of equilibrium equations or:

$$3n + r = 3j$$

A beam is statically indeterminate, as shown in Figure 3.5, when:

$$3n + r > 3j$$

Figure 3.5

In this case:

$$3j = 3 \times 2$$
$$= 6$$

and:

$$3n + r = 3 \times 1 + 4$$
$$= 7$$
$$> 3j$$

A rigid frame is unstable, as shown in Figure 3.6, when:

$3n + r < 3j$

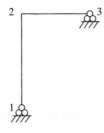

Figure 3.6

In this case:

$$3j = 3 \times 3$$
$$= 9$$

and:

$$3n + r = 3 \times 2 + 2$$
$$= 8$$
$$< 3j$$

(b) Beam or rigid frame with internal hinges or rollers

The introduction of an internal hinge in a beam or rigid frame provides an additional equation of equilibrium at the hinge of $M = 0$. In effect, a moment release has been introduced in the member.

The introduction of a horizontal, internal roller provides two additional equations of equilibrium at the roller of $M = 0$ and $H = 0$. In effect, a moment release and a release of horizontal restraint have been introduced in the member. Thus, a beam or frame with internal hinges or rollers is determinate when:

$$3n + r = 3j + h + 2s$$

where n is the number of members, j is the number of nodes in the rigid frame before the introduction of hinges, r is the number of external restraints, h is the number of internal hinges, and s is the number of rollers introduced.

The compound beam shown in Figure 3.7 is determinate since:

$$3n + r = 3 \times 3 + 5$$
$$= 14$$

Figure 3.7

and:

$$3j + h + 2s = 3 \times 4 + 2 + 0$$
$$= 14$$
$$= 3n + r$$

(c) Rigid frame with internal hinges at a node

The introduction of a hinge into i of the n members meeting at a node in a rigid frame produces i releases. The introduction of a hinge into all n members produces $(n - 1)$ releases.

Thus, a rigid frame with hinges at the nodes is determinate when:

$$3n + r = 3j + c$$

where n is the number of members, j is the number of nodes in the rigid frame, r is the number of external restraints, and c is the number of releases introduced.

As shown in Figure 3.8, for four members meeting at a rigid node there are three unknown moments. The introduction of a hinge into one of the members produces one release, the introduction of a hinge into two members produces two releases, and the introduction of a hinge into all four members produces three releases.

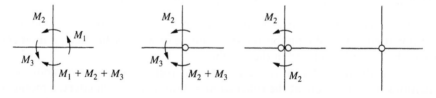

Figure 3.8

The rigid frame shown in Figure 3.9 is determinate since:

$$3n + r = 3 \times 3 + 4$$
$$= 13$$

and:

$$3j + c = 3 \times 4 + 1$$
$$= 13$$
$$= 3n + r$$

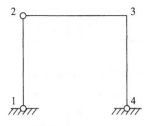

Figure 3.9

3.4 Beams

Beams are normally subject to transverse loads only, and roller and hinge supports are typically represented by vertical arrows. Typical examples of beams are shown in Figure 3.10. Beams may be analyzed using the equations of static equilibrium and the method of sections, as illustrated in Section 3.1. Alternatively, the principle of virtual work may be utilized to provide a simple and convenient solution.

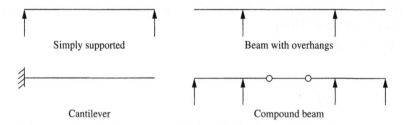

Figure 3.10

The principle of virtual work may be defined as follows: If a structure in equilibrium under a system of applied forces is subjected to a system of displacements compatible with the external restraints and the geometry of the structure, the total work done by the applied forces during these external displacements equals the work done by the internal forces, corresponding to the applied forces, during the internal deformations corresponding to the external displacements.

The expression "virtual work" signifies that the work done is the product of a real loading system and imaginary displacements or an imaginary loading system and real displacements.

For the simply supported beam shown in Figure 3.11 (i), the support reaction V_2, caused by the applied load W, may be determined by the principle of virtual work. As shown at (ii), a unit virtual displacement of $\delta = 1$ is imposed

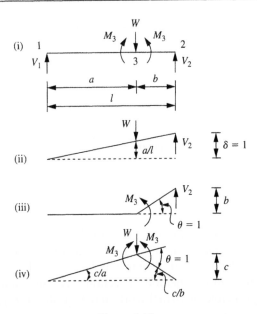

Figure 3.11

on end 2 in the direction of V_2 and the internal work done equated to the external work done. Then:

$$V_2 \times 1 = W \times a/l$$

and:

$$V_2 = Wa/l$$

Similarly, as shown at (iii), the bending moment produced at point 3 by the applied load may be determined by cutting the beam at 3 and imposing a unit virtual angular discontinuity of $\theta = 1$. Equating internal work done to external work done gives:

$$M_3 \times 1 = V_2 \times b$$

and:

$$M_3 = Wab/l$$

Alternatively, after cutting the beam at 3 and imposing a unit virtual angular discontinuity, the ends may be clamped together to produce the deformed shape shown in (iv). Equating internal work done to external work done gives:

$$M_3(c/a + c/b) = W \times c$$

and:

$$M_3 = Wab/l$$

Example 3.3

Use the virtual work method to determine the support reactions and significant bending moments for the compound beam shown in Figure 3.12 due to the applied loads indicated. Hence, draw the shear force and the bending moment diagrams for the beam.

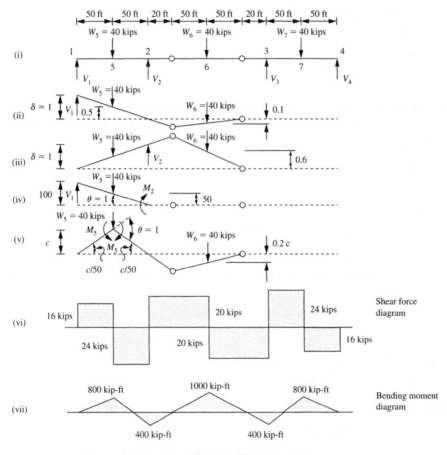

Figure 3.12

Solution

As shown at (ii), a unit virtual displacement of $\delta = 1$ is imposed on end 1 in the direction of V_1 and the internal work done equated to the external work done. Then:

$$V_1 \times 1 = W_5 \times 0.5 - W_6 \times 0.1$$

and:

$$V_1 = 40 \times 0.4$$
$$= 16 \text{ kips}$$

As shown at (iii), a unit virtual displacement of $\delta = 1$ is imposed at support 2 in the direction of V_2 and the internal work done equated to the external work done. Then:

$$V_2 \times 1 = W_5 \times 0.5 + W_6 \times 0.6$$

and:

$$V_2 = 40 \times 1.1$$
$$= 44 \text{ kips}$$

As shown at (iv), the beam is cut at 2 and a unit virtual angular discontinuity of $\theta = 1$ is imposed. The internal work done is equated to the external work done to give:

$$M_2 \times 1 = W_5 \times 50 - V_1 \times 100$$

and:

$$M_2 = 40 \times 50 - 16 \times 100$$
$$= 400 \text{ kip-ft ... tension in the top fibers}$$

As shown at (v), the beam is cut at 5, and a unit virtual angular discontinuity of $\theta = 1$ is imposed. The internal work done is equated to the external work done to give:

$$M_5(c/50 + c/50) = W_5 \times c - W_6 \times 0.2c$$

and:

$$M_5 = 25 \times 40 \times 0.8$$
$$= 800 \text{ kip-ft ... tension in the bottom fibers}$$

Similarly:

$$M_6 = 25 \times 40$$
$$= 1000 \text{ kip-ft ... tension in the bottom fibers}$$

The shear force and bending moment diagrams are shown at (vi) and (vii). The bending moment is drawn on the compression side of the beam.

3.5 Rigid frames

The support reactions of rigid frame structures may be determined using the equations of static equilibrium, and the internal forces in the members from a free body diagram of the individual members. The internal forces on a member are most conveniently indicated as acting from the node on the member: i.e., as support reactions at the node.

The vertical reaction at support 4 of the frame shown in Figure 3.13 is obtained by considering moment equilibrium about support 1. Hence:

$$M_1 = 0$$
$$= lV_4 - Wh$$

and:

$$V_4 = Wh/l \text{ ... upward}$$

Resolving forces vertically gives:

$$V_1 = -V_4$$
$$= -Wh/l \text{ ... downward}$$

Resolving forces horizontally gives:

$$H_1 = -W \text{ ... to the left}$$

and the support reactions are shown at (ii). The deformed shape of the frame is shown at (iii).

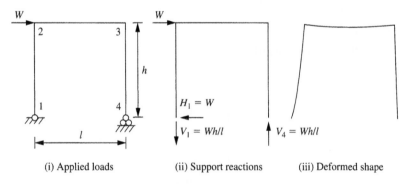

(i) Applied loads   (ii) Support reactions   (iii) Deformed shape

Figure 3.13

The internal forces in member 12 are determined from a free body diagram of the member, as shown in Figure 3.14. Resolving forces vertically gives the internal force acting at node 2 as:

$$V_2 = -V_1$$
$$= Wh/l \text{ ... upward}$$

Considering moment equilibrium about node 2 gives the internal moment at node 2 as:

$$M_{21} = -hH_1$$
$$= Wh \text{ ... counter-clockwise}$$

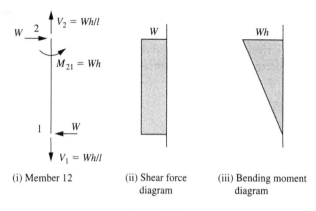

(i) Member 12   (ii) Shear force   (iii) Bending moment
                        diagram   diagram

Figure 3.14

Hence, member 12 is subject to a tensile force of $P_{12} = Wh/l$, the shear force diagram is shown at (ii) and the bending moment diagram at (iii) with the moment drawn on the compression side of the member.

The internal forces in member 34 are determined from a free body diagram of the member, as shown in Figure 3.15. Resolving forces vertically gives the internal force acting at node 3 as:

$$V_3 = -V_4$$
$$= -Wh/l \dots \text{downward}$$

Member 34

Figure 3.15

Considering moment equilibrium about node 3 gives the internal moment at node 3 as:

$$M_{34} = 0$$

Resolving forces horizontally gives the internal force acting at node 3 as:

$$H_3 = 0$$

Hence, member 34 is subject to a compressive force of $P_{34} = Wh/l$, and both the shear force and the bending moment are zero.

The internal forces in member 23 are determined from a free body diagram of the member, as shown in Figure 3.16. Resolving forces vertically gives the internal force acting at node 2 as:

$$V_2 = -V_3$$
$$= -Wh/l \ldots \text{downward}$$

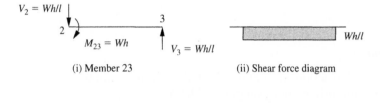

(i) Member 23   (ii) Shear force diagram

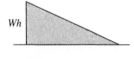

(iii) Bending moment diagram

Figure 3.16

Considering moment equilibrium about node 2 gives the internal moment at node 2 as:

$$M_{23} = -lV_3$$
$$= -Wh \ldots \text{clockwise}$$

Resolving forces horizontally gives the internal force acting at node 2 as:

$$H_2 = 0$$

Hence, member 23 has no axial force. The shear force diagram is shown at (ii) and the bending moment diagram at (iii), with the moment drawn on the compression side of the member.

Example 3.4

Determine the support reactions and member forces in the rigid frame shown in Figure 3.17 due to the applied loads indicated. Hence, draw the shear force and the bending moment diagrams for the members.

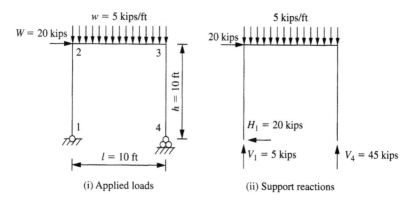

(i) Applied loads   (ii) Support reactions

Figure 3.17

Solution

The vertical reaction at support 4 of the frame is obtained by considering moment equilibrium about support 1. Hence:

$$M_1 = 0$$
$$= lV_4 - Wh - wl^2/2$$
$$= 10V_4 - 20 \times 10 - 5 \times 100/2$$

and:

$$V_4 = 45 \text{ kips ... upward}$$

Resolving forces vertically gives:

$$V_1 = -V_4 + wl$$
$$= -45 + 5 \times 10$$
$$= 5 \text{ kips ... upward}$$

Resolving forces horizontally gives:

$$H_1 = -W$$
$$= 20 \text{ kips ... to the left}$$

and the support reactions are shown at (ii).

The internal forces in member 12 are determined from a free body diagram of the member, as shown in Figure 3.18. Resolving forces vertically gives the internal force acting at node 2 as:

$$V_2 = -V_1$$
$$= -5 \text{ kips ... downward}$$

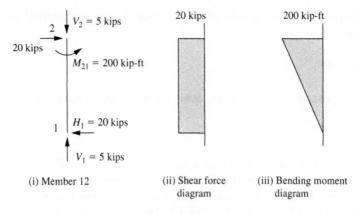

(i) Member 12   (ii) Shear force diagram   (iii) Bending moment diagram

Figure 3.18

Considering moment equilibrium about node 2 gives the internal moment at node 2 as:

$$M_{21} = -hH_1$$
$$= -10 \times -20$$
$$= 200 \, \text{kip-ft} \ldots \text{counter-clockwise}$$

Hence, member 12 is subject to a compressive force of $P_{12} = 5 \, \text{kips}$. The shear force diagram is shown at (ii) and the bending moment diagram at (iii), with the moment drawn on the compression side of the member.

The internal forces in member 34 are determined from a free body diagram of the member, as shown in Figure 3.19. Resolving forces vertically gives the internal force acting at node 3 as:

$$V_3 = -V_4$$
$$= -45 \, \text{kips} \ldots \text{downward}$$

Member 34

Figure 3.19

Considering moment equilibrium about node 3 gives the internal moment at node 3 as:

$$M_{34} = 0$$

Resolving forces horizontally gives the internal force acting at node 3 as:

$$H_3 = 0$$

Hence, member 34 is subject to a compressive force of $P_{34} = 45$ kips, and both the shear force and the bending moment are zero.

The internal forces in member 23 are determined from a free body diagram of the member, as shown in Figure 3.20. Resolving forces vertically gives the internal force acting at node 2 as:

$$\begin{aligned}
V_2 &= -V_3 + wl \\
&= -45 + 5 \times 10 \\
&= 5 \text{ kips ... upward}
\end{aligned}$$

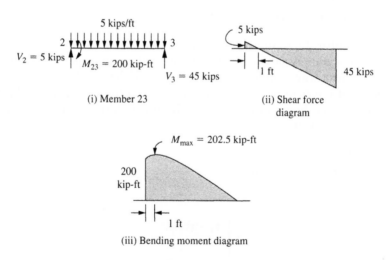

(i) Member 23

(ii) Shear force diagram

(iii) Bending moment diagram

Figure 3.20

Considering moment equilibrium about node 2 gives the internal moment at node 2 as:

$$\begin{aligned}
M_{23} &= -lV_3 + wl^2/2 \\
&= -10 \times 45 + 5 \times 100/2 \\
&= -200 \text{ kip-ft ... clockwise}
\end{aligned}$$

Resolving forces horizontally gives the internal force acting at node 2 as:

$$H_2 = 0$$

Hence, member 23 has no axial force. The shear force diagram is shown at (ii) and the bending moment diagram at (iii), with the moment drawn on the compression side of the member.

3.6 Three-hinged arch

The three-hinged arch has hinged supports at each abutment that provide four external restraints. The introduction of another hinge in the arch member provides a moment release. The extra hinge provides an additional equation of equilibrium that, together with the three basic equations of equilibrium, makes the solution of the arch possible. Typical examples of three-hinged arches are shown in Figure 3.21.

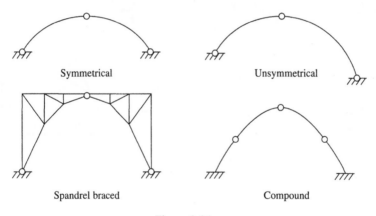

Figure 3.21

For a three-hinged arch subjected to vertical loads only, the horizontal support reactions at the arch springings are equal and opposite and act inward. The vertical support reactions at the arch springings are equal to those of a simply supported beam of identical length with identical loads.

For the symmetrical three-hinged arch shown in Figure 3.22 with a vertical applied load W, the unknown horizontal thrust at the springings is H, and the unknown vertical reactions are V_1 and V_2. The vertical reaction at support 2 is obtained by considering moment equilibrium about support 1. Hence:

$$M_1 = 0$$
$$= lV_2 - Wa$$

and:

$$V_2 = Wa/l$$

Resolving forces vertically gives:

$$V_1 = W - V_2$$
$$= W(1 - a/l)$$

These values for V_1 and V_2 are identical to the reactions of a simply supported beam of the same span as the arch with the same applied load W.

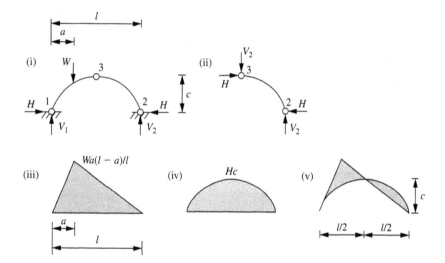

Figure 3.22

The horizontal thrust at the springings is determined from a free body diagram of the right half of the arch, as shown at (ii). Considering moment equilibrium about the crown hinge at 3:

$$M_3 = 0$$
$$= lV_2/2 - Hc$$

and:

$$H = lV_2/2c$$

This value for H is identical to the bending moment at the center of a simply supported beam of the same length with the same applied load W multiplied by $1/c$.

The bending moment in the arch at any point a distance x from the left support is given by the expressions:

$$M_x = V_1 x - Hy \text{ ... for } x \leq a$$
$$= V_2(l - x) - Hy \text{ ... for } a < x \leq l$$

where y is the height of the arch a distance x from the left support.

At $x = l/2$, $y = c$, and the bending moment at the crown hinge is:

$$M_{l/2} = 0$$

The expressions for bending moment may be considered as the superposition of the bending moment of a simply supported beam of the same span with the same applied load W plus the bending moment due to the horizontal thrust H. The bending moment due to the applied load on a simply supported beam is shown at (iii) and the bending moment due to the horizontal thrust is shown at (iv); this is identical to the shape of the arch. Since the bending moment is zero at the crown hinge, the combined bending moment for the arch is obtained by adjusting the scale of the free bending moment to give an ordinate of magnitude c at $x = l/2$ and superimposing this on a drawing of the arch. This is shown at (v), drawn on the compression side of the arch. In the case of a three-hinged parabolic arch with a uniformly distributed applied load, no bending moment is produced in the arch rib.

Example 3.5

Determine the support reactions in the parabolic unsymmetrical three-hinged arch shown in Figure 3.23 due to the applied load indicated. Hence, draw the bending moment diagram for the arch.

Solution

The equation of the arch rib is:

$$y = x(l - x)/l$$

The reactions at support 2 are obtained by considering moment equilibrium about support 1. Hence:

$$M_1 = 0$$
$$= 15V_2 + 3.75H_2 - 5 \times 40 \sin 45° - 3.75 \times 40 \cos 45°$$

and:

$$= 15V_2 + 3.75H_2 - 247.49$$

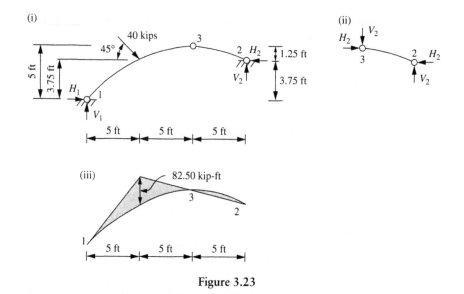

Figure 3.23

Resolving forces vertically gives:

$$0 = V_1 + V_2 - 40 \sin 45°$$
$$= V_1 + V_2 - 28.28$$

Resolving forces horizontally gives:

$$0 = H_1 - H_2 + 40 \cos 45°$$
$$= H_1 - H_2 + 28.28$$

Considering moment equilibrium about the crown hinge at 3 for the free body diagram of section 23 of the arch shown at (ii) gives:

$$M_3 = 0$$
$$= 5V_2 - 1.25H_2$$

Solving these equations simultaneously gives:

$$H_1 = 4.72$$
$$H_2 = 33.00$$
$$V_1 = 20.03$$
$$V_2 = 8.25$$

The maximum moment occurs at the location of the applied load and is:

$$M_{max} = 10V_2$$
$$= 10 \times 8.25$$
$$= 82.50 \text{ kip-ft}$$

The bending moment is shown at (iii), drawn on the compression side of the arch.

Supplementary problems

S3.1 Determine whether the beams shown in Figure S3.1 are determinate, indeterminate, or unstable.

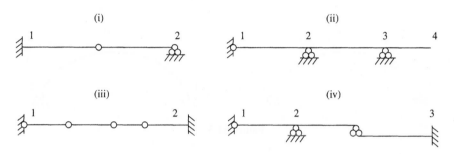

Figure S3.1

S3.2 Determine whether the rigid frames shown in Figure S3.2 are determinate, indeterminate, or unstable.

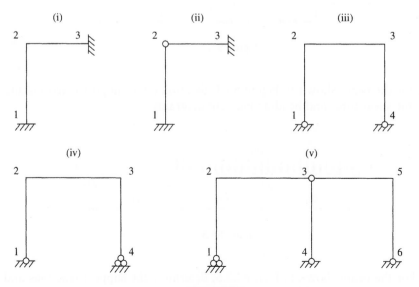

Figure S3.2

S3.3 Determine whether the arch structures shown in Figure S3.3 are determinate, indeterminate or unstable.

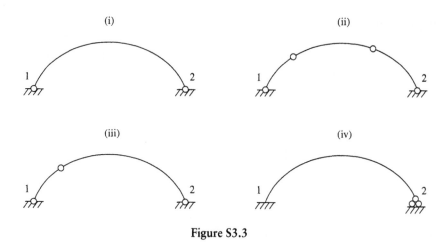

Figure S3.3

S3.4 For the beam shown in Figure S3.4, determine the support reactions and draw the shear force and bending moment diagrams.

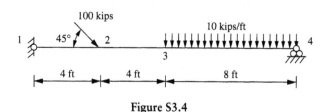

Figure S3.4

S3.5 For the beam shown in Figure S3.5, determine the support reactions and draw the shear force and bending moment diagrams.

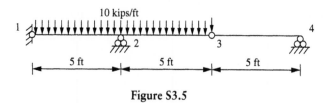

Figure S3.5

S3.6 For the beam shown in Figure S3.6, determine the support reactions and draw the shear force and bending moment diagrams.

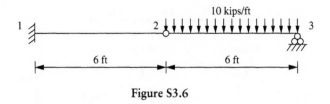

Figure S3.6

S3.7 For the frame shown in Figure S3.7, determine the support reactions and draw the shear force and bending moment diagrams.

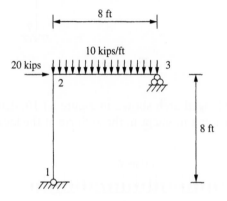

Figure S3.7

S3.8 For the frame shown in Figure S3.8, determine the support reactions and draw the shear force and bending moment diagrams.

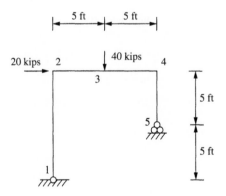

Figure S3.8

S3.9 For the frame shown in Figure S3.9, determine the support reactions and draw the shear force and bending moment diagrams.

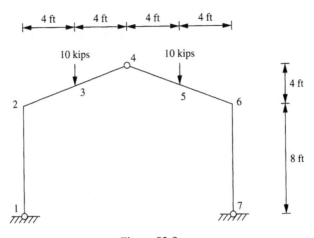

Figure S3.9

S3.10 For the three-hinged arch shown in Figure S3.10, determine the support reactions and the bending moment in the arch rib at the location $x = 8\,\text{ft}$.

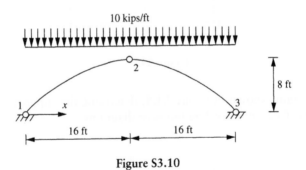

Figure S3.10

4 Elastic deformations

Notation

A	cross–sectional area of a member
c	rise of an arch
C	constant of integration
E	modulus of elasticity
F	force
G	modulus of torsional rigidity
h	height of a rigid frame
H	horizontal force
I	moment of inertia
l	length of member
m	bending moment in a member due to a unit virtual load
M	bending moment in a member due to the applied loads
M'	bending moment in a conjugate member due to the elastic load
P	axial force in a member due to the applied load
q	shear force in a member due to a unit virtual load
Q	shear force in a member due to the applied load
Q'	shear force in a conjugate member due to the elastic load
R	redundant force in a member, radius of curvature
u	axial force in a member due to a unit virtual load
V	vertical force
w	intensity of applied distributed load on a member
w'	intensity of elastic load on a conjugate member, expressed as M/EI
W	concentrated load, applied load on a member, expressed as $\int w \, dx$
W'	elastic load on a conjugate member, expressed as $\int M \, dx/EI$
x	horizontal deflection
y	vertical deflection
δ	deflection due to the applied load
δx	element of length of a member
$\delta\theta$	relative rotation between two sections in a member due to the applied loads
θ	rotation due to the applied loads
μ	form factor in shear

4.1   Deflection of beams

(a) Macaulay's method

Macaulay's method provides a simple and convenient method for determining the deflection of beams. It may be used to obtain an expression for the entire elastic curve over the whole length of a beam.

A small element δs of a beam is shown in Figure 4.1 and is assumed to be bent in the shape of an arc of a circle of radius R. The slope of the elastic curve

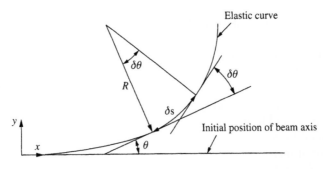

Figure 4.1

at one end of the element is θ. The change in slope of the elastic curve over the length of the element is $\delta\theta$, and the curvature, or rate of change of slope, over the element is:

$$\delta\theta/\delta s = 1/R$$
$$\approx \delta\theta/\delta x \ \dots \text{ positive as shown}$$

The slope of the beam is positive as shown and for small displacements is given by:

$$\theta \approx \tan\theta$$
$$\approx \delta y/\delta x$$

In the limit:

$$d\theta/dx = 1/R$$
$$\theta = dy/dx$$

and:

$$d\theta/dx = d/dx(dy/dx)$$
$$= d^2y/dx^2$$
$$= 1/R$$
$$= M/EI$$

and:

$$EI \ d^2y/dx^2 = M$$

Hence, by setting up an expression for M in terms of the applied loads on a beam and x and integrating this expression twice, an equation is obtained for the deflection of the beam.

Thus, the curvature of the elastic curve is given by the expression:

$$d^2y/dx^2 = 1/R$$
$$= M(x)/EI$$

The slope of the elastic curve is given by the expression:

$$dy/dx = \theta$$
$$= \int M(x)/EI + C_1$$

The deflection of the elastic curve is given by the expression:

$$y = \delta$$
$$= \iint M(x)/EI + C_1 x + C_2$$

where:

$$C_1 = \text{constant of integration}$$
$$C_2 = \text{constant of integration}$$
$$M(x) = \text{bending moment at any point in the beam in terms of } x$$
$$R = \text{radius of curvature}$$

The expression for the bending moment at any point in the cantilever shown in Figure 4.2 is:

$$EI \ d^2y/dx^2 = M$$
$$= -W(l - x) \ \dots \ \text{tension in the top fiber of the cantilever}$$
$$= -Wl + Wx$$

Integrating this expression with respect to x gives:

$$EI \ dy/dx = -Wlx + Wx^2/2 + C_1$$

where:

$$C_1 = \text{constant of integration}$$
$$= 0$$

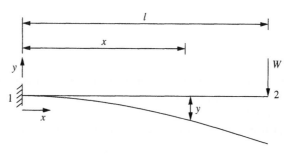

Figure 4.2

since:

$$dy/dx = 0 \text{ at } x = 0$$

Hence, the slope of the elastic curve is given by the expression:

$$EI\,dy/dx = -Wlx + Wx^2/2$$

Integrating this expression with respect to x gives

$$EIy = -Wlx^2/2 + Wx^3/6 + C_2$$

where:

$$C_2 = \text{constant of integration}$$
$$= 0$$

since:

$$y = 0 \text{ at } x = 0$$

Hence, the deflection of the elastic curve is given by the expression:

$$EIy = -Wlx^2/2 + Wx^3/6$$

At $x = l$ the deflection of the cantilever is:

$$y = -Wl^3/2EI + Wl^3/6EI$$
$$= -Wl^3/3EI \ldots \text{ downward}$$

The expression for the bending moment at any point in the cantilever shown in Figure 4.3 is:

$$EI\,d^2y/dx^2 = M$$
$$= -W(a - x) - W[x - a]$$

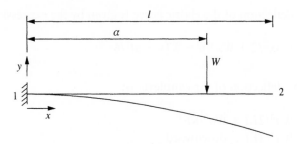

Figure 4.3

The term $[x - a]$ is valid only when positive: i.e., when $x > a$.
Hence:

$$EI \, d^2y/dx^2 = -Wa + Wx - W[x - a]$$

Integrating this expression with respect to x for the terms outside the brackets and with respect to $(x - a)$ for the term in brackets gives:

$$EI \, dy/dx = -Wax + Wx^2/2 - W[x - a]^2/2 + C_1$$

where:

C_1 = constant of integration
  = 0

since:

$dy/dx = 0$ at $x = 0$

Hence, the slope of the elastic curve is given by the expression:

$$EI \, dy/dx = -Wax + Wx^2/2 - W[x - a]^2/2$$

Integrating this expression gives:

$$EIy = -Wax^2/2 + Wx^3/6 - W[x - a]^3/6 + C_2$$

where:

C_2 = constant of integration
  = 0

since:

$y = 0$ at $x = 0$

Hence, the deflection of the elastic curve is given by the expression:

$$EIy = -Wax^2/2 + Wx^3/6 - W[x - a]^3/6$$

At $x = a$ the deflection of the cantilever is:

$$y = -Wa^3/2EI + Wa^3/6EI$$
$$= -Wa^3/3EI \ldots \text{downward}$$

At $x = l$ the deflection of the cantilever is:

$$y = -Wal^2/2EI + Wl^3/6EI - W(l - a)^3/6EI$$
$$= -Wal^2/2EI + Wl^3/6EI - W(l^3 - 3al^2 + 3a^2l - a^3)/6EI$$
$$= -Wa^2(3l - a)/6EI \ldots \text{downward}$$

Example 4.1

For the simply supported beam shown in Figure 4.4, determine the maximum deflection due to the applied load. The flexural rigidity of the beam is $EI = 7 \times 10^6$ kip in^2.

Solution

The expression for the bending moment at any point in the beam shown in Figure 4.4 is:

$$EI \, d^2y/dx^2 = V_1x - W[x - a]$$
$$= 20x - 60[x - 12]$$

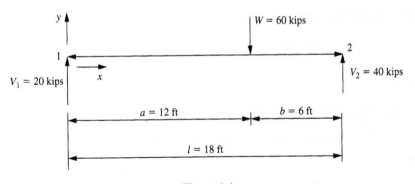

Figure 4.4

Integrating this expression with respect to x for the terms outside the brackets and with respect to $(x - a)$ for the term in brackets gives an expression for the slope of the elastic curve of:

$$EI\ dy/dx = 10x^2 - 30[x - 12]^2 + C_1$$

where:

C_1 = constant of integration

Integrating this expression gives an expression for the deflection of the elastic curve of:

$$EIy = 10x^3/3 - 10[x - 12]^3 + C_1 x + C_2$$

where:

C_2 = constant of integration

At

$x = 0, y = 0$

and:

$C_2 = 0$

At

$x = 18\ \text{ft}, y = 0$

and:

$$0 = 10 \times 18^3/3 - 10(18 - 12)^3 + C_1 \times 18$$

then

$C_1 = -960\ \text{kip ft}^2$

Hence, the deflection of the elastic curve is given by the expression:

$$EIy = 10x^3/3 - 10[x - 12]^3 - 960x$$

The maximum deflection occurs where the slope of the elastic curve is zero and:

$$EI \, dy/dx = 0$$
$$= 10x^2 - 960$$

and:

$$x = 9.80 \text{ ft}$$

The maximum deflection is given by:

$$EIy_{max} = 10 \times 9.8^3/3 - 960 \times 9.8$$
$$= -6273$$
$$y_{max} = -6273 \times 144/(7 \times 10^6)$$
$$= -0.129 \text{ ft}$$
$$= -1.55 \text{ in} \dots \text{ downward}$$

(b) Virtual work method

The virtual work, or unit-load, method may be used to obtain the displacement of a single point in a beam. The principle may be defined as follows: if a structure in equilibrium under a system of applied forces is subjected to a system of displacements compatible with the external restraints and the geometry of the structure, the total work done by the applied forces during these external displacements equals the work done by the internal forces, corresponding to the applied forces, during the internal deformations, corresponding to the external displacements. The expression "virtual work" signifies that the work done is the product of a real loading system and imaginary displacements or an imaginary loading system and real displacements.

To the cantilever shown in Figure 4.5 (i), the external loads W are gradually applied. This results in the deflection of any point 3 a distance δ, while each load moves a distance y in its line of action. The loading produces a bending moment M and a relative rotation $\delta\theta$ to the ends of the element shown at (ii). From the principle of conservation of energy and ignoring the effects of axial and shear forces, the external work done during the application of the loads must equal the internal energy stored in the beam.

Then:

$$\sum Wy/2 = \sum M \delta\theta/2 \dots \tag{1}$$

To the unloaded structure a unit virtual load is applied at 3 in the direction of δ as shown at (iii). This results in a bending moment m in the element.

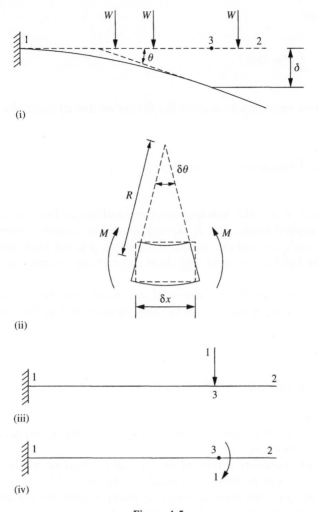

(i)

(ii)

(iii)

(iv)

Figure 4.5

Now, while the virtual load is still in position, the real loads W are gradually applied to the structure. Again equating external work and internal energy:

$$\sum Wy/2 + 1 \times \delta = \sum M\delta\theta/2 + \sum m\delta\theta \dots \tag{2}$$

Subtracting expression (1) from expression (2):

$$\begin{aligned}
1 \times \delta &= \sum m\delta\theta \\
&= \sum m\delta x/R \\
&= \sum m\delta x/EI
\end{aligned}$$

In the limit:

$$1 \times \delta = \int Mm\,dx/EI$$

If it becomes necessary to include the deflection due to shear, the expression becomes:

$$1 \times \delta = \int Mm\,dx/EI + \int Vv\,dx/\mu AG$$

where M and V are the bending moment and shear force at any section due to the applied loads, and I, G, and A are the second moment of area, the rigidity modulus, and the area of the section; μ is the form factor; and m and v are the bending moment and shear force at any section due to the unit virtual load.

In a similar manner, the rotation θ of any point 3 of the structure may be obtained by applying a unit virtual bending moment at 3 in the direction of θ, as shown at (iv).

Then:

$$1 \times \theta = \int Mm\,dx/EI + \int Vv\,dx/\mu AG$$

where m and v are the bending moment and shear force at any section due to the unit virtual moment.

For a beam, moments produced by the virtual load or moment are considered positive, and moments produced by the applied loads, which are of opposite sense, are considered negative. A positive value for the displacement indicates that the displacement is in the same direction as the virtual force or moment.

The deflection and slope at the free end of the cantilever shown in Figure 4.6 may be obtained by the virtual work method. Taking the origin of coordinates at point 3, the expression for the bending moment due to the applied load is obtained from Figure 4.6 (ii) as:

$$M = Wx$$

A unit vertical load is applied at the end of the cantilever and the function m derived from (iii) as:

$$m = l - a + x$$

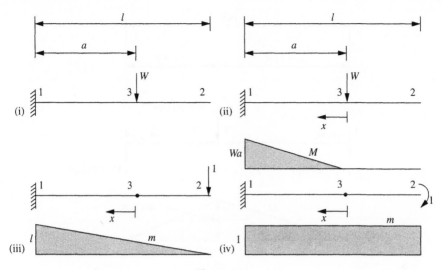

Figure 4.6

The vertical deflection at 2 is given by:

$$1 \times \delta = \int_0^a Mm \, dx/EI$$
$$= W \int_0^a x(l - a + x) \, dx/EI$$
$$= W[x^2l/2 - x^2a/2 + x^3/3]_0^a/EI$$
$$= W(a^2l/2 - a^3/2 + a^3/3)/EI$$
$$= Wa^2(3l - a)/6EI \dots \text{downward}$$

To determine the slope at the end of the cantilever, a unit clockwise rotation is applied at the end of the cantilever and the function m derived from (iv) as:

$$m = 1$$

The slope at 2 is given by:

$$1 \times \theta = \int_0^a Mm \, dx/EI$$
$$= W \int_0^a x \, dx/EI$$
$$= Wa^2/2EI \dots \text{clockwise}$$

Example 4.2

For the simply supported beam shown in Figure 4.7, determine the deflection at the location of the applied load. The flexural rigidity of the beam is $EI = 7 \times 10^6$ kip in^2.

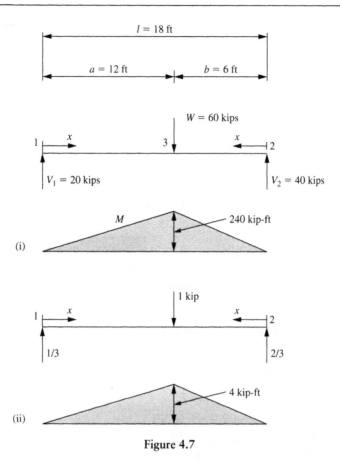

Figure 4.7

Solution

Taking the origin of coordinates at end 1 and end 2 in turn, the expressions for the bending moment due to the applied load are obtained from Figure 4.7(i) as:

$$M = V_1 x$$
$$= 20x$$

and:

$$M = V_2 x$$
$$= 40x$$

A unit vertical load is applied at 3 and the corresponding functions for m derived from (ii) as:

$$m = x/3$$

and:

$$m = 2x/3$$

The vertical deflection at 3 is given by:

$$1 \times \delta = \int_0^a Mm\,dx/EI + \int_0^b Mm\,dx/EI$$
$$= 20 \int_0^a x^2 \, dx/3EI + 80 \int_0^b x^2 \, dx/3EI$$
$$= 20[x^3/9EI]_0^{12} + 80[x^3/9EI]_0^6$$
$$= 3840/EI + 1920/EI$$
$$= 5760 \times 144/(7 \times 10^6)$$
$$= 0.118 \text{ ft}$$
$$= 1.422 \text{ in} \dots \text{ downward}$$

As an alternative to evaluating the integrals in the virtual work method, advantage may be taken of the volume integration method. For a straight prismatic member:

$$\int Mm\,dx/EI = \int Mm\,dx \times 1/EI.$$

The function m is always either constant along the length of the member or varies linearly. The function M may vary linearly for real concentrated loads or parabolically for real distributed loads. Thus, $\int Mm\,dx$ may be regarded as the volume of a solid with a cross-section defined by the function M and a height defined by the function m. The volume of this solid is given by the area of cross-section multiplied by the height of the solid at the centroid of the cross-section.

Commonly occurring values of $\int Mm\,dx$ are provided in Part 2, Chapter 2, Table 2.3 for various types of functions M and m.

Example 4.3

Determine the slope at the free end of the cantilever shown in Figure 4.8 using the volume integration method.

Solution

The functions M and m derived from Figure 4.8 (i) and (ii) and the solid defined by these functions are shown at (iii). The slope at the end of the cantilever is given by the volume of this solid × 1/EI, which is:

$$\theta = Wa \times a/2 \times 1 \times 1/EI$$
$$= Wa^2/2EI$$

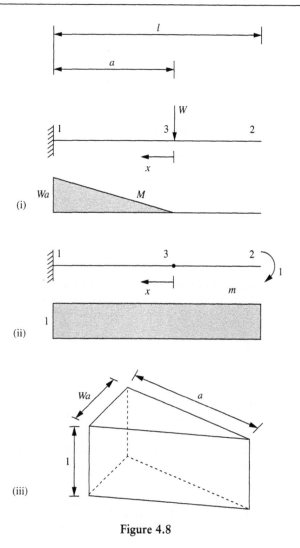

Figure 4.8

Alternatively, the value of $\int Mm\,dx/EI$ may be obtained directly from Part 2, Table 2.3, and is given by:

$$\theta = a \times 1 \times Wa/2EI$$
$$= Wa^2/2EI$$

4.2 Deflection of rigid frames

(a) Virtual work method

The virtual work method may be applied to rigid frames to obtain the displacement at a specific point on the frame. Integration is carried out over all members of the frame and the values summed to provide the required displacement.

To determine the horizontal deflection of node 2 for the frame shown in Figure 4.9 (i), use is made of the bending moment diagrams produced by the applied load and by a unit virtual horizontal load at node 2. These are shown at (ii) and (iv). Member 34 has zero bending moment under both loading cases.

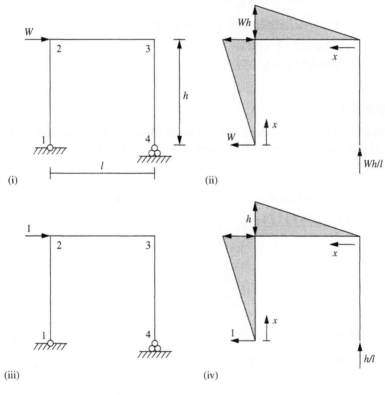

Figure 4.9

With the origin of coordinates as indicated, expressions for the bending moment produced by the applied load are obtained from Figure 4.9 (ii) as:

$$M = Wx \text{ ... member 12}$$

and:

$$M = Whx/l \text{ ... member 23}$$

A unit horizontal load is applied at node 2, as shown at (iii), and the corresponding functions for m derived from (iv) as:

$$m = x \text{ ... member 12}$$

and:

$m = hx/l$ ... member 23

The horizontal deflection at node 2 is obtained by summing the values for members 12 and 23 to give:

$$1 \times \delta = \int_0^h Mm \, dx/EI + \int_0^l Mm \, dx/EI$$
$$= W \int_0^h x^2 \, dx/EI + W \int_0^l h^2 x^2 \, dx/l^2 EI$$
$$= W[x^3/3EI]_0^h + W[h^2 x^3/3l^2 EI]_0^l$$
$$= W(h^3/3EI + h^2 l/3EI)$$

Example 4.4

Determine the horizontal deflection of node 2 for the frame shown in Figure 4.10. The flexural rigidity of the members is $EI = 30 \times 10^6$ kip in^2.

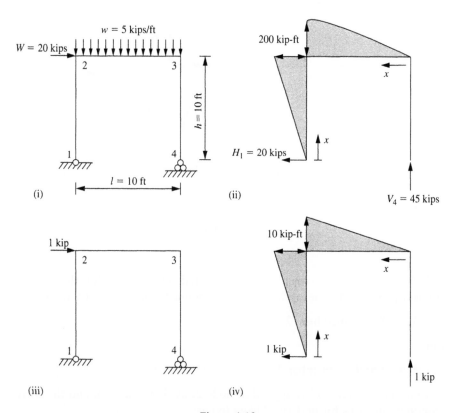

Figure 4.10

Solution

The bending moments produced in the members of the frame by the applied loads shown in Figure 4.10 (i) are obtained from (ii). With the origin of coordinates as indicated, the expressions for the moments are:

$$M = H_1 x \ \dots \text{ member } 12$$
$$= 20x$$

and:

$$M = V_4 x - wx^2/2 \ \dots \text{ member } 23$$
$$= 45x - 5x^2/2$$

A unit horizontal load is applied at node 2, as shown at (iii), and the corresponding functions for m derived from (iv) as:

$$m = x \ \dots \text{ member } 12$$

and:

$$m = hx/l \ \dots \text{ member } 23$$
$$= x$$

No moments are produced in member 34, and the horizontal deflection at node 2 is obtained by summing the values for members 12 and 23 to give:

$$1 \times \delta = \int_0^h Mm\,dx/EI + \int_0^l Mm\,dx/EI$$
$$= 20 \int_0^{10} x^2\,dx/EI + 5\int_0^{10}(9x^2 - x^3/2)dx/EI$$
$$= 20[x^3/3EI]_0^{10} + 5[3x^3/EI - x^4/8EI]_0^{10}$$
$$= 15,417 \times 144/(30 \times 10^6)$$
$$= 0.074 \text{ ft}$$
$$= 0.89 \text{ in}$$

(b) Conjugate beam method

The conjugate beam method may be used to obtain an expression for the entire elastic curve over the whole of a rigid frame or beam.

The beam shown in Figure 4.11 (i) is subjected to a distributed applied load of intensity w, positive when acting upward as indicated. The shear force at any section is given by the area under the load intensity curve as:

$$Q = \int w\,dx$$

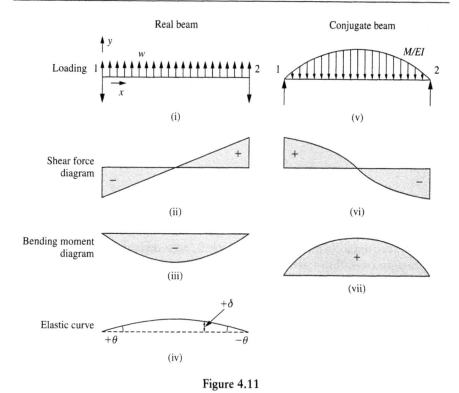

Figure 4.11

with shear force upward on the left of a section regarded as positive. The shear force diagram is shown at (ii) and is negative at end 1 and positive at end 2.

The bending moment at any section is given by the area under the shear force curve as:

$$M = \int Q \, dx$$
$$= \iint w \, dx$$

with bending moment producing tension in the bottom fiber regarded as positive. The bending moment diagram is shown at (iii) and is negative as indicated.

In addition, the curvature of the elastic curve at any section is given by:

$$d^2 y / dx^2 = d\theta / dx$$
$$= 1/R$$
$$= M/EI$$

and the slope and deflection at any section are given by:

$$dy/dx = \theta$$
$$= \int M\,dx/EI$$
$$y = \delta$$
$$= \int \theta\,dx$$
$$= \iint M\,dx/EI$$

with x positive to the right and y positive upward. The elastic curve is shown at (iv) with deflections positive (i.e., upward) as indicated and with a positive slope (i.e., counterclockwise rotation) at end 1 and a negative slope (i.e., clockwise rotation) at end 2.

An analogous beam known as the conjugate beam and of the same length as the real beam, as shown in Figure 4.11 (v), is subjected to an applied loading of intensity:

$$w' = M/EI$$

where M is the bending moment in the actual beam at any section and $\int M\,dx/EI$ is known as the elastic load, W'. The elastic load acts in a positive direction (upward) when the bending moment in the real beam is positive. The loading diagram is shown at (v) and is negative as indicated.

Then, the shear at any section in the conjugate beam is given by:

$$Q' = \int w'\,dx$$
$$= \int M\,dx/EI$$
$$= \theta$$

where θ is the slope at the corresponding section in the real beam. The shear force diagram is shown at (vi) and is positive at end 1 and negative at end 2. Hence, as shown at (iv), the slope of the elastic curve is positive at end 1 and negative at end 2.

The bending moment at any section in the conjugate beam is given by:

$$M' = \int Q'\,dx$$
$$= \iint M\,dx/EI$$
$$= \delta$$

where δ is the deflection at the corresponding section in the real beam. The bending moment diagram is shown at (vii) and is positive as indicated. Hence, as shown at (iv), the deflection of the elastic curve is positive.

Thus, the slope and deflection at any section in the real beam are given by the shear and bending moment at the corresponding section in the conjugate beam, and the elastic curve of the real beam is given by the bending moment diagram of the conjugate beam. The end slope and end deflection of the real beam are given by the end reaction and end moment of the conjugate beam. The maximum deflection in the real beam occurs at the position of zero shear in the conjugate beam.

In the case of frames, the elastic load applied to the conjugate frame is positive (i.e., acts vertically upward) when the outside fiber of the real frame is in compression. Then, the displacement of the real frame at any section is perpendicular to the lever arm used to determine the moment in the conjugate frame and is outward when a positive bending moment occurs at the corresponding section of the conjugate frame.

The restraints of the conjugate structure must be consistent with the displacements of the real structure. Details of the necessary restraints in the conjugate structure are provided in Part 2, Chapter 4, Table 4.1.

At a simple end support in a real structure, there is a rotation but no deflection. Thus, the corresponding restraints in the conjugate structure must be a shear force and a zero moment, which are produced by a simple end support in the conjugate structure.

At a fixed end in a real structure, there is neither a rotation nor a deflection. Thus, there must be no restraint at the corresponding point in the conjugate structure, which must be a free end.

At a free end in a real structure, there is both a rotation and a deflection. Thus, the corresponding restraints in the conjugate structure are a shear force and a bending moment, which are produced at a fixed end.

At an interior support in a real structure, there is no deflection and a smooth change in slope. Thus, there can be no moment and no reaction at the corresponding point in the conjugate structure, which must be an unsupported hinge.

At an interior hinge in a real structure, there is a deflection and an abrupt change of slope. Thus, the corresponding restraints in the conjugate structure are a moment and a reaction, which are produced by an interior support.

Example 4.5

Determine the rotation of nodes 1 and 4 and the horizontal deflection of node 2 for the frame shown in Figure 4.12. The flexural rigidity of the members is $EI = 30 \times 10^6 \, \text{kip in}^2$.

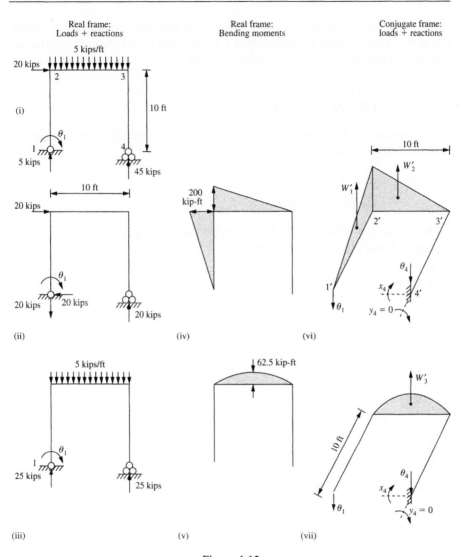

Figure 4.12

Solution

It is convenient to utilize the principle of superposition and consider the distributed load and the lateral load separately.

The lateral load is applied to the frame at (ii), which results in the bending moment diagram, drawn on the compression side of the members, shown at (iv). The elastic loads are applied to the conjugate frame at (vi) and are given by:

$$W'_1 = W'_2$$
$$= 0.5 \times 10 \times 200/EI$$
$$= 1000/EI \dots \text{upward}$$

Since the vertical deflection at 4 is zero, the bending moment at 4' about axis 3'4' is zero. Hence, the rotation at 1 is obtained by taking moments in the conjugate frame about axis 3'4' and is given by:

$$\theta_1 = -(20/3 \times W'_2 + 10 \times W'_1)/10$$
$$= -1667/EI \ldots \text{clockwise}$$

The rotation at 4 is obtained by resolving forces vertically in the conjugate frame and is given by:

$$\theta_4 = -W'_2 - W'_1 - \theta_1$$
$$= -333/EI \ldots \text{clockwise}$$

The horizontal deflection at 2 is given by the bending moment at 2' in the conjugate frame about axis 2'3', which is:

$$\delta_2 = -10\theta_1 - 10/3 \times W'_1$$
$$= 13,336/EI \ldots \text{to the right}$$

The distributed load is applied to the frame at (iii), which results in the bending moment diagram, drawn on the compression side of the members, shown at (v). The elastic load is applied to the conjugate frame at (vii) and is given by:

$$W'_3 = 0.667 \times 10 \times 62.5/EI$$
$$= 416.9/EI \ldots \text{upward}$$

Since the vertical deflection at 4 is zero, the bending moment at 4' about axis 3'4' is zero. Hence, the rotation at 1 is obtained by taking moments in the conjugate frame about axis 3'4' and is given by:

$$\theta_1 = -(10/2 \times W'_3)/10$$
$$= -208.4/EI \ldots \text{clockwise}$$

The rotation at 4 is obtained by resolving forces vertically in the conjugate frame and is given by:

$$\theta_4 = -W'_3 - \theta_1$$
$$= -208.4/EI \ldots \text{clockwise}$$

The horizontal deflection at 2 is given by the bending moment in the conjugate frame about axis 2'3', which is:

$$\delta_2 = -10\theta_1$$
$$= 2084/EI \ldots \text{to the right}$$

The total horizontal deflection at 2 is obtained by summing the individual values calculated for the lateral load and the distributed load and is given by:

$$\delta_2 = 13,336/EI + 2084/EI$$
$$= 15420 \times 144/(30 \times 10^6)$$
$$= 0.074 \text{ ft}$$
$$= 0.89 \text{ in} \dots \text{to the right}$$

The total rotation at 1 is obtained by summing the individual values for the lateral load and the distributed load and is given by:

$$\theta_1 = -1667/EI - 208.4/EI$$
$$= -1875.4 \times 144/(30 \times 10^6)$$
$$= -0.0090 \text{ rad} \dots \text{clockwise}$$

The total rotation at 4 is obtained by summing the individual values for the lateral load and the distributed load and is given by:

$$\theta_4 = -333/EI - 208.4/EI$$
$$= -541.4 \times 144/(30 \times 10^6)$$
$$= -0.00260 \text{ rad} \dots \text{clockwise}$$

4.3 Deflection of pin-jointed frames

The virtual work method may be applied to pin-jointed frames to obtain the displacement at a node on the frame.

To the pin-jointed frame shown in Figure 4.13 (i), the external loads W are gradually applied. This results in the deflection of any node 3 a distance δ,

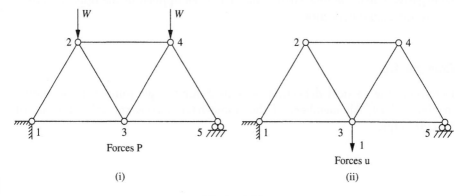

Figure 4.13

while each load moves a distance y in its line of action. The loading produces an internal force P and an extension δl in any member of the frame. The external work done during the application of the loads must equal the internal energy stored in the structure, from the principle of conservation of energy. Then:

$$\sum Wy/2 = \sum P\,\delta l/2 \ \ldots \ (1)$$

To the unloaded structure a unit virtual load is applied at node 3 in the direction of δ, as shown in Figure 4.13 (ii). This results in a force u in any member.

Now, while the virtual load is still in position, the real loads W are gradually applied to the structure. Again equating external work and internal energy:

$$\sum Wy/2 + 1 \times \delta = \sum P\,\delta l/2 + \sum u \delta l \ \ldots \ (2)$$

Subtracting expression (1) from expression (2):

$$1 \times \delta = \sum u\,\delta l$$
$$= \sum Pul/AE$$

where P is the internal force in a member due to the applied loads and l, A and E are its length, area, and modulus of elasticity, and u is the internal force in a member due to the unit virtual load.

For a pin-jointed frame, tensile forces are considered positive and compressive forces negative. Increase in the length of a member is considered positive and decrease in length negative. The unit virtual load is applied to the frame in the anticipated direction of the deflection. If the assumed direction is correct, the deflection obtained will have a positive value. The deflection obtained will be negative when the unit virtual load has been applied in the opposite direction to the actual deflection.

Example 4.6

Determine the vertical deflection of node 5 for the pin-jointed frame shown in Figure 4.14. All members of the frame have a constant value for AE of 100,000 kips.

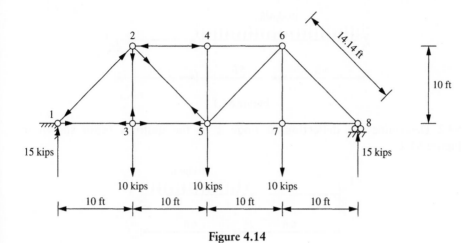

Figure 4.14

Solution

Because of the symmetry in the structure and loading, only half of the members need be considered. The member forces P due to the applied loads and u due to the unit virtual load are tabulated in Table 4.1.

Table 4.1 Determination of forces and displacements in Example 4.6

Member	P	l	u	Pul
12	−21.21	14.14	−0.707	212
23	10.00	10.00	0	0
13	15.00	10.00	0.500	75
45	0.00	10.00	0	0
24	−20.00	10.00	−1.000	200
25	7.07	14.14	0.707	71
35	15.00	10.00	0.500	75
Total				633

The vertical deflection is given by:

$$\delta_5 = \sum Pul/AE$$
$$= 2 \times 633 \times 12/100,000$$
$$= 0.15 \text{ in downward}$$

Supplementary problems

S4.1 Determine the rotations at nodes 1 and 2 and the deflection at node 3 of the uniform beam shown in Figure S4.1.

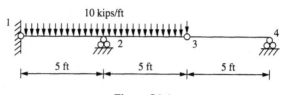

Figure S4.1

S4.2 Determine the deflection at node 2 of the uniform beam shown in Figure S4.2.

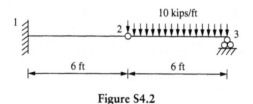

Figure S4.2

S4.3 Determine the rotation at node 1 and the deflection at node 2 of the uniform beam shown in Figure S4.3.

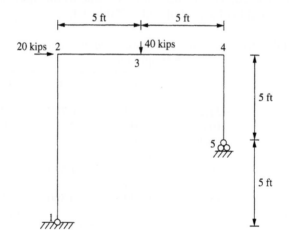

Figure S4.3

S4.4 Determine the equation of the elastic curve for the uniform beam shown in Figure S4.4. Determine the location of the maximum deflection in span 12 and the magnitude of the maximum deflection. Determine the deflection at node 3.

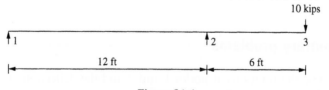

Figure S4.4

S4.5 Determine the deflection at node 3 of the uniform beam shown in Figure S4.5.

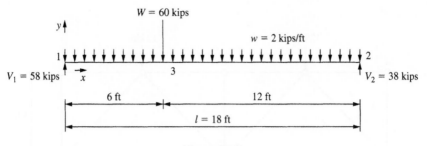

Figure S4.5

S4.6 Determine the deflection at node 3 of the pin-jointed truss shown in Figure S4.6.

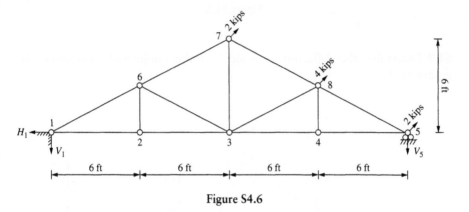

Figure S4.6

S4.7 Determine the deflection at node 4 of the pin-jointed truss shown in Figure S4.7.

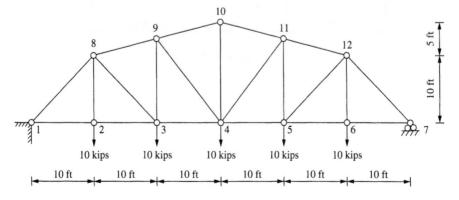

Figure S4.7

S4.8 Determine the deflection at node 2 of the pin-jointed truss shown in Figure S4.8.

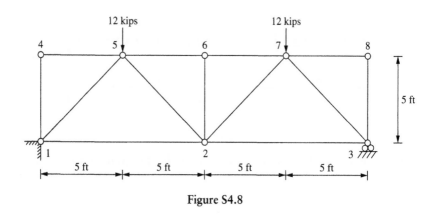

Figure S4.8

S4.9 Determine the deflection at node 3 of the pin-jointed truss shown in Figure S4.9.

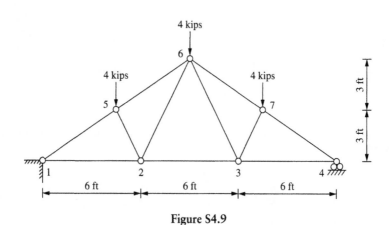

Figure S4.9

S4.10 Determine the deflection at node 2 of the pin-jointed truss shown in Figure S4.10.

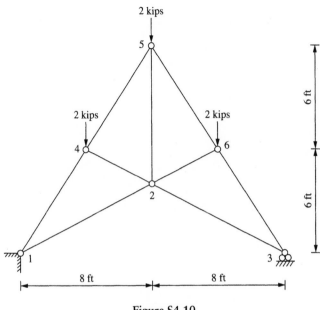

Figure S4.10

Problem 3.11

5 Influence lines

Notation

A	cross-sectional area of a member
E	modulus of elasticity
I	second moment of area of a member
l	length of a member
M	bending moment in a member due to the applied loads
P	axial force in a member due to the applied loads
Q	shear force in a member due to the applied loads
V	vertical reaction
W	applied load
δ	deflection
θ	rotation

5.1 Introduction

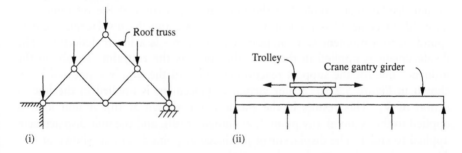

Figure 5.1

The maximum design force in a member of a structure subjected to a system of stationary loads is readily determined. The static loads are applied to the structure as shown in Figure 5.1 (i) and the member forces calculated. However, a member such as a bridge girder or a crane gantry girder are subjected to moving loads, and the maximum design force in the member depends on the location of the moving load. As shown in Figure 5.1 (ii), this involves the trial and error positioning of the crane loads on the girder to determine the most critical location. Alternatively, an influence line may be utilized to determine the location of the moving load that produces the maximum design force in a member.

5.2 Construction of influence lines

An influence line for a member is a graph representing the variation in shear, moment or force in the member due to a unit load traversing a structure. The construction of an influence line may be obtained by the application of Müller-Breslau's principle.

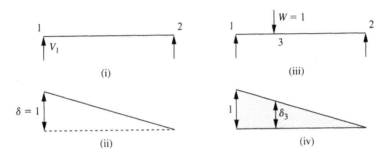

Figure 5.2

In accordance with Müller-Breslau's principle, the influence line for any constraint in a structure is the elastic curve produced by the corresponding unit virtual displacement applied at the point of application of the restraint. The term "displacement" is used in its general sense, and the displacement corresponding to a moment is a rotation and to a force is a linear deflection. The displacement is applied in the same direction as the restraint. To obtain the influence line for the support reaction at end 1 of the simply supported beam shown in Figure 5.2 (i), a unit virtual displacement is applied in the line of action of V_1. This results in the elastic curve shown at (ii). A unit load is applied to the beam at any point 3, as shown at (iii), and the unit displacement applied to end 1. The displacement produced at point 3 is δ_3, as shown at (iv). Then, applying the virtual work principle:

$$V_1 \times (\delta = 1) = (W = 1) \times \delta_3$$
$$V_1 = \delta_3$$

and the elastic curve is the influence line for V_1.

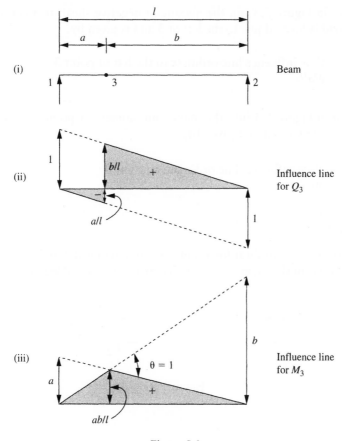

Figure 5.3

Similarly, as shown in Figure 5.3 (ii), the influence line for shear at point 3 is obtained by cutting the beam at 3 and displacing the cut ends a unit distance apart. The influence line for bending moment at 3 is produced by cutting the beam at 3 and imposing a unit virtual rotation, as shown at (iii).

5.3 Maximum effects

(a) Single concentrated load

A concentrated load W produces the maximum positive shear at point 3 in the beam shown in Figure 5.3 (i) when the load is located just to the right of 3. The shear is given by:

$$Q_{max} = W \times \text{influence line ordinate to the right of point 3}$$
$$= Wb/l$$

As shown in Figure 5.3 (ii), the maximum negative shear at point 3 occurs when the load is located just to the left of 3 and is given by:

$$Q_{min} = W \times \text{influence line ordinate to the left of point 3}$$
$$= Wa/l$$

As shown in Figure 5.3 (iii), the maximum moment at point 3 occurs when the load is located at 3 and is given by:

$$M_{max} = W \times \text{influence line ordinate at point 3}$$
$$= Wab/l$$

Example 5.1

Determine the maximum shear force and maximum moment at point 3 in the simply supported beam shown in Figure 5.4 due to a concentrated load of 10 kips.

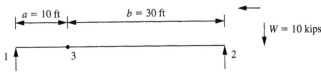

Figure 5.4

Solution

The influence line for shear is shown in Figure 5.3 (ii), and the maximum shear at point 3 occurs when the 10 kip load is immediately to the right of 3. The maximum shear is;

$$Q = Wb/l$$
$$= 10 \times 30/40$$
$$= 7.5 \text{ kips}$$

The influence line for moment is shown in Figure 5.3 (iii), and the maximum moment at point 3 occurs when the 10 kip load is at 3. The maximum moment is;

$$M = Wab/l$$
$$= 10 \times 10 \times 30/40$$
$$= 75 \text{ kip-ft}$$

(b) Uniformly distributed load

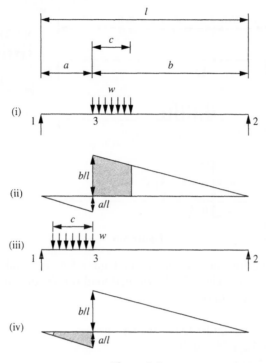

Figure 5.5

A uniformly distributed load w of length c is applied to the beam, as shown in Figure 5.5(i). This produces the maximum positive shear at point 3 when the distributed load is located just to the right of 3. As shown in (ii), the shear is given by;

$$Q_{max} = w \times \text{area under the influence line}$$

As shown at (iii) and (iv), the maximum negative shear at point 3 is given by;

$$Q_{min} = w \times \text{area under the influence line}$$

Example 5.2

Determine the maximum shear force at point 3 in the simply supported beam shown in Figure 5.6 due to a distributed load of 2 kips/ft over a length of 10 ft.

Solution

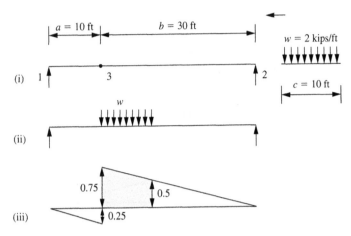

Figure 5.6

The influence line for shear is shown in Figure 5.6 (iii), and the maximum shear at point 3 occurs when the distributed load is immediately to the right of 3, as shown at (ii). The maximum shear is;

$$Q_{max} = w \times \text{area under the influence line}$$
$$= 2 \times 10(0.75 + 0.5)/2$$
$$= 12.5 \text{ kips}$$

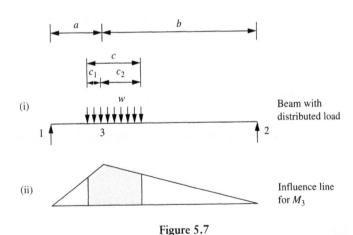

Figure 5.7

The maximum moment at point 3 is produced when point 3 divides the distributed load in the same ratio as it divides the span. As shown in Figure 5.7 (i), this occurs when:

$$c_1/c_2 = a/b$$

As shown at (ii), the maximum moment at point 3 is given by:

M_{max} = w × area under the influence line

Example 5.3

Determine the maximum bending moment at point 3 in the simply supported beam shown in Figure 5.8 due to a distributed load of 2 kips/ft over a length of 10 ft.

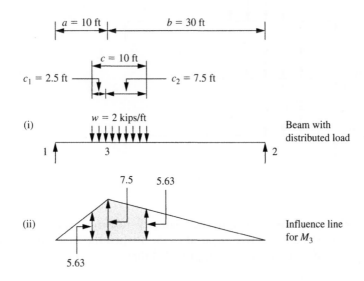

Figure 5.8

Solution

The maximum moment at point 3 is produced when point 3 divides the distributed load in the same ratio as it divides the span. As shown in Figure 5.8 (i), this occurs when:

c_1/c_2 = a/b
      = 10/30

and: c_1 = 2.5 ft
     c_2 = 7.5 ft

As shown at (ii), the maximum moment at point 3 is given by:

M_{max} = w × area under the influence line
      = 2[2.5(5.63 + 7.5)/2 + 7.5(5.63 + 7.5)/2]
      = 2 × 10(5.63 + 7.5)/2
      = 131 kip-ft

(c) Train of wheel loads

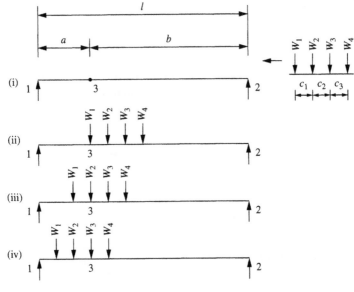

Figure 5.9

A train of wheel loads is applied to the beam, as shown in Figure 5.9 (i). This produces the maximum positive shear at point 3 when the first load is located just to the right of 3, as shown in (ii), provided that:

$$W_1/c_1 > \sum W/l$$

where: $\sum W$ = total weight of the wheel loads
l = span length

The maximum positive shear at point 3 occurs when the second load is located just to the right of 3, as shown in (iii), provided that:

$$W_2/c_2 > \sum W/l$$

The maximum positive shear at point 3 occurs when the third load is located just to the right of 3, as shown in (iv), provided that:

$$W_3/c_3 > \sum W/l$$

and so on.

Example 5.4

Determine the maximum shear at point 3 in the simply supported beam shown in Figure 5.10 due to the train of wheel loads indicated.

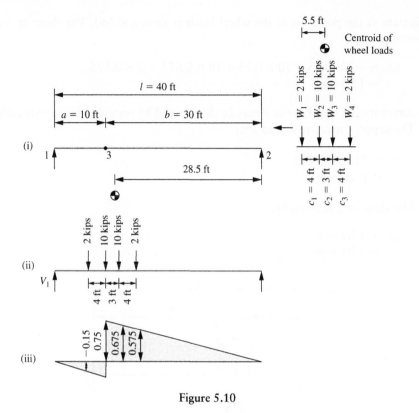

Figure 5.10

Solution

The ratio of total weight of wheel loads to span length is:

$$\sum W/l = 24/40$$
$$= 0.6 \text{ kips/ft}$$

Placing the first wheel at point 3 gives:

$$W_1/c_1 = 2/4$$
$$= 0.5 \cdots <0.6$$

Placing the second wheel at point 3, as shown in (ii), gives:

$$W_2/c_2 = 10/3$$
$$= 3.3 \cdots >0.6, \text{ governs}$$

Hence, the maximum shear occurs at point 3 when the second load is located immediately to the right of 3; the ordinates of the influence line diagram at the positions of the wheel loads is shown at (iii). The shear at 3 is given by:

$$Q_3 = -2 \times 0.15 + 10 \times 0.75 + 10 \times 0.675 + 2 \times 0.575$$
$$= 15.1 \text{ kips}$$

Alternatively, the shear at 3 may be determined by resolving forces vertically. The support reaction at end 1 is:

$$V_1 = 24 \times 28.5/40$$
$$= 17.1 \text{ kips}$$

The shear at 3 is given by:

$$Q_3 = 17.1 - 2$$
$$= 15.1 \text{ kips}$$

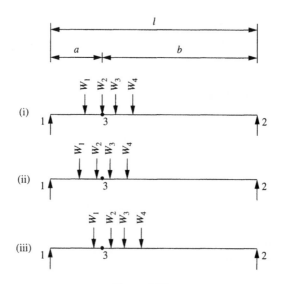

Figure 5.11

A train of wheel loads is applied to the beam, as shown in Figure 5.11 (i). This produces the maximum bending moment at a specific point 3 when a specified load is located at 3, as shown in (ii), such that if the load is moved to the left of 3, the intensity of loading on section 13 is greater than on section 23, but if it moves to the right of point 3, the intensity of loading on section 23 is greater than on section 13. The first requirement is shown at (ii) and is:

$$\sum W_L/a > \sum W_R/b$$

where: $\sum W_L = W_1 + W_2$
$\quad\quad\quad \sum W_R = W_3 + W_4$

The second requirement is shown at (iii) and is:

$$\sum W_R / b > \sum W_L / a$$

where: $\sum W_L = W_1$
$\quad\quad \sum W_R = W_2 + W_3 + W_4$

Example 5.5

Determine the maximum bending moment at point 3 in the simply supported beam shown in Figure 5.12 due to the train of wheel loads indicated.

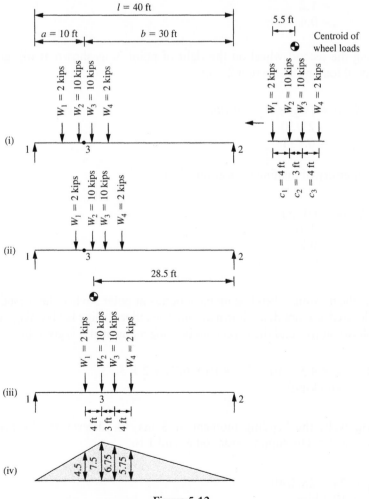

Figure 5.12

Solution

Placing the second wheel on the left of point 3, as shown at (i), gives an intensity of loading on section 23 of:

$$\sum W_R/b = (W_3 + W_4)/b$$
$$= 12/30$$
$$= 0.4$$

and an intensity of loading on section 13 of:

$$\sum W_L/a = (W_1 + W_2)/a$$
$$= 12/10$$
$$= 1.2$$
$$> 0.4$$

Placing the second wheel on the right of point 3, as shown at (ii), gives an intensity of loading on section 23 of:

$$\sum W_R/b = (W_2 + W_3 + W_4)/b$$
$$= 22/30$$
$$= 0.733$$

and an intensity of loading on section 13 of:

$$\sum W_L/a = (W_1)/a$$
$$= 2/10$$
$$= 0.2$$
$$< 0.733$$

Hence the maximum bending moment occurs at point 3 when the second wheel load is located at point 3, as shown at (iii). The influence line for bending moment at 3 is shown at (iv), and the maximum bending moment at 3 is given by:

$$M_3 = 2 \times 4.5 + 10 \times 7.5 + 10 \times 6.75 + 2 \times 5.75$$
$$= 163 \text{ kip-ft}$$

Alternatively, the bending moment at 3 may be determined by resolving forces vertically. The support reaction at end 1 is:

$$V_1 = 24 \times 28.5/40$$
$$= 17.1 \text{ kips}$$

The moment at 3 is given by:

$$M_3 = 17.1 \times 10 - 2 \times 4$$
$$= 163 \text{ kip-ft}$$

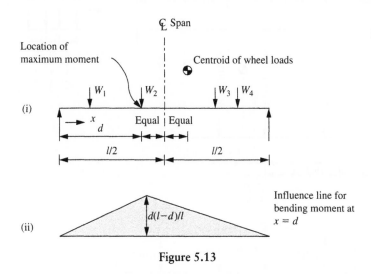

Figure 5.13

The maximum bending moment produced by a train of wheel loads in a simply supported beam always occurs under one of the wheels and does not necessarily occur at midspan. As shown in Figure 5.13 (i), a train of wheel loads produces the maximum possible bending moment in a simply supported beam under one of the wheels when the center of the span bisects the distance between this wheel and the centroid of the train. The maximum moment usually occurs under one of the wheels adjacent to the centroid of the train. The influence line for bending moment at the location of the maximum moment is shown at (ii).

Example 5.6

Determine the maximum possible bending moment in the simply supported beam shown in Figure 5.14 due to the train of wheel loads indicated.

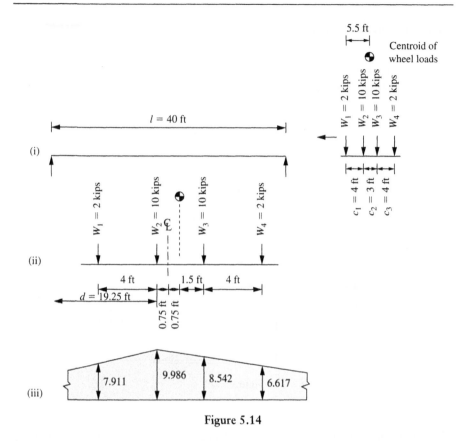

Figure 5.14

Solution

Placing the train of wheel loads as indicated in Figure 5.14 (ii) produces the maximum bending moment under the second wheel load. Hence, the maximum bending moment occurs under the second wheel load when it is located at a distance of 0.75 ft left of midspan, as shown at (ii). The influence line for bending moment at the location of the second wheel load is shown at (iii), and the maximum bending moment at this location is given by:

$$M_{max} = 2 \times 7.911 + 10 \times 9.986 + 10 \times 8.542 + 2 \times 6.617$$
$$= 214.336 \text{ kip-ft}$$

Alternatively, the bending moment may be determined by resolving forces vertically. The support reaction at end 1 is:

$$V_1 = 24 \times 19.25/40$$
$$= 11.55 \text{ kips}$$

The maximum moment is given by:

$$M_{max} = 11.55 \times 19.25 - 2 \times 4$$
$$= 214.338 \text{ kip-ft}$$

Because of symmetry in the train of wheel loads, this moment also occurs under the third wheel load when it is located at a distance of 0.75 ft right of midspan.

When the second or third wheel load is located at midspan, the moment at midspan is:

$$M = 214.00 \text{ kip-ft}$$

and this is the maximum bending moment that occurs at midspan.

(d) Envelope of maximum effects

To design a member, it is necessary to determine the maximum bending moment and shear force that can occur at all sections of the member. Diagrams indicating maximum values are known as envelope diagrams and are determined using influence lines at selected points along the member.

Example 5.7

Construct the maximum possible bending moment envelope in the simply supported beam shown in Figure 5.15 due to a concentrated load of 100 kips.

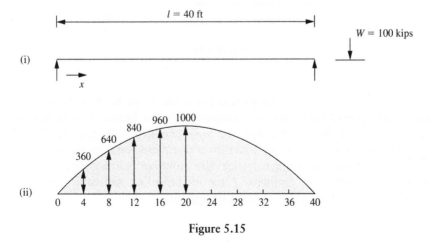

Figure 5.15

Table 5.1

x ft	Influence line ordinate	M_{max}
4	$4 \times 36/40 = 3.6$	360
8	$8 \times 32/40 = 6.4$	640
12	$12 \times 28/40 = 8.4$	840
16	$16 \times 24/40 = 9.6$	960
20	$20 \times 20/40 = 10.0$	1000

Solution

The maximum bending moment occurs at any section when the concentrated load is located at the section. Values of the moment are calculated in Table 5.1, and the bending moment envelope is shown in Figure 5.15 (ii).

5.4 Pin-jointed truss

(a) Stringers and cross beams

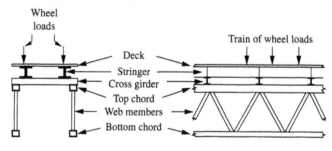

Figure 5.16

As moving loads traverse a pin-jointed truss, the loads are transferred to the truss panel points by a system of stringers and cross beams. This is shown in Figure 5.16 for a deck bridge with the loads applied to the top chord of the truss. The moving load is transferred from one panel point to the next as the load moves across the stringer. Hence, the influence line for axial force in a member is completed by connecting the influence line ordinates at the panel points on either side of a panel with a straight line.

(b) Influence lines for a Warren truss

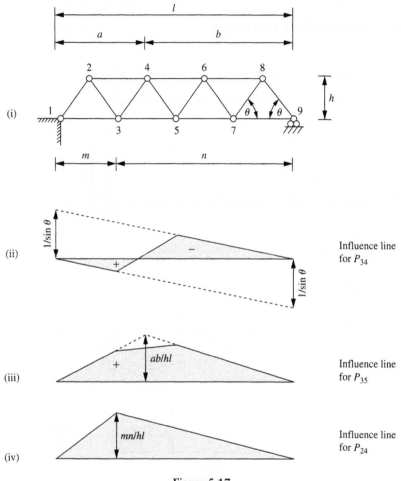

Figure 5.17

Figure 5.17 (i) shows a Warren truss with the loads from the stringers applied to the bottom panel points. The influence lines for axial force in members 34, 35, and 24 are obtained by taking a section through these three members and considering the relevant free body diagrams.

The influence line for axial force in web member 34 is obtained by multiplying the influence line for shear force in panel 34 by $1/\sin \theta$ and is shown at (ii). Positive sense of the influence line indicates tension in member 34. Because of the effect of the stringers, the influence line between nodes 3 and 5 is obtained by connecting the ordinates at 3 and 5 with a straight line.

The influence line for axial force in bottom chord member 35 is obtained by multiplying the influence line for moment at node 4 by $1/h$ and is shown at (iii). The influence line is positive, indicating tension in member 35.

The influence line for axial force in top chord member 24 is obtained by multiplying the influence line for moment at node 3 by $1/h$.

(c) Influence lines for a Pratt truss

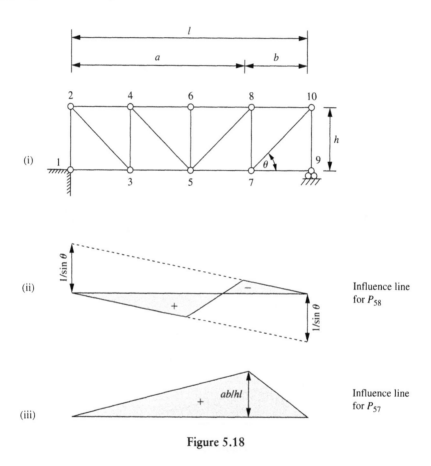

Figure 5.18

Figure 5.18 (i) shows a Pratt truss with the loads from the stringers applied to the bottom panel points. The influence line for axial force in web member 58 is obtained by multiplying the influence line for shear force in panel 58 by $1/\sin\theta$ and is shown at (ii). Positive sense of the influence line indicates tension in member 58. The influence line for axial force in member 78 is identical in shape and of opposite sign.

The influence line for axial force in bottom chord member 57 is obtained by multiplying the influence line for moment at node 8 by $1/h$ and is shown at (iii). The influence line is positive, indicating tension in member 57.

The influence line for axial force in top chord member 810 is obtained by multiplying the influence line for moment at node 7 by $1/h$. This influence line is identical in shape to the influence line for P_{57} and of opposite sign.

(d) Influence lines for a bowstring truss

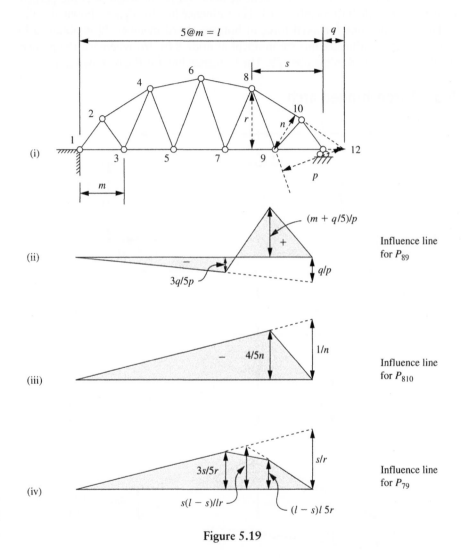

Figure 5.19

Figure 5.19 (i) shows a bowstring truss with the loads from the stringers applied to the bottom panel points. The influence lines for axial force in members 89, 810, and 79 are obtained by taking a section through these three members and considering the relevant free body diagrams.

The influence line for axial force in web member 89 is obtained by multiplying the influence line for moment at point 12 by $1/p$, where p is the perpendicular from point 12 to the line of action of member 89. The influence line for P_{89} is shown at (ii).

The influence line for axial force in top chord member 810 is obtained by multiplying the influence line for moment at node 9 by $1/n$, where n is the perpendicular from node 9 to member 810. The influence line for P_{810} is shown at (iii).

The influence line for axial force in bottom chord member 79 is obtained by multiplying the influence line for moment at node 8 by $1/r$, where r is the perpendicular from node 8 to member 79. The influence line for P_{79} is shown at (iv).

5.5 Three-hinged arch

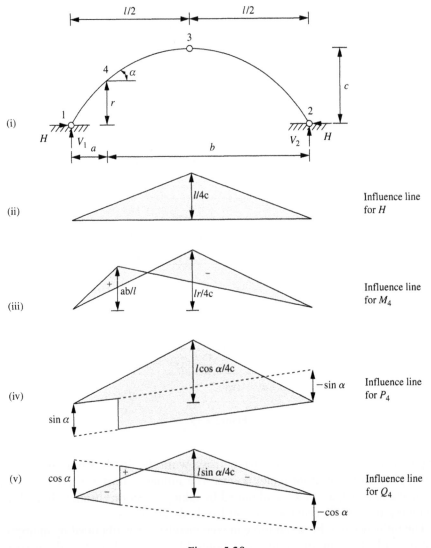

Figure 5.20

The horizontal thrust at the springings of a three-hinged arch is equal to the bending moment at the center of a simply supported beam of the same length multiplied by $1/c$. Hence, the influence line for horizontal thrust is the influence line for the free bending moment multiplied by $1/c$ and is shown in Figure 5.20 (ii).

The influence line for bending moment at point 4 is the influence line for free bending moment at 4 minus the horizontal thrust multiplied by r and is given by:

$$M_4 = (M_s)_4 - Hr$$

The influence line is shown at (iii).
The influence line for thrust at point 4 is given by:

$$P_4 = H \cos \alpha - V_2 \sin \alpha \ldots \text{ unit load from 1 to 4}$$
$$= H \cos \alpha + V_2 \sin \alpha \ldots \text{ unit load from 4 to 2}$$

The influence line is shown at (iv).
The influence line for shear at point 4 is given by:

$$Q_4 = -H \sin \alpha - V_2 \cos \alpha \ldots \text{ unit load from 1 to 4}$$
$$= -H \sin \alpha + V_1 \cos \alpha \ldots \text{ unit load from 4 to 2}$$

The influence line is shown at (v).

Supplementary problems

S5.1 Construct the influence lines for V_2 and M_2 for the beam shown in Figure S5.1. Determine the maximum value of M_2 due to a distributed load of 2 kips/ft over a length of 60 ft.

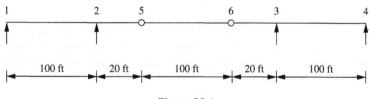

Figure S5.1

S5.2 Construct the influence line for V_2 for the beam shown in Figure S5.2. Determine the maximum value of V_2 due to distributed load of 10 kips/ft over a length of 6 ft.

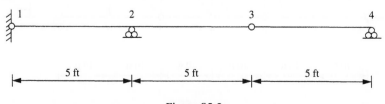

Figure S5.2

S5.3 Construct the influence line for V_4 for the beam shown in Figure S5.3. Determine the maximum value of V_4 due to a train of three wheel loads each of 3 kips at 2 ft on center.

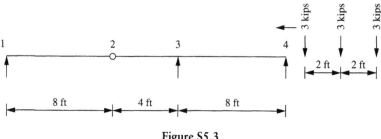

Figure S5.3

S5.4 Figure S5.4 shows a truss with the loads from the stringers applied to the bottom panel points. Construct the influence line for axial force in member 24. Determine the maximum value of P_{24} due to a concentrated load of 5 kips.

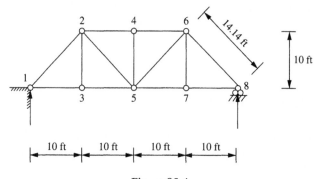

Figure S5.4

S5.5 Figure S5.5 shows a truss with the loads from the stringers applied to the bottom panel points. Construct the influence line for axial force in member 45. Determine the maximum value of P_{45} due to concentrated load of 10 kips.

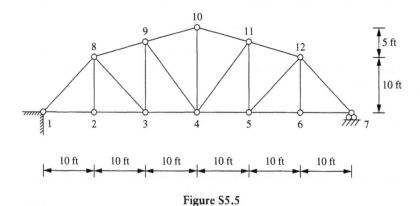

Figure S5.5

S5.6 Figure S5.6 shows a truss with the loads from the stringers applied to the bottom panel points. Construct the influence line for axial force in member 1718. Determine the maximum value of P_{1718} due to concentrated load of 20 kips.

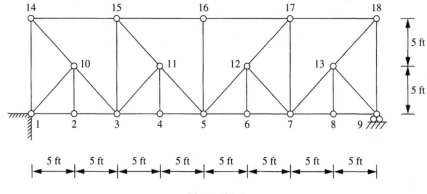

Figure S5.6

6 Space frames

Notation

F	force
F_x	force component along the x-axis
F_y	force component along the y-axis
F_z	force component along the z-axis
f_i	angle in a triangle opposite side F_i
H	horizontal force
l	length of member
M	bending moment
M_x	bending moment about the x-axis
M_y	bending moment about the y-axis
M_z	bending moment about the z-axis
P	axial force in a member
R	support reaction
V	vertical force
W_{LL}	concentrated live load
w_{DL}	distributed dead load
θ	angle of inclination

6.1 Introduction

The design of a building is generally accomplished by considering the structure as an assemblage of planar frames, each of which is designed as an independent two-dimensional frame. In some instances, however, it is necessary to consider the building as a whole and design it as a three-dimensional structure.

The sign convention shown in Figure 6.1 may be adopted for a three-dimensional structure acted on by a generalized system of forces. A space structure is illustrated in Figure 6.2 (i). The plan view of the structure is in the xz plane, as shown in Figure 6.2 (ii), and the elevation of the structure is in the xy plane, as shown in Figure 6.2 (iii).

Displacements in a space structure may occur in six directions, a displacement in the x, y, and z directions and a rotation about the x-, y-, and z-axes. The sign convention for displacements is shown in Figure 6.3 (i). A total of six displacement components define the restraint conditions at support 1 of the

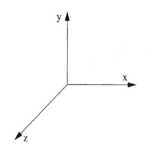

Figure 6.1

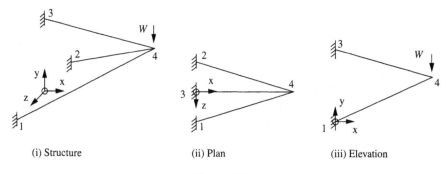

(i) Structure   (ii) Plan   (iii) Elevation

Figure 6.2

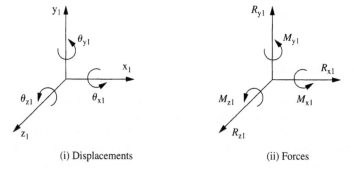

(i) Displacements   (ii) Forces

Figure 6.3

structure shown in Figure 6.2. The arrows indicate the positive directions of the displacement components, and, using the right-hand screw system, rotations are considered positive when acting clockwise as viewed from the origin. Similarly, a total of six force components, as shown in Figure 6.3 (ii), define the support reactions at support 1 of the structure in Figure 6.2.

6.2 Conditions of equilibrium

For a three-dimensional structure, six conditions of static equilibrium may be obtained at any point in the structure and at each support. These six equations of statics are:

$$\Sigma F_x = 0, \ \Sigma F_y = 0, \ \Sigma F_z = 0$$
$$\Sigma M_x = 0, \ \Sigma M_y = 0, \ \Sigma M_z = 0$$

A three-dimensional structure is externally determinate when six external restraints are applied to the structure, since these restraints may be determined by the available six equations of static equilibrium.

The cranked cantilever of Figure 6.4 (i) has an applied load W at the free end that has the components W_x, W_y, and W_z as shown. The magnitude of W is given by:

$$W = (W_x^2 + W_y^2 + W_z^2)^{0.5}$$

and the three direction cosines of the applied load are given by:

$$\cos \theta_x = W_x/W$$
$$\cos \theta_y = W_y/W$$
$$\cos \theta_z = W_z/W$$

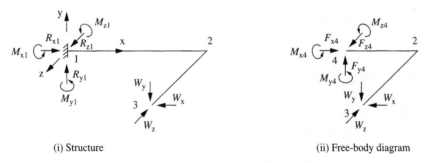

(i) Structure   (ii) Free-body diagram

Figure 6.4

As shown in Figure 6.4 (i), the cranked cantilever is statically determinate since six restraints are provided at the fixed end.

Similarly, six member stresses may be determined at a section cut through the structure, at any point 4, as shown by the free-body diagram shown in Figure 6.4 (ii).

Example 6.1

The cranked cantilever shown in Figure 6.4 (i) has an applied load at the free end with components $W_x = -10\,\text{kips}$, $W_y = -15\,\text{kips}$, and $W_z = -20\,\text{kips}$. The relevant lengths of the cantilever are $l_{12} = 12$ feet and $l_{23} = 6$ feet. Determine the magnitude of the reactions at support 1.

Solution

Applying the equilibrium equations, with the origin of the coordinates at support 1, gives:

Resolving along the x-axis:

$$R_{x1} + W_x = 0$$
$$R_{x1} = -W_x$$
$$= 10 \text{ kips}$$

Resolving along the y-axis:

$$R_{y1} + W_y = 0$$
$$R_{y1} = -W_y$$
$$= 15 \text{ kips}$$

Resolving along the z-axis:

$$R_{z1} + W_z = 0$$
$$R_{z1} = -W_z$$
$$= 20 \text{ kips}$$

Taking moments about the x-axis:

$$M_{x1} + W_y l_{23} = 0$$
$$M_{x1} = -W_y l_{23}$$
$$= -15 \times 6$$
$$= -90 \text{ kip-ft}$$

Taking moments about the y-axis:

$$M_{y1} + W_x l_{23} + W_z l_{12} = 0$$
$$M_{y1} = -W_x l_{23} - W_z l_{12}$$
$$= 10 \times 6 - 20 \times 12$$
$$= -180 \text{ kip-ft}$$

Taking moments about the z-axis:

$$M_{z1} + W_y l_{12} = 0$$
$$M_{z1} = -W_y l_{12}$$
$$= 15 \times 12$$
$$= 180 \text{ kip-ft}$$

6.3   Pin-jointed space frames

In a pin-jointed three-dimensional space frame with j nodes, including the supports, $3j$ equations of equilibrium may be derived, since each node provides the relationships:

$$\sum F_x = 0, \quad \sum F_y = 0, \quad \sum F_z = 0$$

If the frame has n members and r external restraints, the number of unknowns is $(n + r)$. In a pin-jointed three-dimensional space frame the frame is statically determinate when:

$$(n + r) = 3j$$

The frame is indeterminate when:

$$(n + r) > 3j$$

Example 6.2

The pin-jointed space frame shown in Figure 6.5 consists of nine members. The supports consist of a fixed pin at node 1, providing three restraints as shown, and rollers at nodes 2, 3, and 4, providing only vertical restraint. Determine if the structure is statically determinate.

Solution

The total number of external restraints is:

$$r = 3 + 3 \times 1$$
$$= 6$$

Hence, the structure is stable and determinate externally.
The total number of members is:

$$n = 9$$

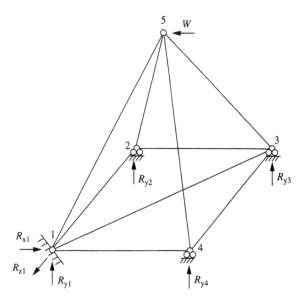

Figure 6.5

The total number of nodes is

$$j = 5$$
$$(n + r) = 9 + 6$$
$$= 15$$
$$3j = 3 \times 5$$
$$= 15$$

Hence:

$$(n + r) = 3j \ ...$$

and the structure is statically determinate.

6.4 Member forces

The member forces in a pin-jointed space frame may be obtained by resolution of forces at the nodes. Figure 6.6 shows a total of i members, 01, 02 ... 0i with a common node 0. The force in member 0i is P_{0i}, and the three direction cosines of member 0i are $\cos\theta_{x0i}$, $\cos\theta_{y0i}$, and $\cos\theta_{z0i}$. The force components of member 0i along the three coordinate axes are:

$$P_{x0i} = P_{0i} \cos\theta_{x0i}$$
$$P_{y0i} = P_{0i} \cos\theta_{y0i}$$
$$P_{z0i} = P_{0i} \cos\theta_{z0i}$$

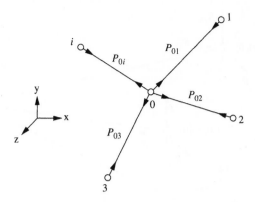

Figure 6.6

The length of member $0i$ is l_{0i}, and the projections of member $0i$ along the three coordinate axes are:

$$x_{0i} = l_{0i} \cos\theta_{x0i}$$
$$y_{0i} = l_{0i} \cos\theta_{y0i}$$
$$z_{0i} = l_{0i} \cos\theta_{z0i}$$

Assuming that no external force is applied at node 0, resolving along the three coordinate axes gives:

$$\sum P_{x0i} = 0$$
$$\sum P_{y0i} = 0$$
$$\sum P_{z0i} = 0$$

and

$$P_{0i}/l_{0i} = P_{x0i}/x_{0i} = P_{y0i}/y_{0i} = P_{z0i}/z_0 i$$

Example 6.3

The pin-jointed space frame shown in Figure 6.7 consists of three members. The supports consist of fixed pins at nodes 1, 2, and 3, each providing three restraints as shown. Determine the member forces produced by the 100 kip vertical load applied at node 4.

Solution

The total number of external restraints is:

$$r = 3 \times 3$$
$$= 9$$

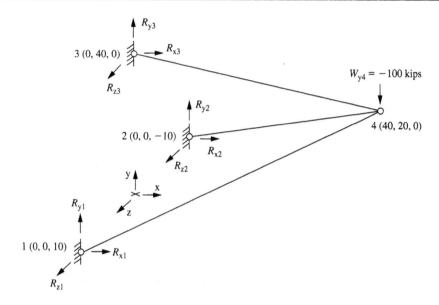

Figure 6.7

Hence, the structure is stable.
The total number of members is:

$n = 3$

The total number of nodes is:

$$j = 4$$
$$(n + r) = 3 + 9$$
$$= 12$$
$$3j = 3 \times 4$$
$$= 12$$

Hence:

$(n + r) = 3j$ ... the structure is statically determinate

And the structure is statically determinate.
The lengths of the members are:

$$l_{14} = (x_{14}^2 + y_{14}^2 + z_{14}^2)^{0.5}$$
$$= (40^2 + 20^2 + 10^2)^{0.5}$$
$$= 45.83 \text{ ft}$$
$$l_{24} = 45.83 \text{ ft}$$
$$l_{34} = (x_{34}^2 + y_{34}^2 + z_{34}^2)^{0.5}$$
$$= (40^2 + 20^2 + 0^2)^{0.5}$$
$$= 44.72 \text{ ft}$$

The direction cosines are:

$$\cos \theta_{x14} = x_{14}/l_{14}$$
$$= 40/45.83$$
$$= 0.873$$
$$\cos \theta_{y14} = y_{14}/l_{14}$$
$$= 20/45.83$$
$$= 0.436$$
$$\cos \theta_{z14} = z_{14}/l_{14}$$
$$= -10/45.83$$
$$= -0.218$$
$$\cos \theta_{x34} = x_{34}/l_{34}$$
$$= 40/44.72$$
$$= 0.894$$
$$\cos \theta_{y34} = y_{34}/l_{34}$$
$$= -20/44.72$$
$$= -0.447$$
$$\cos \theta_{z34} = z_{34}/l_{34}$$
$$= 0/44.72$$
$$= 0$$

Because of the symmetry of the structure and the loading, the forces in members 14 and 24 are identical. Hence:

$$P_{14} = P_{24}$$

Resolving along the x-axis at node 4 gives:

$$2P_{x14} + P_{x34} = 0$$
$$2P_{14} \cos \theta_{x14} + P_{34} \cos \theta_{x34} = 0$$
$$1.746P_{14} + 0.894P_{34} = 0$$

Resolving along the y-axis at node 4 gives:

$$2P_{y14} + P_{y34} = -W_{y4}$$
$$2P_{14} \cos \theta_{y14} + P_{34} \cos \theta_{y34} = -W_{y4}$$
$$0.872P_{14} - 0.447P_{34} = 100 \text{ kips}$$

Hence:

$$P_{34} = -111.86 \text{ kips ... tension}$$
$$P_{14} = +57.21 \text{ kips ... compression}$$

Supplementary problems

S6.1 The cranked cantilever shown in Figure S6.1 has a load of 100 kips applied at the free end. Determine the magnitude of the reactions at support 1.

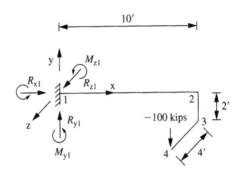

Figure S6.1

S6.2 The pin-jointed space frame shown in Figure S6.2 consists of three members. The supports consist of fixed pins at nodes 1, 2, and 3, each providing three restraints as shown. Determine the member forces produced by the 100 kip vertical load applied at node 4.

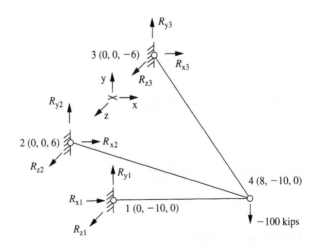

Figure S6.2

S6.3 The pin-jointed space frame shown in Figure S6.3 consists of three members. The supports consist of fixed pins at nodes 1, 2, and 3, each providing three restraints. Determine the member forces produced by the 100 kip vertical load applied at node 4.

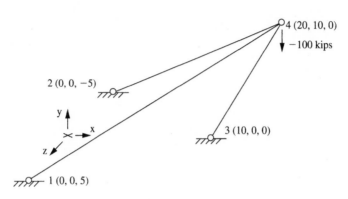

Figure S6.3

S6.4 The pin-jointed space frame shown in Figure S6.4 consists of three members. The supports consist of fixed pins at nodes 1, 2, and 3, each providing three restraints. Determine the member forces produced by the 100 kip vertical load applied at node 4.

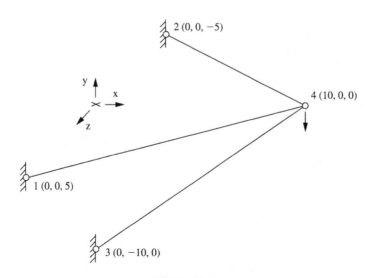

Figure S6.4

50.3 The pin-jointed space frame shown in Figure 50.3 consists of three members. The supports consist of fixed pins at nodes 1, 2, and 3. Each providing three reactions. Determine the member forces produced by the 100 kN vertical load applied at node 4.

Figure 50.3

50.4 The pin-jointed space tower shown in Figure 50.4 consists of three members. The supports consist of fixed pins at nodes 1, 2, and 3, each providing three reactions. Determine the member forces produced by the 100 kN vertical load applied at node 4.

Figure 50.4

Answers to supplementary problems part 1

Chapter 1

S1.1 V_1 = 14 kips
        H_1 = 10 kips
        M_1 = 100 kip-ft
        V_4 = 6 kips
        H_4 = 0 kips

S1.2 V_1 = –4.8 kips ... downward
        V_2 = 28.8 kips ... upward
        M_2 = 480 kip-ft ... producing tension in the top fiber of the girder

S1.3 V_1 = 5 kips
        H_1 = 5 kips
        M_{32} = 100 kip-ft ... producing tension in the top fiber of the member
        V_7 = 5 kips
        H_7 = 5 kips

S1.4 V_1 = 25 kips
        H_1 = 25 kips
        V_2 = 75 kips
        H_2 = 25 kips
        P_{13} = 35.33 kips ... tension
        P_{23} = 79.04 kips ... compression

S1.5 V_1 = 10 kips
        H_1 = 10 kips
        V_4 = 10 kips
        H_4 = 0 kips
        P_{13} = 14.14 kips ... tension

S1.6 V_1 = 11.67 kips
        H_1 = 10 kips
        V_4 = 8.33 kips
        H_4 = 0 kips
        M_{34} = 166.6 kips ... producing tension in the bottom fiber of the member

S1.7 $V_1 = 30$ kips
        $H_1 = 10$ kips
        $V_6 = 50$ kips
        $H_6 = 0$ kips

S1.8 $V_1 = 10$ kips
        $H_1 = 10$ kips
        $V_3 = 10$ kips
        $H_3 = 10$ kips
        $P_{12} = 14.14$ kips ... compression
        $P_{23} = 10$ kips ... compression

S1.9 $V_1 = 0$ kips
        $H_1 = 10.5$ kips
        $M_1 = 147$ kip-ft

S1.10 $V_1 = 4$ kips
        $H_1 = 0$ kips
        $M_1 = 24$ kip-ft
        $P_{34} = 7.2$ kips ... tension
        $P_{24} = 6.0$ kips ... compression

Chapter 2

S2.1 $V_1 = 2.24$ kips
        $H_1 = 3.57$ kips
        $V_5 = 4.92$ kips
        $P_{34} = 5.37$ kips ... compression
        $P_{38} = 5.0$ kips ... tension
        $P_{78} = 4.0$ kips ... tension

S2.2 $P_{45} = 32$ kips ... tension
        $P_{411} = 3.2$ kips ... compression
        $P_{1011} = 30.93$ kips ... compression

S2.3 $P_{23} = 17.78$ kips ... tension
        $P_{27} = 21.89$ kips ... tension
        $P_{67} = 33.33$ kips ... compression

S2.4 $P_{12} = 12$ kips ... tension
        $P_{14} = 0$ kips
        $P_{15} = 16.97$ kips ... compression

$P_{45} = 0$ kips
$P_{56} = 12$ kips ... compression
$P_{25} = 0$ kips
$P_{26} = 0$ kips

S2.5 $P_{12} = 9$ kips ... tension
$P_{15} = 10.82$ kips ... compression
$P_{23} = 6$ kips ... tension
$P_{56} = 9.02$ kips ... compression
$P_{25} = 3.35$ kips ... compression
$P_{26} = 3.35$ kips ... tension

S2.6 P_{12} $= 5$ kips ... tension
P_{114} $= 30$ kips ... compression
P_{110} $= 7.07$ kips ... compression
P_{23} $= 5$ kips ... tension
P_{310} $= 35.35$ kips ... tension
P_{315} $= 10$ kips ... compression
$P_{1014} = 42.43$ kips ... tension
$P_{1415} = 30$ kips ... compression

S2.7 $P_{23} = 6.67$ kips ... tension
$P_{27} = 2.10$ kips ... compression
$P_{37} = 0$ kips
$P_{78} = 6.67$ kips ... compression
$P_{67} = 6.72$ kips ... compression

S2.8 $P_{12} = 3.35$ kips ... tension
$P_{14} = 5.41$ kips ... compression
$P_{24} = 1.12$ kips ... compression
$P_{45} = 3.60$ kips ... compression
$P_{25} = 4$ kips ... tension

S2.9 $P_{23} = 13.34$ kips ... tension
$P_{26} = 2.5$ kips ... compression
$P_{27} = 3$ kips ... tension
$P_{67} = 16.77$ kips ... compression
$P_{37} = 0$ kips

S2.10 $P_{49} = 0$ kips
$P_{59} = 28.28$ kips ... compression
$P_{89} = 22.36$ kips ... compression
$P_{78} = 20$ kips ... compression

Chapter 3

S3.1 (i)   Determinate
      (ii) Indeterminate
      (iii) Unstable
      (iv) Indeterminate

S3.2 (i)   Indeterminate
      (ii) Indeterminate
      (iii) Indeterminate
      (iv) Determinate
      (v)   Determinate

S3.3 (i)   Indeterminate
      (ii) Unstable
      (iii) Determinate
      (iv) Indeterminate

S3.4 Support reactions:

$$V_1 = 73.03 \text{ kips}$$
$$H_1 = -70.71 \text{ kips}$$
$$V_4 = 77.68 \text{ kips}$$

Shears:

$$Q_{21} = 73.03 \text{ kips}$$
$$Q_{23} = 2.32 \text{ kips}$$
$$Q_{32} = 2.32 \text{ kips}$$
$$Q_{34} = 2.32 \text{ kips}$$
$$Q_{43} = -77.68 \text{ kips}$$

Moments:

$$M_{21} = 292.12 \text{ kip-ft} \ldots \text{compression in top fiber}$$
$$M_{32} = 301.40 \text{ kip-ft} \ldots \text{compression in top fiber}$$
$$M_{max} = 301.67 \text{ kip-ft} \ldots \text{at } x = 8.23 \text{ ft}$$

S3.5 Support reactions:

$$V_1 = 0 \text{ kips}$$
$$H_1 = 0 \text{ kips}$$
$$V_2 = 100 \text{ kips}$$
$$V_4 = 0 \text{ kips}$$

Shears:

$$Q_{21} = -50 \text{ kips}$$
$$Q_{23} = 50 \text{ kips}$$
$$Q_{32} = 0 \text{ kips}$$

$Q_{34} = 0$ kips
$Q_{43} = 0$ kips

Moments:

$M_{21} = 125$ kip-ft ... compression in bottom fiber
$M_{32} = 0$ kip-ft
$M_{34} = 0$ kip-ft

S3.6 Support reactions:

$V_1 = 30$ kips
$H_1 = 0$ kips
$M_1 = 180$ kip-ft
$V_3 = 30$ kips

Shears:

$Q_{21} = 30$ kips
$Q_{23} = 30$ kips
$Q_{32} = -30$ kips

Moments:

$M_{12} = 180$ kip-ft ... compression in bottom fiber
$M_{21} = 0$ kip-ft
$M_{32} = 0$ kip-ft
$M_{max} = 45$ kip-ft ... at $x = 9$ ft, compression in top fiber

S3.7 Support reactions:

$V_1 = 20$ kips
$H_1 = -20$ kips
$M_1 = 0$ kip-ft
$V_3 = 60$ kips

Shears:

$Q_{12} = 20$ kips
$Q_{21} = 20$ kips
$Q_{23} = 20$ kips
$Q_{32} = -60$ kips

Moments:

$M_{21} = 160$ kip-ft ... compression in outer fiber
$M_{23} = 160$ kip-ft ... compression in top fiber
$M_{32} = 0$ kip-ft
$M_{max} = 180$ kip-ft ... at $x = 2$ ft, compression in top fiber

S3.8 Support reactions:

$V_1 = 0$ kips
$H_1 = -20$ kips
$M_1 = 0$ kip-ft

$$V_5 = 40 \text{ kips}$$
$$H_5 = 0 \text{ kips}$$

Shears:

$$Q_{12} = 20 \text{ kips}$$
$$Q_{21} = 20 \text{ kips}$$
$$Q_{23} = 0 \text{ kips}$$
$$Q_{32} = 0 \text{ kips}$$
$$Q_{34} = -40 \text{ kips}$$
$$Q_{43} = -40 \text{ kips}$$
$$Q_{45} = 0 \text{ kips}$$
$$Q_{54} = 0 \text{ kips}$$

Moments:

$M_{21} = 200 \text{ kip-ft} \ldots$ compression in outer fiber
$M_{23} = 200 \text{ kip-ft} \ldots$ compression in top fiber
$M_{32} = 200 \text{ kip-ft} \ldots$ compression in top fiber
$M_{34} = 200 \text{ kip-ft} \ldots$ compression in top fiber
$M_{43} = 0 \text{ kip-ft}$
$M_{45} = 0 \text{ kip-ft}$
$M_{54} = 0 \text{ kip-ft}$

S3.9 Support reactions:

$$V_1 = -10 \text{ kips}$$
$$H_1 = 3.33 \text{ kips}$$
$$M_1 = 0 \text{ kip-ft}$$
$$V_7 = -10 \text{ kips}$$
$$H_7 = -3.33 \text{ kips}$$

Shears:

$$Q_{12} = -3.33 \text{ kips}$$
$$Q_{21} = -3.33 \text{ kips}$$
$$Q_{23} = 7.45 \text{ kips}$$
$$Q_{32} = 7.45 \text{ kips}$$
$$Q_{34} = -1.49 \text{ kips}$$
$$Q_{43} = -1.49 \text{ kips}$$

Moments:

$M_{21} = 26.64 \text{ kip-ft} \ldots$ compression in inner fiber
$M_{23} = 26.64 \text{ kip-ft} \ldots$ compression in bottom fiber
$M_{32} = 6.70 \text{ kip-ft} \ldots$ compression in top fiber
$M_{34} = 6.70 \text{ kip-ft} \ldots$ compression in top fiber
$M_{43} = 0 \text{ kip-ft}$

S3.10 Support reactions:

$$V_1 = -160 \text{ kips}$$
$$H_1 = 160 \text{ kips}$$

$M_1 = 0\,\text{kip-ft}$
$V_3 = -160\,\text{kips}$
$H_3 = -160\,\text{kips}$
$M_2 = 0\,\text{kip-ft}$
At $x = 8$ ft the bending moment in the arch rib is
$M = 0\,\text{kip-ft}$

Chapter 4

S4.1 $\theta_1 = 52/EI$ rad
$\theta_2 = 156/EI$ rad
$y_3 = 1560/EI$ ft

S4.2 $y_2 = 2160/EI$ ft

S4.3 $\theta_1 = 1917/EI$ rad
$x_2 = 15837/EI$ ft

S4.4 The equation of the elastic curve is
$y = -2.5x^3/3EI + 2.5[x - 12]^3 + 120x$
The location of the maximum deflection in span 12 is
$x = 6.93$ ft
The maximum deflection in span 12 is
$y = 554/EI$ ft
The deflection at node 3 is
$y_3 = -2158/EI$

S4.5 $y_3 = 8136/EI$ ft

S4.6 $y_3 = 247/EA$ ft

S4.7 $y_4 = 3062/EA$ ft

S4.8 $y_2 = 410/EA$ ft

S4.9 $y_3 = 317/EA$ ft

S4.10 $y_2 = 199/EA$ ft

Chapter 5

S5.1 $M_2 = 1800\,\text{kip-ft}$

S5.2 $V_2 = 96\,\text{kips}$

S5.3 $V_3 = 6.75\,\text{kips}$

S5.4 $P_{24} = 5\,\text{kips} \ldots$ compression

S5.5 $P_{45} = 10.67\,\text{kips} \ldots$ tension

S5.6 $P_{1718} = 15\,\text{kips} \ldots$ compression

Chapter 6

S6.1 $M_{x1} = 400\,\text{kip-ft}$
$M_{z1} = 1000\,\text{kip-ft}$
$R_{y1} = 100\,\text{kips}$

S6.2 $P_{14} = 80\,\text{kips}$
$P_{24} = 70.71\,\text{kips}$
$P_{34} = 70.71\,\text{kips}$

S6.3 $P_{14} = 114.55\,\text{kips}$
$P_{24} = 114.55\,\text{kips}$
$P_{34} = 282.80\,\text{kips}$

S6.4 $P_{14} = 55.90\,\text{kips}$
$P_{24} = 55.90\,\text{kips}$
$P_{34} = 141.40\,\text{kips}$

Part Two

Analysis of Indeterminate Structures

PART TWO

Analysis of Indeterminate Structures

1 Statical indeterminacy

Notation

c number of releases introduced in a structure
d degree of indeterminacy
h number of internal hinges introduced in a structure
H horizontal reaction
j number of joints
M bending moment
n number of members
r number of external restraints
s number of internal rollers introduced in a structure
V vertical reaction

1.1   Introduction

A structure is in equilibrium with a system of applied loads when the resultant force in any direction and the resultant moment about any point are zero. For a two-dimensional plane structure, three equations of static equilibrium may be obtained:

$$\Sigma H = 0$$
$$\Sigma V = 0$$
$$\Sigma M = 0$$

where H and V are the resolved components in the horizontal and vertical directions of all forces and M is the resultant moment about any point.

A statically determinate structure is one in which all member forces and external reactions may be determined by applying the equations of equilibrium.

An indeterminate or redundant structure is one that possesses more unknown member forces and reactions than available equations of equilibrium. To determine the member forces and reactions, additional equations must be obtained from conditions of geometrical compatibility. The number of unknowns, in excess of the available equations of equilibrium, is the degree of indeterminacy, and the unknown forces and reactions are the redundants. The redundants may be removed from the structure, leaving a stable, determinate structure, which is known as the cut-back structure. External redundants are redundants that exist among the external reactions. Internal redundants are redundants that exist among the member forces.

Several methods have been proposed[1–5] for evaluating the indeterminacy of a structure.

1.2   Indeterminacy in pin-jointed frames

In a pin-jointed frame, external reactions are provided by either roller supports or hinge supports, as shown in Figure 1.1 (i) and (ii). The roller support provides only one degree of restraint in the vertical direction, and both horizontal and rotational displacements can occur. The hinge support provides two degrees of restraint in the vertical and horizontal directions, and only rotational displacement can occur. The magnitudes of the external restraints may be obtained from the three equations of equilibrium. Thus, a structure is externally indeterminate when it possesses more than three external restraints and unstable when it possesses fewer than three.

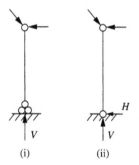

Figure 1.1

Figure 1.2 (i) and (ii) shows pin-jointed frames that have three degrees of restraint and are stable and determinate. Figure 1.3 (i) shows a pin-jointed frame that has four degrees of restraint and is one degree indeterminate. The

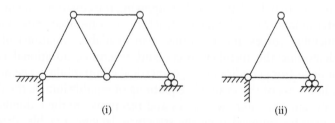

Figure 1.2

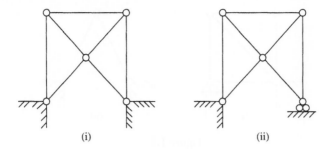

Figure 1.3

cut-back structure is shown in Figure 1.3 (ii). Figure 1.4 (i) shows a pin-jointed frame that is two degrees indeterminate; the cut-back structure is shown in Figure 1.4 (ii).

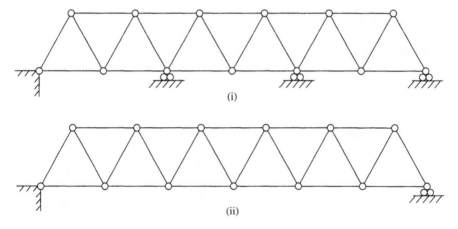

Figure 1.4

In a pin-jointed frame with j joints, including the supports, $2j$ equations of equilibrium may be obtained, since at each joint:

$$\Sigma H = 0 \quad \text{and} \quad \Sigma V = 0$$

Each member of the frame is subjected to an axial force, and if the frame has n members and r external restraints, the number of unknowns is $(n + r)$. Thus, the degree of indeterminacy is:

$$D = n + r - 2j$$

Figure 1.5 (i) and (ii) shows pin-jointed frames that are determinate. For frame (i):

$$D = 5 + 3 - (2 \times 4) = 0$$

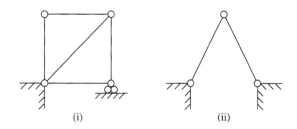

Figure 1.5

and for (ii):

$$D = 2 + 4 - (2 \times 3) = 0$$

Figure 1.6 (i) and (ii) shows pin-jointed frames that are indeterminate. For frame (i)

$$D = 10 + 3 - (2 \times 6) = 1$$

and for (ii):

$$D = 11 + 4 - (2 \times 6) = 3$$

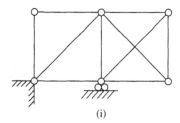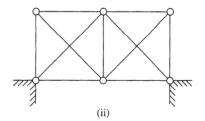

Figure 1.6

1.3 Indeterminacy in rigid frames

In addition to roller and hinge supports a rigid frame may be provided with a rigid support, shown in Figure 1.7, which provides three degrees of restraint.

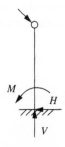

Figure 1.7

In a rigid frame with j joints, including the supports, $3j$ equations of equilibrium may be obtained, since at each joint:

$$\Sigma H = 0, \quad \Sigma V = 0, \quad \Sigma M = 0$$

Each member of the frame is subjected to three forces, axial and shear forces and a moment, and the number of unknowns is $(3n + r)$. Thus, the degree of indeterminacy is:

$$D = 3n + r - 3j$$

The degree of indeterminacy of the arches shown in Figure 1.8 is

arch (i) $D = 3 + 4 - (3 \times 2) = 1$
arch (ii) $D = 3 + 4 - (3 \times 2) = 1$
arch (iii) $D = 3 + 5 - (3 \times 2) = 2$
arch (iv) $D = 3 + 6 - (3 \times 2) = 3$
arch (v) $D = 3 + 3 - (3 \times 2) = 0$

and arch (v) is the cut-back structure for (i), (ii), (iii), and (iv).

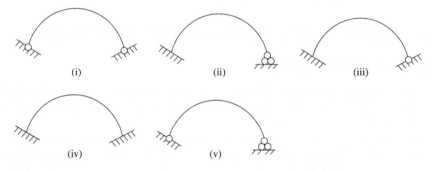

Figure 1.8

The degree of indeterminacy of the portal frame shown in Figure 1.9 (i) is:

$$D = (3 \times 3) + 6 - (3 \times 4) = 3$$

and the cut-back structure is obtained by introducing three releases, as at (ii), (iii), or (iv).

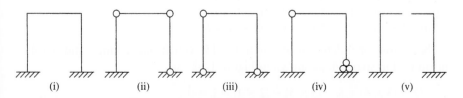

Figure 1.9

The redundants in the portal frame may also be regarded as the axial and shear forces and the moment in the beam, and the cut-back structure is obtained by cutting the beam as shown at (v). Similarly in a multibay, multistory frame the degree of indeterminacy equals 3 × the number of beams.

For the rigid frame shown in Figure 1.10, the degree of indeterminacy is:

$$D = 3 \times 6$$
$$= 18$$

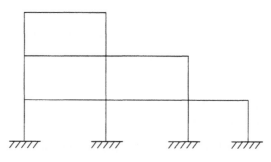

Figure 1.10

1.4   Indeterminacy in rigid frames with internal hinges

(a) Hinges within a member

The introduction of an internal hinge in a rigid frame provides an additional equation of equilibrium at the hinge of $M = 0$. In effect, a moment release has been introduced in the member.

The introduction of a horizontal, internal roller provides two additional equations of equilibrium at the roller of $M = 0$ and $H = 0$. In effect, a moment release and a release of horizontal restraint have been introduced in the member. Thus, the degree of indeterminacy is:

$$D = 3n + r - 3j - h - 2s$$

where n is the number of members, j is the number of joints in the rigid frame before the introduction of hinges, r is the number of external restraints, h is the number of internal hinges, and s is the number of rollers introduced.

The degree of indeterminacy of the frames shown in Figure 1.11 is:

(i) $D = (3 \times 3) + 4 - (3 \times 4) - 1 = 0$
(ii) $D = (3 \times 1) + 4 - (3 \times 2) - 1 = 0$
(iii) $D = (3 \times 7) + 12 - (3 \times 8) - 3 = 6$

The degree of indeterminacy of beam 15, which has a hinge and a roller introduced in span 34, as shown in Figure 1.12, is:

$$D = (3 \times 4) + 6 - (3 \times 5) - (2 \times 1) - 1 = 0$$

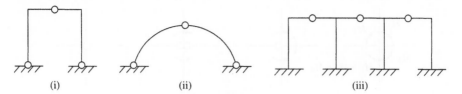

Figure 1.11

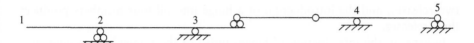

Figure 1.12

(b) Hinges at a joint

For two members meeting at a rigid joint there is one unknown moment, as shown in Figure 1.13, and the introduction of a hinge is equivalent to producing one release.

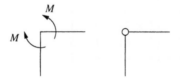

Figure 1.13

For three members meeting at a rigid joint there are two unknown moments, as shown in Figure 1.14, and the introduction of a hinge into one of the members produces one release; the introduction of a hinge into all three members produces two releases.

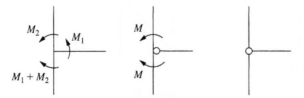

Figure 1.14

For four members meeting at a rigid joint there are three unknown moments, as shown in Figure 1.15; the introduction of a hinge into one of the members produces one release, the introduction of a hinge into two members produces

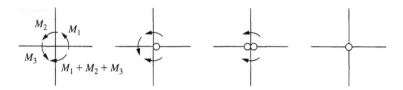

Figure 1.15

two releases, and the introduction of a hinge into all four members produces three releases.

In general, the introduction of hinges into i of the n members meeting at a rigid joint produces i releases. The introduction of a hinge into all n members produces $(n - 1)$ releases.

Thus, the degree of indeterminacy is given by:

$$D = 3n + r - 3j - c$$

where c is the number of releases introduced.

The degree of indeterminacy of the frames shown in Figure 1.16 is:

(i) $D = (3 \times 5) + 3 - (3 \times 4) - 5 = 1$
(ii) $D = (3 \times 6) + 3 - (3 \times 6) - 2 = 1$
(iii) $D = (3 \times 2) + 3 - (3 \times 2) - 2 = 1$
(iv) $D = (3 \times 5) + 9 - (3 \times 6) - 1 = 5$
(v) $D = (3 \times 4) + 5 - (3 \times 4) - 4 = 1$

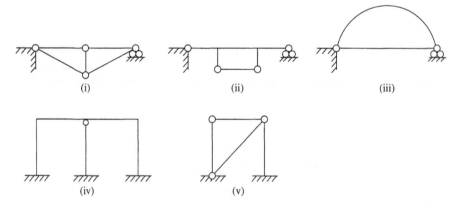

Figure 1.16

Supplementary problems

S1.1 Determine the degree of indeterminacy of the braced beam shown in Figure S1.1.

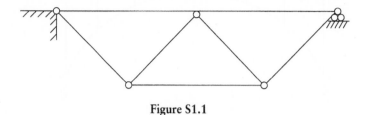

Figure S1.1

S1.2 Determine the degree of indeterminacy of the tied arch shown in Figure S1.2.

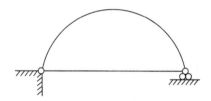

Figure S1.2

S1.3 Determine the degree of indeterminacy of the rigid-jointed frame shown in Figure S1.3.

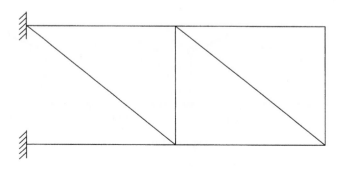

Figure S1.3

S1.4 Determine the degree of indeterminacy of the open spandrel arch shown in Figure S1.4.

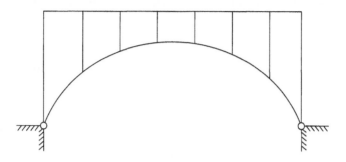

Figure S1.4

S1.5 Determine the degree of indeterminacy of the frames shown in Figure S1.5.

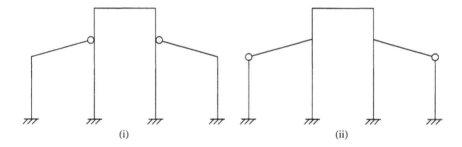

(i)   (ii)

Figure S1.5

S1.6 Determine the degree of indeterminacy of the gable frames shown in Figure S1.6.

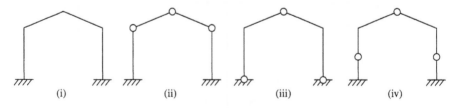

(i)   (ii)   (iii)   (iv)

Figure S1.6

S1.7 Determine the degree of indeterminacy of the bridge structures shown in Figure S1.7.

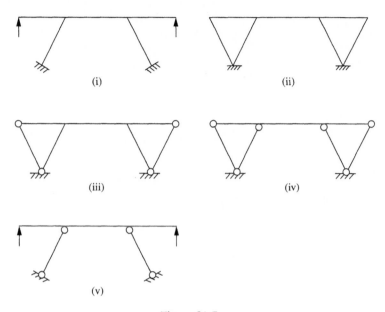

Figure S1.7

References

1. Matheson, J. A. L. Degree of redundancy of plane frameworks. Civil Eng. and Pub. Works. Rev. 52. June 1957. pp. 655–656.
2. Henderson, J. C. De C. and Bickley, W. G. Statical indeterminacy of a structure. Aircraft Engineering. 27. December 1955. pp. 400–402.
3. Rockey, K. C. and Preece, B. W. The degree of redundancy of structures. Civil Eng. and Pub. Works. Rev. 56. December 1961. pp. 1593–1596.
4. Di Maggio, F. C. Statical indeterminacy and stability of structures. Proc. Am. Soc. Civil Eng. 89 (ST3). June 1963. pp. 63–75.
5. Rangasami, K. S. and Mallick, S. K. Degrees of freedom of plane and space frames. The Structural Engineer. 44. March 1966. pp. 109–111 and September 1966. pp. 310–312.

2 Virtual work methods

Notation

A	cross–sectional area of a member
E	Young's modulus
G	modulus of torsional rigidity
H	horizontal reaction
I	second moment of area of a member
I_o	second moment of area of an arch at its crown
l	length of a member
m	bending moment in a member due to a unit virtual load
M	bending moment in a member due to the applied loads
P	axial force in a member due to the applied loads
q	shear force in a member due to a unit virtual load
Q	shear force in a member due to the applied loads
u	axial force in a member due to a unit virtual load
V	vertical reaction
W	applied load
x	horizontal deflection
y	displacement of an applied load in its line of action; vertical deflection
δ	deflection due to the applied load
δl	extension of a member, lack of fit of a member
δx	element of length of a member
$\delta \theta$	relative rotation between two sections in a member due to the applied loads
θ	rotation due to the applied loads
μ	form factor in shear
ϕ	shear deformation due to the applied loads

2.1 Introduction

The principle of virtual work provides the most useful means of obtaining the displacement of a single point in a structure. In conjunction with the principles of superposition and geometrical compatibility, the values of the redundants in indeterminate structures may then be evaluated.

The principle may be defined as follows: if a structure in equilibrium under a system of applied forces is subjected to a system of displacements compatible with the external restraints and the geometry of the structure, the total work done by the applied forces during these external displacements equals the work done by the internal forces, corresponding to the applied forces, during the internal deformations, corresponding to the external displacements. The expression "virtual work" signifies that the work done is the product of a real loading system and imaginary displacements or an imaginary loading system and real displacements. Thus, in Chapters 2 and 3 and Sections 6.2, 6.4, 6.5, 9.8, and 10.3 displacements are obtained by considering virtual forces undergoing real displacements, while in Sections 7.8–7.14, 9.5–9.8, and 11.3 equilibrium relationships are obtained by considering real forces undergoing virtual displacements.

A rigorous proof of the principle based on equations of equilibrium has been given by Di Maggio[1]. A derivation of the virtual-work expressions for linear structures is given in the following section.

2.2 Virtual-work relationships

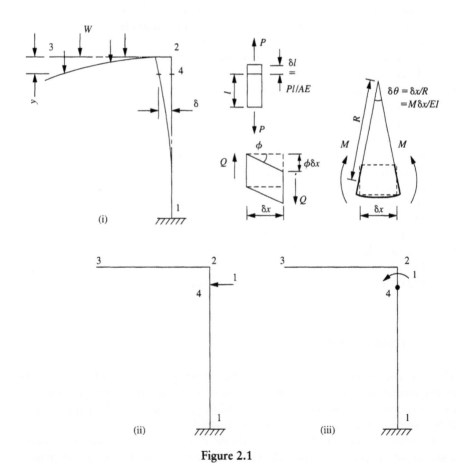

Figure 2.1

To the structure 123 shown in Figure 2.1 (i), the external loads W are gradually applied. This results in the deflection of any point 4 a distance δ while each load moves a distance y in its line of action. The loading produces an internal force P and an extension δl in any element of the structure, a bending moment M and a relative rotation $\delta\theta$ to the ends of any element, a shear force Q, and a shear strain ϕ. The external work done during the application of the loads must equal the internal energy stored in the structure from the principle of conservation of energy.

Then:

$$\sum Wy/2 = \sum P\delta l/2 + \sum M\delta\theta/2 + \sum Q\phi\ \delta x/2 \tag{1}$$

To the unloaded structure a unit virtual load is applied at 4 in the direction of δ as shown in Figure 2.1 (ii). This results in a force u, a bending moment m, and a shear force q in any element.

Now, while the virtual load is still in position, the real loads W are gradually applied to the structure. Again, equating external work and internal energy:

$$\sum Wy/2 + 1 \times \delta = \sum P\ \delta l/2 + \sum M\ \delta\theta/2 + \sum Q\phi\ \delta x/2 + \sum u\delta l \\ + \sum m\ \delta\theta + \sum q\phi\ \delta x \tag{2}$$

Subtracting expression (1) from expression (2):

$$1 \times \delta = \sum u\ \delta l + \sum m\ \delta\theta + \sum q\phi\ \delta x$$

For pin-jointed frameworks, with the loading applied at the joints, only the first term on the right-hand side of the expression is applicable.

Then:

$$1 \times \delta = \sum u\ \delta l \\ = \sum Pul/AE$$

where P is the internal force in a member due to the applied loads and l; A and E are its length, area, and modulus of elasticity; and u is the internal force in a member due to the unit virtual load.

For rigid frames, only the last two terms on the right-hand side of the expression are significant.

Then:

$$1 \times \delta = \sum m\ \delta\theta + \sum q\phi\ \delta x \\ = \int Mm\ dx/EI + \int Qq\ dx/\mu AG$$

where M and Q are the bending moment and shear force at any section due to the applied loads and I, G and A are the second moment of area, the rigidity modulus, and the area of the section; μ is the form factor; and m and q are the bending moment and shear force at any section due to the unit virtual load.

Usually the term representing the deflection due to shear can be neglected, and the expression reduces to:

$$1 \times \delta = \int Mm \ dx/EI$$

In a similar manner, the rotation θ of any point 4 of the structure may be obtained by applying a unit virtual bending moment at 4 in the direction of θ. Then:

$$1 \times \theta = \int Mm \ dx/EI + \int Qq \ dx/\mu AG$$

where m and q are the bending moment and shear force at any section due to the unit virtual moment.

2.3   Sign convention

For a pin-jointed frame, tensile forces are considered positive and compressive forces negative. Increase in the length of a member is considered positive and decrease in length negative. The unit virtual load is applied to the frame in the anticipated direction of the deflection. If the assumed direction is correct, the deflection obtained will have a positive value. The deflection obtained will be negative when the unit virtual load has been applied in the opposite direction to the actual deflection.

For a rigid frame, moments produced by the virtual load or moment are considered positive, and moments produced by the applied loads, which are of opposite sense, are considered negative. A positive value for the displacement indicates that the displacement is in the same direction as the virtual force or moment.

2.4   Illustrative examples

Example 2.1

Determine the horizontal and vertical deflection of point 4 of the pin-jointed frame shown in Figure 2.2. All members have a cross-sectional area of $8 \, in^2$ and a modulus of elasticity of $29,000 \, kips/in^2$.

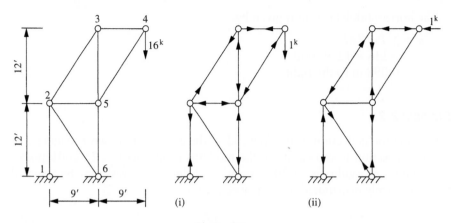

Figure 2.2

Solution

Member forces u_1 due to a vertical unit load at 4 are obtained from (i) and are tabulated in Table 2.1. The stresses in each member, P/A, due to the real load of 16 kips are given by:

$$P/A = 16u_1 \times 1/8$$
$$= 2u_1$$

Member forces u_2 due to a horizontal unit load at 4 are obtained from (ii) and are tabulated in Table 2.1.

Table 2.1 Determination of forces and displacements in Example 2.1

Member	P/A	l	u_1	u_2	Pu_1l/A	Pu_2l/A
12	2.0	12	1.00	−2.67	24.00	−64.0
23	2.5	15	1.25	−1.67	46.88	−62.5
34	1.5	9	0.75	−1.00	10.12	−13.5
45	−2.5	15	−1.25	0	46.88	0
56	−4.0	12	−2.00	1.33	96.00	−64.0
53	−2.0	12	−1.00	1.33	24.00	−32.0
52	−1.5	9	−0.75	0	10.12	0
26	0	15	0	1.67	0	0
Total					258.0	−236.0

The vertical deflection is given by:

$$v = \sum Pu_1l/AE$$
$$= 258 \times 12/29{,}000$$
$$= 0.107 \text{ in downward}$$

The horizontal deflection is given by:

$$h = \sum Pu_2l/AE$$
$$= -236 \times 12/29,000$$
$$= 0.098 \text{ in to the right}$$

Example 2.2

Determine the deflection at the free end of the cantilever shown in Figure 2.3. The cross-section is shown at (i), the modulus of elasticity is 29,000 kips/in², the modulus of rigidity is 11,200 kips/in², and the shear stress may be assumed to be uniformly distributed over the web area.

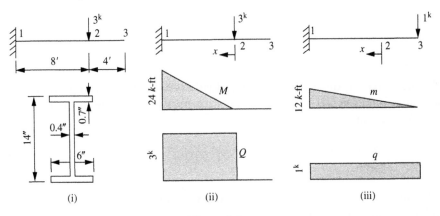

Figure 2.3

Solution

The origin of coordinates is taken at 2 and the functions M and Q derived from (ii) as:

$$M = 3x$$
and $Q = 3$

A unit vertical load is applied at 3 and the functions m and q derived from (iii) as:

$$m = 4 + x$$
and $q = 1$.

The moment of inertia and the area of the web are given by:

$$I = 6 \times (14)^3/12 - 5.6 \times (12.6)^3/12$$
$$= 440 \text{ in}^4$$
$$A = 0.4(14 - 1.4)$$
$$= 5.04 \text{ in}^2$$

Since M and Q are zero over the length 23, the vertical deflection at 3 is given by:

$$1 \times \delta = \int_0^8 Mm \ dx/EI + \int_0^8 Qq \ dx/AG$$
$$= 3 \times 12^3 \int_0^8 (4x + x^2) \ dx/(29,000 \times 440)$$
$$+ 3 \times 12 \int_0^8 dx/(11,200 \times 5.04)$$
$$= 0.121 + 0.005$$
$$= 0.126 \text{ in}$$

Example 2.3

Determine the vertical deflection of point 4 of the pin-jointed frame shown in Figure 2.2 if members 12 and 23 are made 0.1 in too short and members 56 and 53 are made 0.1 in too long.

Solution

Member forces u_1 due to a vertical downward unit load at 4 have already been determined in Example 2.1 and are tabulated in Table 2.2.

Table 2.2 Determination of forces and displacements in Example 2.3

Member	δl	u_1	$u_1 \delta l$
12	−0.1	1.0	−0.1
23	−0.1	1.25	−0.125
53	0.1	−1.0	−0.1
56	0.1	−2.0	−0.2
Total			−0.525

The vertical deflection is given by:

$$v = \sum u_1 \delta l$$
$$= -0.525$$
$$= 0.525 \text{ in upward}$$

Example 2.4

Determine the deflection at the free end of the cantilever shown in Figure 2.4. The moment of inertia has a constant value I over the length 23 and increases linearly from I at 2 to $2I$ at 1.

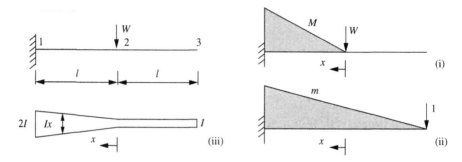

Figure 2.4

Solution

The origin of coordinates is taken at 2 and the function M derived from (i) as:

$M = Wx$.

A unit vertical load is applied at 3 and the function m derived from (ii) as:

$m = l + x$.

The moment of inertia is given by:

$I_x = I + Ix/l$.

Since the value of M is zero over the length 23, the vertical deflection at 3 is given by:

$$1 \times \delta = \int_0^l Mm \, dx/EI$$
$$= \int_0^l Wx(l + x) \, dx/EI(1 + x/l)$$
$$= W \int_0^l lx \, dx/EI$$
$$= Wl^3/2EI$$

2.5 Volume integration

For straight prismatic members, $\int Mm \, dx/EI = \int Mm \, dx \times 1/EI$. The function m is always either constant along the length of the member or varies linearly. The function M may vary linearly for real concentrated loads or parabolically for real distributed loads. Thus, $\int Mm \, dx$ may be regarded as the volume of a solid with a cross-section defined by the function M and a height defined by the function m. The volume of this solid is given by the area of cross-section multiplied by the height of the solid at the centroid of the cross-section.

The value of $\int Mm \, dx$ has been tabulated[2] for various types of functions M and m, and common cases are given in Table 2.3.

Table 2.3 Volume integrals

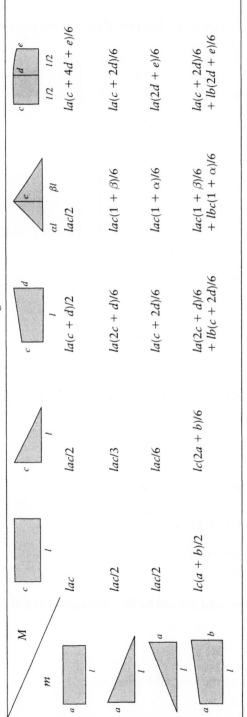

M / m	(rectangle) c, l	(triangle) c, l	(trapezoid) c, d, l	(triangle) αl, βl, c	(parabola) c, d, e, $1/2$, $1/2$
(rectangle) a, l	lac	$lac/2$	$la(c+d)/2$	$lac/2$	$la(c+4d+e)/6$
(triangle) a, l	$lac/2$	$lac/3$	$la(2c+d)/6$	$lac(1+\beta)/6$	$la(c+2d)/6$
(triangle) a, l	$lac/2$	$lac/6$	$la(c+2d)/6$	$lac(1+\alpha)/6$	$la(2d+e)/6$
(trapezoid) a, b, l	$lc(a+b)/2$	$lc(2a+b)/6$	$la(2c+d)/6$ $+ lb(c+2d)/6$	$lac(1+\beta)/6$ $+ lbc(1+\alpha)/6$	$la(c+2d)/6$ $+ lb(2d+e)/6$

Example 2.5

Determine the deflection at the free end of the cantilever shown in Figure 2.5.

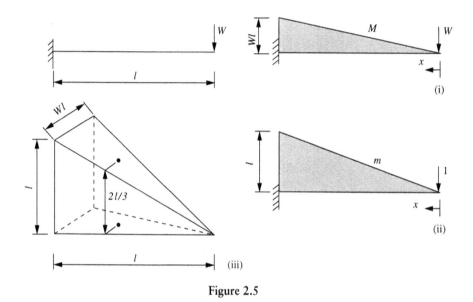

Figure 2.5

Solution

The origin of coordinates is taken at the free end and the functions M and m derived from (i) and (ii) as:

$$M = Wx$$
and $m = x$

The deflection at the free end is given by:

$$1 \times \delta = \int_0^l Mm \, dx/EI$$
$$= W \int_0^l x^2 \, dx/EI$$
$$= Wl^3/3EI$$

Alternatively, the solid defined by the functions M and m is shown at (iii); its volume is:

$$Wl^2/2 \times 2l/3 = Wl^3/3$$

and the deflection at the free end is given by:

$$\delta = Wl^3/3EI.$$

Alternatively, from Table 2.3, the value of $\int Mm\ \mathrm{d}x/EI$ is given by:

$\delta = lac/3EI$
$\quad = Wl^3/3$

Example 2.6

Determine the rotation at the free end of the cantilever shown in Figure 2.6.

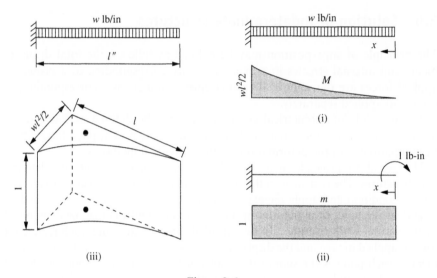

Figure 2.6

Solution

The functions M and m are derived from (i) and (ii) as:

$M = wx^2/2$
and $m = 1$

The rotation at the free end is given by:

$$1 \times \theta = \int_0^l Mm\ \mathrm{d}x/EI$$
$$= w \int_0^l x^2\ \mathrm{d}x/2EI$$
$$= wl^3/6EI.$$

Alternatively, the solid defined by these functions is shown at (iii); its volume is:

$Wl^2 \times l/6 = wl^3/6$

and the rotation at the free end is given by:

$\theta = wl^3/6EI.$

From Table 2.3, the value of $\int Mm\ dx/EI$ is given by:

$\theta = l(c + 4d + e)/6EI$
$\quad = l(wl^2/2 + wl^2/2 + 0)/6EI$
$\quad = wl^3/6EI$

2.6 Solution of indeterminate structures

The principle of superposition may be defined as follows: the total displacements and internal stresses in a linear structure corresponding to a system of applied forces are the sum of the displacements and stresses corresponding to each load applied separately.

The principle of geometrical compatibility may be defined as follows: the displacement of any point in a structure due to a system of applied forces must be compatible with the deformations of the individual members.

The two above principles may be used to evaluate the redundants in indeterminate structures. The first stage in the analysis is to cut back the structure to a determinate condition and apply the external loads. The displacements corresponding to and at the point of application of the removed redundants may be determined by the virtual-work relations. To the unloaded cut-back structure, each redundant force is applied in turn and the displacements again determined. The total displacement at each point is the sum of the displacements due to the applied loads and the redundants and must be compatible with the deformations of the individual members. Thus, a series of compatibility equations is obtained equal in number to the number of redundants. These equations are solved simultaneously to obtain the redundants and the remaining forces obtained from equations of equilibrium.

Example 2.7

Determine the reaction in the prop of the propped cantilever shown in Figure 2.7 (a) when the prop is firm and rigid, (b) when the prop is rigid and settles an amount y, and (c) when the prop is elastic.

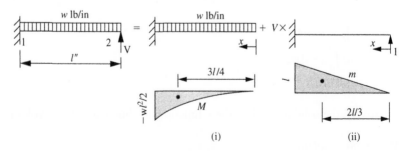

Figure 2.7

Solution

(a) The structure is one degree redundant, and the reaction in the prop is selected as the redundant and removed as shown at (i). The deflection of the free end of the cantilever in the line of action of V is:

$$\delta_2' = -wl^3/6 \times 3l/4EI$$
$$= -wl^4/8EI$$

To the cut-back structure, the redundant V is applied as shown at (ii). The deflection of the free end of the cantilever in the line of action of V is:

$$\delta_2'' = Vl^2/2 \times 2l/3EI$$
$$= Vl^3/3EI$$

The total deflection of 2 in the original structure is:

$$\delta_2 = \delta_2' + \delta_2''$$
$$= 0$$

Thus:

$$-wl^4/8EI + Vl^3/3EI = 0$$

and

$$V = 3wl/8$$

(b) The total deflection of 2 in the original structure is:

$$\delta_2 = \delta_2' + \delta_2''$$
$$= -y$$

Thus:

$$-wl^4/8EI + Vl^3/3EI = -y$$

and

$$V = 3wl/8 - 3EIy/l^3$$

(c) The total deflection of 2 in the original structure is:

$$\delta_2 = \delta_2' + \delta_2''$$
$$= -VL/AE$$

where L, A, and E are the length, cross-section, and modulus of elasticity of the prop.

Thus:

$$-wl^4/8EI + Vl^3/3EI = -VL/AE$$

and

$$V = wl^4/8EI(l^3/3EI + L/AE)$$

Example 2.8

The parabolic arch shown in Figure 2.8 has a second moment of area that varies directly as the secant of the slope of the arch rib. The value of the thrust
required to restore the arch to its original span is H_1, and the value of the
thrust required to reduce the deflection of 2 to zero is H_2. Determine the ratio
of H_2 to H_1, neglecting the effects of axial and shearing forces.

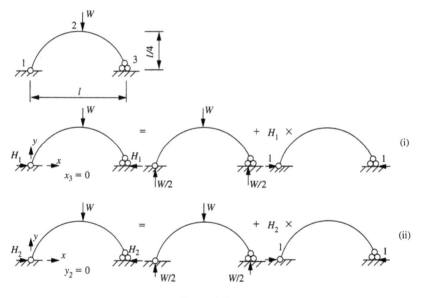

Figure 2.8

Solution

The equation of the arch axis, taking the origin of coordinates at 1, is:

$$y = x(l - x)/l$$

The second moment of area at any section is given by:

$$I = I_o \sec \alpha$$
$$= I_o \, ds/dx$$

where I_o is the second moment of area at the crown.

The horizontal deflection at 3 due to W is obtained by considering a virtual unit load applied horizontally inwards at 3. Then from (i) the deflection due to W is obtained by integrating over the length of the arch:

$$
\begin{aligned}
1 \times x_3' &= \int Mm \ ds/EI \\
&= \int Mm \ dx/EI_o \\
&= -2\int_0^{l/2} Wxy \ dx/2EI_o \\
&= -W\int_0^{l/2} (lx^2 - x^3) \ dx/lEI_o \\
&= -5Wl^3/192EI_o
\end{aligned}
$$

The horizontal deflection at 3 due to H_1 is:

$$
\begin{aligned}
1 \times x_3'' &= 2\int_0^{l/2} H_1 y^2 \ dx/EI_o \\
&= 2H_1 \int_0^{l/2} (l^2 x^2 - 2lx^3 + x^4) \ dx/l^2 EI_o \\
&= H_1 l^3/30EI_o
\end{aligned}
$$

The total horizontal deflection of 3 in the original structure is:

$$
\begin{aligned}
x_3 &= x_3' + x_3'' \\
&= 0
\end{aligned}
$$

Thus,

$$
H_l = 25W/32
$$

The vertical deflection of 2 due to W is obtained by considering a virtual unit load applied vertically upwards at 2. Then, from (ii):

$$
\begin{aligned}
1 \times y_2' &= -2\int_0^{l/2} Wx^2 \ dx/4EI_o \\
&= -W\int_0^{l/2} x^2 \ dx/2EI_o \\
&= -Wl^3/48
\end{aligned}
$$

The vertical deflection at 2 due to H_2 is:

$$
\begin{aligned}
1 \times y_2'' &= 2\int_0^{l/2} H_2 xy \ dx/EI_o \\
&= H_2 \int_0^{l/2} (lx^2 - x^3) \ dx/l^2 EI_o \\
&= 5H_2 l^3/192EI_o
\end{aligned}
$$

The total vertical deflection of 2 in the original structure is:

$$y_2 = y_2' + y_2'' = 0$$

Thus,

$$H_2 = 4W/5$$

and

$$H_2/H_1 = 128/125$$
$$= 1.024$$

Example 2.9

The parabolic arch shown in Figure 2.9 has a second moment of area that varies directly as the secant of the slope of the arch rib. Determine the bending moment at 2, neglecting the effects of axial and shearing forces.

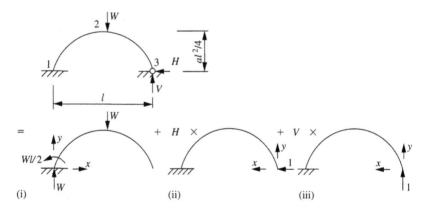

Figure 2.9

Solution

The redundant forces consist of the reactions H and V at 3, and the cut-back structure is a curved cantilever.

The equation of the arch axis is:

$$y = ax(l - x)$$

The horizontal deflection of 3 due to condition (i) is obtained by considering a virtual-unit load applied horizontally inwards at 3. Then, over the span from $x = 0$ to $x = l/2$:

$$m = y$$

and

$$M = Wl/2 - Wx$$

Thus the horizontal deflection at 3 is:

$$\begin{aligned}
x_3' &= W \int_0^{l/2} (l/2 - x) y \, dx/EI_o \\
&= Wa \int_0^{l/2} (l^2 x/2 - 3lx^2/2 + x^3) \, dx/EI_o \\
&= Wal^4/64EI_o
\end{aligned}$$

The horizontal deflection of 3 due to condition (ii) is:

$$\begin{aligned}
x_3'' &= H \int_0^l y^2 \, dx/EI_o \\
&= Ha^2 \int_0^l (l^2 x^2 - 2lx^3 + x^4) \, dx/EI_o \\
&= Ha^2 l^5/30EI_o
\end{aligned}$$

The horizontal deflection of 3 due to condition (iii) is:

$$\begin{aligned}
x_3''' &= -V \int_0^l xy \, dx/EI_o \\
&= -Va \int_0^l (lx^2 - x^3) \, dx/EI_o \\
&= Val^4/12EI_o
\end{aligned}$$

The total horizontal deflection of 3 in the original structure is:

$$x_s = x_3' + x_3'' + x_3''' = 0$$

Thus:

$$W/32 - V/6 + Hal/15 = 0 \tag{1}$$

The vertical deflection of 3 due to condition (i) is obtained by considering a virtual-unit load applied vertically upwards at 3. Then, over the span from $x = 0$ to $x = l/2$:

$$m = l - x$$

and

$$M = Wx - Wl/2$$

Thus, the vertical deflection of 3 is:

$$
\begin{aligned}
y_3' &= W \int_0^{l/2} (x - l/2)(l - x) \, dx/EI_o \\
&= -W \int_0^{l/2} (l^2/2 - 3lx/2 + x^2) \, dx/EI_o \\
&= -5Wl^3/48EI_o
\end{aligned}
$$

The vertical deflection of 3 due to condition (ii) is:

$$
\begin{aligned}
y_3'' &= -H \int_0^l yx \, dx/EI_o \\
&= -Hal^4/12EI_o
\end{aligned}
$$

The vertical deflection of 3 due to condition (iii) is:

$$
\begin{aligned}
y_3''' &= V \int_0^l x^2 \, dx/EI_o \\
&= Vl^3/3EI_o
\end{aligned}
$$

The total vertical deflection of 3 in the original structure is:

$$
y_s = y_3' + y_3'' + y_3''' = 0
$$

Thus,

$$
-5W/16 + V - Hal/4 = 0 \tag{2}
$$

Solving (1) and (2) simultaneously:

$$
H = 5W/6al
$$

and

$$
V = 25W/48
$$

Thus, the bending moment at 2 is:

$$
\begin{aligned}
M_2 &= Vl/2 - Hal^2/4 \\
&= 5Wl/96
\end{aligned}
$$

Supplementary problems

S2.1 Determine the deflection at the free end of the cantilever shown in Figure S2.1.

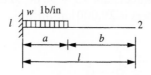

Figure S2.1

S2.2 Determine the vertical deflection at the center of the pin-jointed truss shown in Figure S2.2 if the top chord members are all 0.1 percent in excess of the required length.

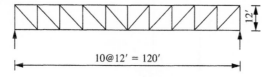

Figure S2.2

S2.3 The pin-jointed truss shown in Figure S2.3 has a vertical load of 20 kips applied at panel point 2. Determine the resulting vertical deflection at panel points 2 and 3. What additional vertical load W must be applied at panel point 3 to increase the deflection at panel point 2 by 50 percent? All members have a cross-sectional area of 2 in^2 and a modulus of elasticity of 29,000 kips/in^2.

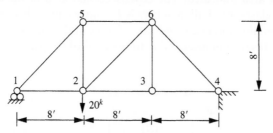

Figure S2.3

S2.4 A cantilever circular arch rib with a uniform section is shown in Figure S2.4. A horizontal force H is applied to the free end of the rib so that end 2 can deflect only vertically when the vertical load V is applied at 2. Determine the ratio of H/V.

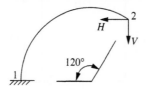

Figure S2.4

S2.5 The frame shown in Figure S2.5 has a vertical load of 10 kips applied at 5. Determine the resulting vertical deflection at node 4. The cross-sectional areas of the members are 13 = 10 in², 24 = 5 in², and 34 = 4 in². The second moment of areas of the members are 13 = 90 in⁴ and 24 = 50 in⁴. The modulus of elasticity of all members is 29,000 kips/in².

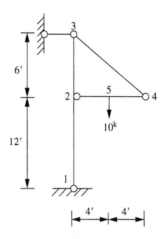

Figure S2.5

S2.6 The two-story, single-bay frame shown in Figure S2.6 has the relative second moments of area indicated. The cross-sectional area of the columns is A. The modulus of elasticity of all members is E, and the modulus of rigidity is G. Determine the bending moments and shear forces in the members and calculate the horizontal deflection of node 3 due to the load of 4 kips.

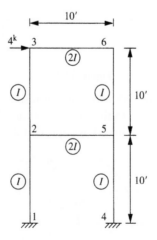

Figure S2.6

S2.7 The parabolic arch shown in Figure S2.7 has a second moment of area that varies directly as the secant of the slope of the arch rib and a temperature coefficient of thermal expansion of α per °F. Neglecting the effects of axial and shearing forces, determine the support reactions at the fixed end produced by a temperature rise of t °F.

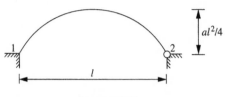

$al^2/4$

Figure S2.7

References

1. Di Maggio, I. F. Principle of virtual work in structural analysis. Proc. Am. Soc. Civil Eng. 86 (ST 11). Nov. 1960. pp. 65–78.
2. Ghose, S. The volume integration method of structural analysis. The Structural Engineer. 42. March 1964. pp. 95–100.

3 Indeterminate pin-jointed frames

Notation

A cross-sectional area of a member
A_t cross-sectional area of the tie of a tied arch
E Young's modulus
H horizontal reaction
I second moment of area of a member
I_o second moment of area of an arch at its crown
l length of a member; span of an arch
m bending moment in a member due to a unit virtual load applied to the cut-back structure
M bending moment in a member due to the external loads applied to the cut-back structure
P axial force in a member due to the external loads applied to the cut-back structure
R redundant force in a member
t change in temperature
u axial force in a member due to a unit virtual load applied to the cut-back structure
V vertical reaction
y settlement of a support
α temperature coefficient of expansion
δ deflection
δl lack of fit
δx spread of arch abutments

3.1   Introduction

The solution of indeterminate pin-jointed frames and two-hinged arches may be readily obtained using the principles of superposition and compatibility. In the case of polygonal two-hinged arches, the preferred solution is by the method of column analogy and is dealt with in Section 6.3.

3.2   Frames one degree redundant

The pin-jointed frame shown in Figure 3.1 contains one redundant member 12 with its unknown force R, assumed tensile. The indeterminate frame can be replaced by system (i) plus $R \times$ system (ii).

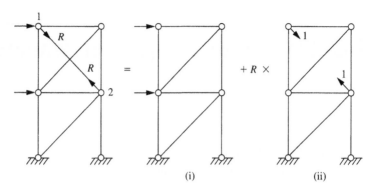

Figure 3.1

The relative inward movement of the points of application of R in system (i) may be determined by the virtual work method by considering R replaced by virtual unit loads. The relative movement is:

$$\delta'_{12} = \sum Pul/AE$$

where P and u are the member forces in the cut-back structure due to the applied loads and the virtual unit loads, respectively.

The relative inward movement of the points of application of R in system (ii) is:

$$\delta''_{12} = \sum u^2 l/AE$$

The relative movement of the points of application of R in the actual structure is outward and consists of the extension in member 12, which is:

$$\delta_{12} = -Rl_{12}/A_{12}E_{12}$$
$$= \delta'_{12} + R\delta''_{12}$$

Thus: $R(l_{12}/A_{12}E_{12} + \sum u^2 l/AE) = - \sum Pul/AE$

If the summations are considered to extend over all the members of the actual structure with $P_{12} = 0$ and $u_{12} = 1$, we obtain:

$$R = - (\sum Pul/AE)/(\sum u^2 l/AE)$$

and the value obtained for R is positive if tensile and negative if compressive.

The actual force in any member is given by $(P + uR)$.

The horizontal deflection of joint 1 of the actual structure may be determined by considering a virtual unit load applied horizontally at 1 in systems (i) and (ii). The horizontal deflection is:

$$\begin{aligned}
\delta_1 &= \delta'_1 + R\delta''_1 \\
&= \sum Pu_1 l/AE + R\sum uu_1 l/AE \\
&= \sum (P + uR)u_1 l/AE
\end{aligned}$$

where u_1 is the force in any member of the cut-back structure due to the horizontal virtual unit load, $(P + uR)$ is the force in any member of the actual structure, and the summation extends over all members of the cut-back structure. In general, to determine the deflection of an indeterminate structure, the unit virtual load may be applied to any cut-back structure that can support it.

Initial inaccuracies in the lengths of members of an indeterminate frame produce forces in the members when they are forced into position. If the member 12 in Figure 3.2 is made too short by an amount $-\delta l$, a tensile force, R, is produced in it on forcing it into position.

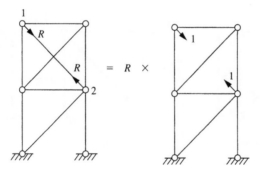

Figure 3.2

Then, the relative movement of points 1 and 2 in the actual structure consists of the extension in member 12 and the initial lack of fit, and:

$$\delta_{12} = -Rl_{12}/A_{12}E_{12} + (-\delta l)$$

There are no applied loads on the structure, thus $\delta'_{12} = 0$ and $\delta''_{12} = \sum u^2 l/AE$ as before.
Thus:

$$\delta_{12} = \delta'_{12} + R\delta''_{12}$$

and:

$$R(l_{12}/A_{12}E_{12} + \sum u^2 l/AE) = -\delta l$$

If the summation is considered to extend over all the members of the actual structure with $u_{12} = 1$, we obtain:

$$R = -\delta l/(\sum u^2 l/AE)$$

where δl is the lack of fit of the member and is positive if too long and negative if too short, and the value obtained for R is positive if tensile and negative if compressive.

The remaining member forces are given by the expression uR.

Example 3.1

All members of the frame shown in Figure 3.3 have a constant value for AE of 10,000 kips. Determine (a) the forces in the members due to the applied load of 10 kips, (b) the horizontal deflection of point 2 due to the applied load of 10 kips, (c) the force in member 24 due to member 24 being 0.1% too long.

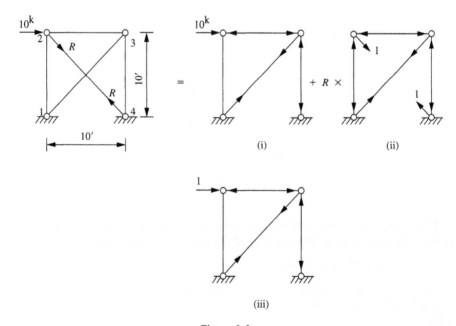

Figure 3.3

Solution

(a) The force R in member 24 is chosen as the redundant, and member forces P and u are obtained from (i) and (ii) and are tabulated in Table 3.1.

Table 3.1 Determination of forces and displacements in Example 3.1

Member	P	l	u	Pul	u^2l	$P + uR$	u_1	$u_1l(P + uR)$
12	0.0	10.0	−0.707	0.0	5.0	5.56	0.0	0.0
23	−10.0	10.0	−0.707	70.7	5.0	−4.44	−1.0	44.4
34	−10.0	10.0	−0.707	70.7	5.0	−4.44	−1.0	44.4
13	14.14	14.14	1.0	200.0	14.14	6.26	1.414	125.2
24	0.0	14.14	1.0	0.0	14.14	−7.88	0.0	0.0
Total				341.4	43.28			214.0

Thus the force in member 24 is:

$$R = -\sum Pul/\sum u^2l$$
$$= -341.4/43.28$$
$$= 7.88 \text{ kips compression}$$

The remaining member forces are given by the expression $(P + uR)$.

(b) The forces u_1 in the cut-back structure due to a virtual unit load applied horizontally at 2 are obtained from (iii) and are tabulated in Table 3.1. The horizontal deflection of 2 is:

$$\delta_2 = \sum (P + uR)u_1l/AE$$
$$= 214 \times 12/10,000$$
$$= 0.257 \text{ in}$$

(c) The force R in member 24 due to the lack of fit is:

$$R = -\delta l/(\sum u^2l/AE)$$
$$= -141.4/43.28$$
$$= 3.27 \text{ kips compression}$$

Example 3.2

Determine the forces in the members of the structure shown in Figure 3.4. All members have the same cross-sectional area and modulus of elasticity.

Solution

The force R in member 13 is chosen as the redundant, and member forces P and u are obtained from (i) and (ii) and are tabulated in Table 3.2.
Thus the force in member 13 is:

$$R = -\sum Pul/\sum u^2l$$
$$= -44.7/21.03$$
$$= 2.13 \text{ kips compression}$$

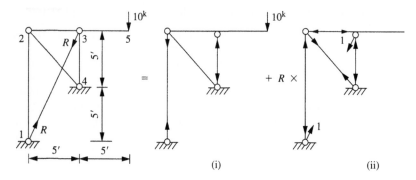

Figure 3.4

Table 3.2 Determination of forces in Example 3.2

Member	P	l	u	Pul	u^2l	P + uR
12	10	10	−0.447	−44.7	2.0	10.95
23	0	5	−0.447	0	1.0	0.95
34	−20	5	−0.894	89.4	4.0	−18.10
24	0	7.07	0.632	0	2.83	−1.35
13	0	11.2	1.000	0	11.2	−2.13
Total				44.7	21.03	

The remaining member forces are given by the expression $(P + uR)$.

3.3 Frames two degrees redundant

The pin-jointed frame shown in Figure 3.5 contains the two redundant members 12 and 34, with unknown forces R_1 and R_2 assumed tensile. The indeterminate frame can be replaced by system (i) plus $R_1 \times$ system (ii) plus $R_2 \times$ system (iii).

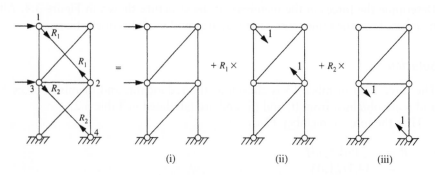

Figure 3.5

The relative movement of points 1 and 2 in the actual structure consists of the extension in member 12 and is:

$$\delta_{12} = -R_{12}l_{12}/A_{12}E_{12}$$

The relative inward movement of points 1 and 2 in systems (i), (ii), and (iii) is:

$$\delta'_{12} + \delta''_{12} + \delta'''_{12} = \sum Pu_1 l/AE + \sum u_1^2 l/AE + \sum u_1 u_2 l/AE$$

where the summations extend over all the members of the cut-back structure. Since:

$$\delta_{12} = \delta'_{12} + R_1 \delta''_{12} + R_2 \delta'''_{12}$$

then:

$$0 = \sum Pu_1 l/AE + R_1 \sum u_1^2 l/AE + R_2 \sum u_1 u_2 l/AE \tag{1}$$

where the summations extend over all the members of the actual structure, $P = 0$ for both redundant members, $u_1 = 1$ for member 12 and 0 for member 34, and $u_2 = 1$ for member 34 and 0 for member 12.

Similarly, by considering the relative movement of points 3 and 4 we obtain:

$$0 = \sum Pu_2 l/AE + R_1 \sum u_1 u_2 l/AE + R_2 \sum u_2^2 l/AE \tag{2}$$

Equations (1) and (2) may be solved simultaneously to obtain the value of the redundant forces R_1 and R_2. The actual force in any member is given by the expression $(P + u_1 R_1 + u_2 R_2)$.

Frames with more than two redundants are best solved by the flexibility matrix method given in Section 10.3.

3.4 Frames redundant externally

The reactions V_1 and V_2 may be considered the external redundants of the frame shown in Figure 3.6. The displacements corresponding to and in the line of action of V_1 and V_2 are:

$$\delta_1 = \sum Pu_1 l/AE + V_1 \sum u_1^2 l/AE + V_2 \sum u_1 u_2 l/AE$$

and:

$$\delta_2 = \sum Pu_2 l/AE + V_1 \sum u_1 u_2 l/AE + V_2 \sum u_2^2 l/AE$$

where P, u_1, and u_2 are the member forces in the cut-back structure due to the applied loads, the virtual unit load corresponding to V_1, and the virtual unit load corresponding to V_2, respectively.

In the case of rigid supports, $\delta_1 = \delta_2 = 0$, and the two equations may be solved simultaneously to obtain V_1 and V_2.

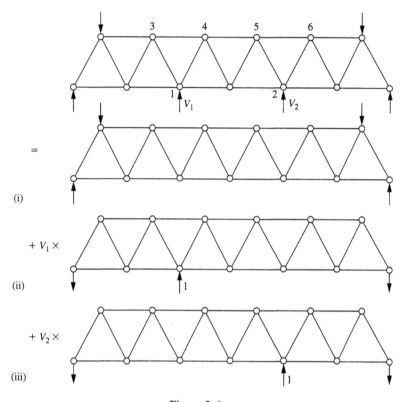

Figure 3.6

In the case of settlement of the supports by amounts y_1 at support 1 and y_2 at support 2, $\delta_1 = -y_1$ and $\delta_2 = -y_2$. Again, the two equations obtained may be solved for V_1 and V_2.

An alternative method of solving the problem is to consider members 34 and 56 as internal redundants. The values of the redundants may be obtained by the method of Section 3.3.

3.5   Frames with axial forces and bending moments

When the loading on pin-jointed frames is applied between the panel points as shown in Figure 3.7, bending moments are produced in some members in addition to axial forces. The force R in member 12 is regarded as the redundant, and the inward movement of the points of application of R in system (i) is:

$$\delta'_{12} = \sum Pul/AE + \sum \int Mm \, dx/EI$$

where M and m are the bending moments at any point in a member of the cut-back structure due to the applied loads and the virtual unit loads corresponding to R, respectively.

The inward movement of the points of application of R in system (ii) is:

$$\delta''_{12} = \Sigma u^2 l/AE + \Sigma \int m^2 \, dx/EI$$

The inward movement of the points of application of R in the actual structure is:

$$\delta_{12} = -Rl_{12}/A_{12}E_{12}$$
$$= \delta'_{12} + R\delta''_{12}$$

Thus $R(\Sigma u^2 l/AE + \Sigma \int m^2 \, dx/EI) = \Sigma Pul/AE + \Sigma \int Mm \, dx/EI$

where the summations extend over all the members of the actual structure with $P_{12} = 0$ and $u_{12} = 1$. In the particular structure shown in Figure 3.7, $\int Mm \, dx/EI = 0$ for all members, and $\int m^2 \, dx/EI$ is applicable only to member 23.

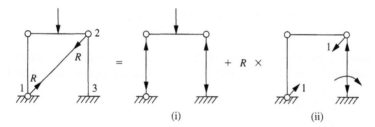

Figure 3.7

Example 3.3

Determine the tension in the member 13 that connects the cantilever to the bracket 234 as shown in Figure 3.8. The cross-sectional areas, second moment of areas, and modulus of elasticity of the members are $A_{13} = 1$ in^2, $A_{34} = A_{23} = 2$ in^2, $I_{12} = 1440$ in^4, $E_{12} = 10,000$ kips/in^2, and $E_{13} = E_{23} = E_{34} = 30,000$ kips/in^2.

Solution

The force in member 13 is considered as the redundant, and the actual structure is replaced by system (i) plus $R \times$ system (ii).

In system (i), member 12 is subjected to a bending moment, and there are no axial forces in the members. Then, over the cantilever from $x = 0$ to $x = 8.66$ ft:

$$M = -15x$$

and:

$$m = x$$

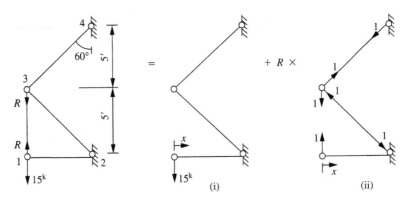

Figure 3.8

and the relative inward movement of 1 and 3 is:

$$\delta'_{13} = \int_1^2 Mm\,dx/EI$$
$$= -15\int_0^{8.66} x^2\,dx/EI$$
$$= -15(8.66)^3 \times (12)^2/\,(3 \times 10{,}000 \times 1440)$$
$$= -\,32.47 \times 10^{-3}\ \text{ft}$$

In system (ii), member 12 is subjected to a bending moment, and members 23 and 34 are subjected to axial forces, and the relative inward movement of 1 and 3 is:

$$\delta''_{13} = \int_1^2 m^2\,dx/\,EI + \sum u^2 l/AE$$
$$= \int_0^{8.66} x^2\,dx/EI + 2 \times (1)^2 \times 10 \times 12^2\,/AE$$
$$= (8.66)^3 \times (12)^2/(3 \times 10{,}000 \times 1440) + 2 \times 10/(2 \times 30{,}000)$$
$$= (2.17 + 0.33) \times 10^{-3}$$
$$= 2.50 \times 10^{-3}\ \text{ft}$$

The extension in member 13 in the actual structure is:

$$\delta_{13} = -R \times 5(1 \times 30{,}000)$$
$$= -R \times 0.167 \times 10^{-3}\ \text{ft}$$

Thus:

$$-0.167R = -32.47 + 2.50R$$

and:

$$R = 12.18\ \text{kips}$$

Example 3.4

Determine the maximum bending moment in the braced beam shown in Figure 3.9. The cross-section areas of the members are $A_{12} = 10$ in^2, $A_{34} = 5$ in^2,

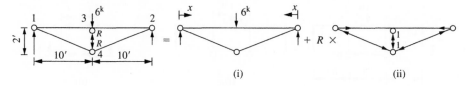

Figure 3.9

$A_{14} = A_{24} = 2.5$ in^2. The second moment of area of member 12 is 1440 in^4, and the modulus of elasticity is constant for all members.

Solution

The force in member 43 is considered the redundant, and the actual structure is replaced by system (i) plus $R \times$ system (ii).

$\int Mm \, dx/EI$ and $\int m^2 \, dx/EI$ are applicable only to member 12 with $M = 3x$ and $m = x/2$, $\Sigma Pul/AE = 0$ and $\Sigma u^2 l/AE$ is applicable to all members:

$$\int_1^2 Mm \, dx/EI = 2 \times 3 \int_0^{10} x^2 \, dx/2EI$$
$$= 1000 \times 12^4/1440E$$
$$= 100 \times 12^2/E$$
$$\int_1^2 m^2 \, dx/EI = 2 \int_0^{10} x^2 \, dx/4EI$$
$$= 16.7 \times 12^2/E$$
$$\Sigma u^2 l/AE = 2 \times (2.55)^2 \times 10.2 \times 12^2/2.5E + (2.50)^2 \times 20$$
$$\times 12^2/10E + 1 \times 2 \times 12^2/5E$$
$$= (53 + 12.5 + 0.4) \times 12^2/E$$
$$= 65.9 \times 12^2/E$$

Thus:

$$-R(16.7 + 65.9) = 100$$

and:

$$R = 1.21 \text{ kips compression}$$

The vertical component of the tensile force in member 14 is 0.605 kips. Thus, the maximum bending moment in the beam at 3 is:

$$M_3 = (3 - 0.605) \times 10$$
$$= 23.95 \text{ kip ft}$$

Example 3.5

Determine the force in member 56 of the frame shown in Figure 3.10. All members have the same cross-sectional area, modulus of elasticity, and second moment of area.

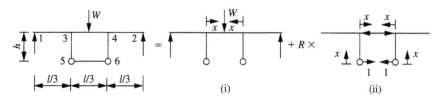

Figure 3.10

Solution

The tensile force in member 56 is considered the redundant, and the actual structure is replaced by system (i) plus $R \times$ system (ii). $\int Mm\, dx/EI$ is applicable only to member 34, with $M = -(Wl/6 + Wx/2)$ and $m = h$. $\int m^2 dx/EI$ is applicable to member 34 and also to members 53 and 64, with $m = x$. $\Sigma Pul/AE = 0$, and $\Sigma u^2 l/AE$ is applicable only to members 34 and 56 with $u = 1$.
Then:

$$
\begin{aligned}
\int_3^4 Mm\, dx/EI &= -2Wh\int_0^{l/6}(l/6 + x/2)dx/EI \\
&= -5Whl^2/72EI \\
\Sigma\int m^2 dx/EI &= \int_3^4 m^2 dx/EI + 2\int_5^3 m^2 dx/EI \\
&= h^2\int_0^{l/3}dx/EI + 2\int_0^h x^2 dx/EI \\
&= h^2 l/3EI + 2h^3/3EI \\
\Sigma u^2 l/AE &= 2 \times 1 \times l/3AE \\
&= 2l/3AE
\end{aligned}
$$

Thus:

$$R(h^2 l/3EI + 2h^3/3EI + 2l/3AE) = 5Whl^2/72EI$$

3.6 Two-hinged arch

The two-hinged arch shown in Figure 3.11 is one degree indeterminate, with the horizontal reaction H considered the external redundant. The value of H may be determined from system (i) and system (ii) by considering the effects of

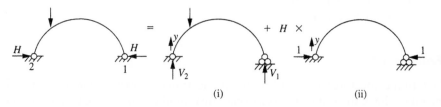

Figure 3.11

moments, axial thrust, and shear on the arch rib. Usually the arch axis closely follows the funicular polygon for the applied loads, and shear effects are small and may be neglected.

The inward movement of 1 in system (i) is:

$$x_1' = \int Mm\, ds/EI + \int Pu\, ds/AE$$

where P and u are the axial thrusts in the cut-back structure due to the applied loads and the virtual unit loads, respectively, and M and m are the moments in the cut-back structure due to the applied loads and the virtual unit loads, respectively.

Thus:

$$x_1' = \int My\, ds/EI + \int V \sin\alpha \cos\alpha\, ds/AE$$
$$\approx \int My\, ds/EI$$

The inward movement of 1 in system (ii) is:

$$x_1'' = \int m^2\, ds/EI + \int u^2\, ds/AE$$
$$= \int y^2\, ds/EI + \int \cos^2\alpha\, ds/AE$$
$$= \int y^2\, ds/EI + \int \cos\alpha\, dx/AE$$
$$= \int y^2\, ds/EI + l/AE$$

where l is the arch span and α is the slope of the arch axis.

When the arch abutments are rigid, the movement of 1 in the actual structure is:

$$x_1 = 0$$

and:

$$x_1 = x_1' + Hx_1''$$

Thus:

$$H = -\int My\, ds/EI/\left(\int y^2\, ds/EI + l/AE\right)$$

When the rise to span ratio of the arch exceeds 0.2, the rib-shortening effects may be neglected and:

$$H = -\int My\, ds/EI/\int y^2\, ds/EI$$

In the case of spread of the abutments by an amount δx:

$$x_1 = -\delta x$$

and:

$$H = -\left(\delta x + \int My\, ds/EI\right)/\int y^2\, ds/EI$$

A change in temperature from that at which the arch was erected causes a change in H. For a rise in temperature of $t°F$ and a coefficient of thermal expansion α:

$$x_1' = -\alpha t l + \int My\, ds/EI$$

and:

$$H = -(\alpha t l + \int My\, ds/EI)/\int y^2\, ds/EI$$

where t is positive for a rise in temperature and negative for a fall in temperature.

In the case of reinforced-concrete arch ribs, the shrinkage of the concrete has a similar effect to a fall in temperature. For concrete with an ultimate shrinkage strain of 300×10^{-6} in/in and a coefficient of thermal expansion of 6×10^{-6} in/in/°F, the shrinkage is equivalent to a fall in temperature of 50°F.

The axial thrust, shear, and bending moment at any section 3 of the arch may be obtained from Figure 3.12 as:

$$P = H\cos\alpha + V\sin\alpha - W\sin\alpha$$
$$Q = H\sin\alpha - V\cos\alpha + W\cos\alpha$$
$$M = -Hy + Vx - Wx_{34}$$

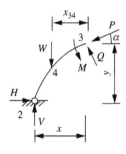

Figure 3.12

Example 3.6

The parabolic arch shown in Figure 3.13 has a second moment of area that varies directly as the secant of the slope of the arch rib. Neglecting the effects of axial and shear forces, determine the horizontal thrust at the supports and the bending moment at a point 20 ft from the left-hand support.

Solution

The equation of the arch axis, taking the origin at 1, is:

$$y = x(80 - x)/160$$

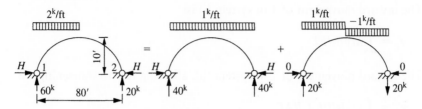

Figure 3.13

The second moment of area at any section is given by:

$$I = I_o \sec \alpha$$
$$= I_o \, ds/dx$$

where I_o is the second moment of area at the crown.
Thus:

$$H = -\int My\,ds/EI / \int y^2 \, ds/EI$$
$$= -\int My\,dx / \int y^2 \, dx$$
$$= -2\int_0^{40}(x^2/2 - 40)y\,dx/2\int_0^{40} y^2 \, dx$$
$$= -160\int_0^{40}(-x^4/2 + 80x^3 - 3200x^2)dx/\int_0^{40}(x^4 - 160x^3 + 6400x^2)dx$$
$$= 80 \text{ kips}$$

At $x = 20$ ft, $y = 7.5$ ft, and the bending moment in the rib is:

$$M = 60 \times 20 - 80 \times 7.5 - 2 \times (20)^2/2$$
$$= 200 \text{ kip-ft with tension on the inside of the arch}$$

3.7   The tied arch

The two-hinged tied arch shown in Figure 3.14 is one degree indeterminate, and the tension in the tie may be considered as the redundant.

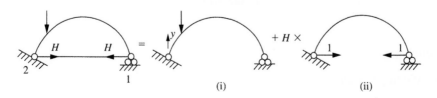

Figure 3.14

The inward movement of 1 in system (i) is:

$$x_1' = \int My\,ds/EI$$

The inward movement of 1 in system (ii), allowing for rib shortening, is:

$$x_1'' = \int y^2\,ds/EI + l/AE$$

The movement of 1 in the actual structure is outward and consists of the extension in the tie, which is given by:

$$x_1 = -Hl/A_tE$$

where A_t is the cross-sectional area of the tie.
Then:

$$x_1 = x_1' + x_1''$$

and:

$$H = -\int My\,ds/EI/(\int y^2\,ds/EI + l/AE + l/A_tE)$$

Example 3.7

The two parabolic arches shown in Figure 3.15 have second movement of areas that vary directly as the secant of the slope of the arch ribs and have the same value at the crown. Neglecting the effects of axial and shear forces, determine the horizontal component of the axial forces in the arches.

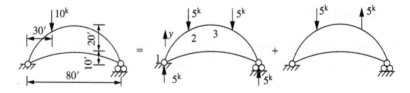

Figure 3.15

Solution

For the upper arch:

$$y = 3x(80 - x)/160$$

and:

$$\int My\,ds/EI = 2\int_1^2 My\,dx/EI_o + 2\int_2^3 My\,dx/EI_o$$
$$= -2\int_0^{30} 5xy\,dx/EI_o - 2\int_{30}^{40} 150y\,dx/EI_o$$
$$= -185{,}000/EI_o$$
$$H\int y^2\,ds/EI = 2H\int_0^{40} y^2\,dx/EI_o$$
$$= 38{,}400H/EI_o$$

For the tie, $y = x(80 - x)/160$, and the extension is:

$$H\int y^2\,ds/EI = 2H\int_0^{40} y^2\,dx/EI_o$$
$$= 4270H/EI_o$$

Thus:

$$H = 185{,}000/(38{,}400 + 4270)$$
$$= 4.35 \text{ kips}$$

3.8 Spandrel braced arch

The two-hinged spandrel braced arch shown in Figure 3.16 is one degree indeterminate, and the horizontal reaction at the hinges may be regarded as an
external redundant. Allowing for a rise in temperature $t°F$ and a spread of the
abutments by an amount δx, the reaction is:

$$H = (\alpha tl - \delta x - \sum Pul/AE)/(\sum u^2 l/AE)$$

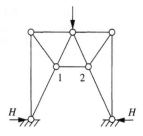

Figure 3.16

An alternative procedure is to regard member 12 as an internal redundant
and obtain a solution by the method of Section 3.2.

Example 3.8

Determine the member forces in the two-hinged spandrel braced arch shown in Figure 3.17. All members have the same cross-sectional area and the same modulus of elasticity.

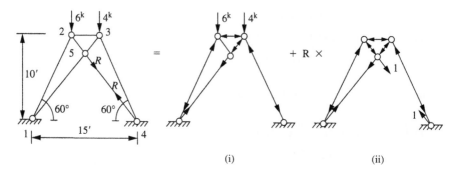

(i)   (ii)

Figure 3.17

Solution

The force R in member 54 is chosen as the redundant, and member forces P and u are obtained from (i) and (ii) and are tabulated in Table 3.3.

Table 3.3 Determination of forces in Example 3.8

Member	P	l	u	Pul	u^2l	$P + uR$
12	−6.94	11.56	−0.85	68.2	8.36	−4.30
23	−3.47	3.44	−1.10	13.1	4.16	−0.05
34	−5.53	11.56	−0.85	54.3	8.36	−2.89
51	1.06	11.05	1.0	11.7	11.05	−2.06
53	1.06	2.55	1.0	2.7	2.55	−2.06
52	0	2.55	1.0	0.0	2.55	−3.11
54	0	11.05	1.0	0.0	11.05	−3.11
Total				150.0	48.08	

Thus:

$$R = -\ \sum Pul / \sum u^2l$$
$$= -150 / 48.08$$
$$= 3.11 \text{ kips compression}$$

The remaining member forces are given by the expression $(P + uR)$.

Supplementary problems

S3.1 All members of the pin-jointed frame shown in Figure S3.1 have a constant value for *AE* of 60,000 kips. Member 24 is fabricated 1/8 in too long. Determine the resultant force in member 24 due to the lack of fit and the applied load of 20 kips.

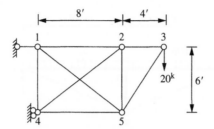

Figure S3.1

S3.2 All members of the pin-jointed frame shown in Figure S3.2 have a constant value for *AE* of 60,000 kips. Member 26 is fabricated 1/20 in too long. Determine the resultant force in members 24 and 26 due to the lack of fit.

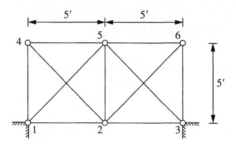

Figure S3.2

S3.3 All members of the frame shown in Figure S3.3 are of uniform section. Determine the force in member 13 due to the applied load *W*.

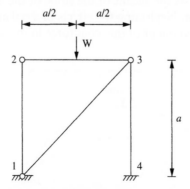

Figure S3.3

S3.4 The trussed beam shown in Figure S3.4 is constructed with a wood beam 12 and strut 34 and 1 in diameter steel tie rods 41 and 42. The cross-sectional area and second moment of area of the beam are 86 in² and 952 in⁴. The cross-sectional area of the strut is 26 in², and the ratio of the modulus of elasticity of steel and wood is 18. Calculate the force in the strut caused by a uniformly distributed load of 1 kip/ft over the beam.

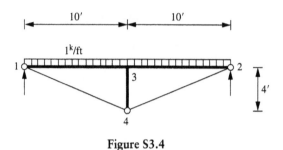

Figure S3.4

S3.5 The tied parabolic arch shown in Figure S3.5 has a flexural rigidity that varies directly as the secant of the slope of the arch rib and has a value of EI_o at the crown. The area and modulus of elasticity of the tie rod are A_t and E_t, and the ratio $EI_o/A_tE_t = 2.5$ ft². Determine the force in the tie rod caused by the 100 kip concentrated load at the crown.

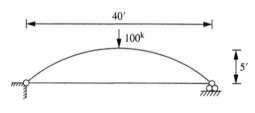

Figure S3.5

S3.6 The two parabolic arches shown in Figure S3.6 have second movement of areas that vary directly as the secant of the slope of the arch ribs and have the same value at the crown. Neglecting the effects of axial and shear forces, determine the horizontal component of the axial force in the arches.

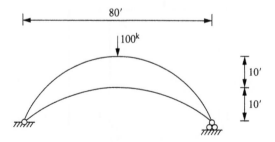

Figure S3.6

S3.7 The two-hinged arch shown in Figure S3.7 has a uniform cross-section throughout. Neglecting the effects of axial and shear forces, determine the horizontal thrust at the supports caused by the 10 kip load located as indicated.

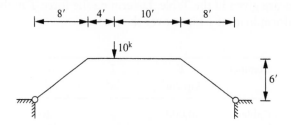

Figure S3.7

S3.8 The parabolic arch rib shown in Figure S3.8 has a flexural rigidity that varies directly as the secant of the slope of the arch rib and has a value of EI_o at the crown. The flexural rigidity of both columns also has a value of EI_o. Determine the horizontal thrust at the supports caused by the concentrated load W at the crown. Neglect the effects of axial and shear force.

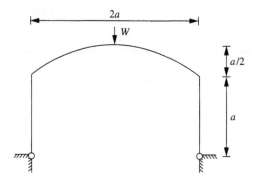

Figure S3.8

S3.9 The cable-stayed bridge shown in Figure S3.9 consists of a continuous main girder 12, supported on rollers where it crosses the rigid piers, and two cables that are continuous over frictionless saddles at the tops of the towers. The modulus of elasticity, cross-sectional areas, and second moments of area of the members are given in the Table. Determine the force T in the cables produced by a uniform load of 1 kip/ft over the girder.

Member	E kips/in^2	I in^4	A in^2
Cable	30,000	–	20
Tower	3000	–	100
Girder	3000	100,000	1000

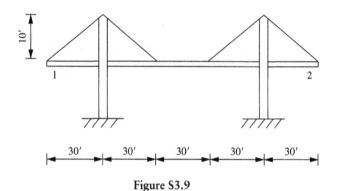

Figure S3.9

4 Conjugate beam methods

Notation

A	cross-sectional area of a member
E	Young's modulus
I	second moment of area of a member
l	length of a member
M	bending moment in a member due to the applied load
M'	bending moment in a conjugate member due to the elastic load
P	axial force in a member due to the applied load
Q	shear force in a member due to the applied load
Q'	shear force in a conjugate member due to the elastic load
R	redundant force in a member
w	intensity of applied load on a member
w'	intensity of elastic load on a conjugate member, M/EI
W	applied load on a member, $\int w\,dx$
W'	elastic load on a conjugate member, $\int M\,dx/EI$
x	horizontal deflection
y	vertical deflection
δ	deflection due to the applied load
δ_{ij}	deflection at i due to a unit load applied at j
δ_{12}	extension produced in member 12 by the applied load
Δ_2	angle change at joint 2 of a triangular frame due to the applied load
ε_{12}	strain produced in member 12 by the applied load
θ	rotation due to the applied load

4.1 Introduction

The conjugate beam method may be used to obtain an expression for the entire deflection curve over the whole of a structure. This, in combination with Müller-Breslau's principle, may then be used to obtain influence lines for the structure.

The method may also be used to determine fixed-end moments and support reactions in continuous beams and frames, though generally the column analogy and moment distribution methods are preferred methods of solution.

4.2   Derivation of the method

The simply supported beam shown in Figure 4.1 is subjected to an applied loading of intensity w, positive when acting upward. The shear force at any section is given by the area under the load intensity curve as:

$$Q = \int w \, dx$$

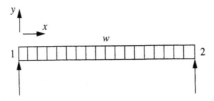

Figure 4.1

with shear force upward on the left of a section regarded as positive. The bending moment at any section is given by the area under the shear force curve as:

$$M = \int Q \, dx$$

with the bending moment producing tension in the bottom fiber regarded as positive. In addition, the curvature at any section is given by:

$$d^2y/dx^2 = 1/R$$
$$= M/EI$$

and the slope and deflection at any section are given by:

$$dy/dx = \theta$$
$$= \int M \, dx/EI$$
$$y = \delta$$
$$= \iint M \, dx/EI$$

with x positive to the right and y positive upward.

An analogous beam, known as the conjugate beam, shown in Figure 4.2, is subjected to an applied loading of intensity:

$$w' = M/EI$$

where M is the bending moment in the actual beam at any section, and $\int M/EI$ is known as the elastic load.

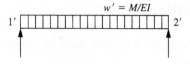

<figure>Figure 4.2</figure>

The shear and bending moment at any section in the conjugate beam are given by:

$$Q' = \int w' \, dx$$
$$= \int M \, dx/EI$$
$$= \theta$$
$$M' = \int Q' \, dx$$
$$= \iint M \, dx/EI$$
$$= \delta$$

where θ and δ are the slope and deflection at any section in the real beam.

Thus, the slope and deflection at any section in the real beam are given by the shear and bending moment at that section in the conjugate beam, and the elastic curve of the real beam is given by the bending moment diagram of the conjugate beam. The end slope and end deflection of the real beam are given by the end reaction and end moment of the conjugate beam. The maximum deflection in the real beam occurs at the position of zero shear in the conjugate beam.

4.3 Sign convention

The conjugate structure consists of the centerlines of the real structure and is placed in a horizontal plane.

In the case of beams, the elastic load applied to the conjugate beam is positive (i.e., acts vertically upward) when the bending moment in the real beam is positive (i.e., tension in the bottom fiber). The deflection of the real beam at any section is positive (i.e., upward) when a positive bending moment occurs at the corresponding section of the conjugate beam. The slope of the real beam is positive when a positive shear force occurs at the corresponding section of the conjugate beam.

In the case of frames, the elastic load applied to the conjugate frame is positive (i.e., acts vertically upward) when the outside fiber of the real frame is in compression. The displacement of the real frame at any section is perpendicular to the lever arm used to determine the moment in the conjugate frame and outward when a positive bending moment occurs at the corresponding section of the conjugate frame.

4.4 Support conditions

The restraints of the conjugate structure must be consistent with the displacements of the real structure.

At a simple end support in the real structure, shown in Table 4.1 (i), there is a rotation but no deflection. Thus, the corresponding restraints in the conjugate structure must be a shear force and a zero moment, which are produced by a simple end support in the conjugate structure. The elastic load on the conjugate beam is upward and is expressed as $W' = Wl^2/8EI$. The slope of the real beam at 1 and the deflection at 3 are:

$$\theta_1 = -W'/2$$
$$= -Wl^2/16EI \text{ ... clockwise}$$
$$\delta_3 = \theta_1 l/2 + W'l/12$$
$$= -Wl^3/48EI \text{ ... downward}$$

At a fixed end in the real structure, shown in Table 4.1 (ii), there is neither a rotation nor a deflection. Thus, there must be no restraint at the corresponding point in the conjugate structure, which must be a free end. The elastic loads on the conjugate beam consist of a downward load due to the fixing moments of $W'_1 = -Ml/EI$ and an upward load due to the free bending moment of $W'_2 = Wl^2/8EI$. For equilibrium of the conjugate beam, $W'_1 = -W'_2$ and $M = Wl/8$, with tension in the top fiber as shown.

At a free end in the real structure, shown in Table 4.1 (iii), there is both a rotation and a deflection. Thus, the corresponding restraints in the conjugate structure are a shear force and a bending moment, which are produced at the fixed end. The elastic load on the conjugate beam is downward and is expressed as $W' = -Wl^2/2EI$. The slope and deflection of the real beam at 2 are:

$$\theta_2 = -W'$$
$$= Wl^2/2EI \text{ ... clockwise}$$
$$\delta_2 = 2W'l/3$$
$$= -Wl^3/3EI \text{ ... downward}$$

At an interior support in the real structure, shown in Table 4.1 (iv), there is no deflection and a smooth change in slope. Thus, there can be no moment and no reaction at the corresponding point in the conjugate structure, which must be an unsupported hinge. The elastic loads on the conjugate structure consist of downward loads due to the fixing moment of $W'_1 = -Ml/2EI$ and upward loads due to the free bending moments of $W'_2 = Wl^2/8EI$. The slope of the real beam at 1 is:

$$\theta_1 = -W'_1 - W'_2$$
$$= Ml/2EI - Wl^2/8EI$$

Table 4.1

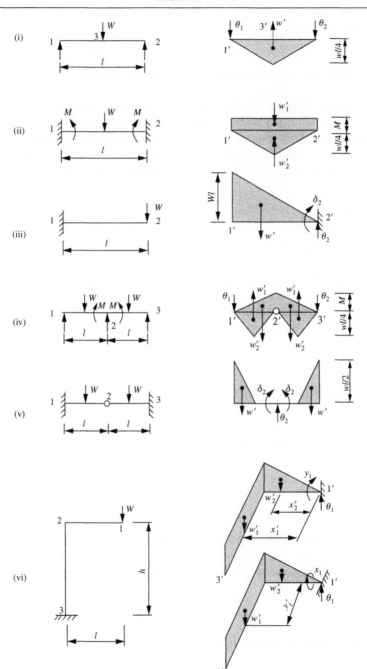

Taking moments about $2'$ for member $1'2'$:

$$0 = l\theta_1 + lW'_1/3 + lW'_2/2$$

And:

$M = 3Wl/16\ldots$ with tension in the top fiber as shown.

At an interior hinge in the real structure, shown in Table 4.1 (v), there is a deflection and an abrupt change of slope. Thus, the corresponding restraints in the conjugate structure are a moment and a reaction, which are produced by an interior support. The elastic loads on the conjugate structure are downward and are expressed as $W' = -Wl^2/8EI$. The slope and deflection of the real beam at 2 are:

$$\theta_2 = -2W'$$
$$= Wl^2/4EI$$
$$\delta_2 = 5lW'/6$$
$$= -5Wl^3/48EI$$

The outside fibers of both members of the real frame, shown in Table 4.1 (vi), are in tension throughout. Thus, the elastic loads on both members of the conjugate frame are downward and are expressed as $W'_1 = -Whl/EI$ and $W'_2 = -Wl^2/2EI$, respectively. The vertical deflection of the real frame at 1 is:

$$y_1 = \sum W'x'$$
$$= W'_1 x'_1 + W'_2 x'_1$$
$$= -Whl^2/EI - Wl^3/3EI$$

and is inward (downward) since the bending moment at $1'$ produces tension in the top fiber. The horizontal deflection of the real frame at 1 is:

$$x_1 = \sum W'y'$$
$$= W'_1 y'_1$$
$$= -Wh^2/2EI$$

The sense of this deflection is obtained by considering the deflection of the real frame at 2. The bending moment at $2'$ produces tension in the top fiber, and thus the deflection of 2 is inward (to the right) and the deflection of 1 must also be to the right.

4.5 Illustrative examples

Example 4.1

Determine the deflection and rotation at the free end of the cantilever shown in Figure 4.3 (i) and derive the equation of the elastic curve.

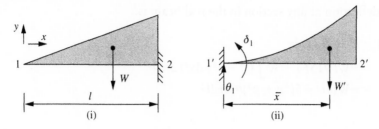

Figure 4.3

Solution

For the real beam, the intensity of loading is:

$$w = -2Wx/l^2$$

the shear is:

$$Q = -Wx^2/l^2$$

and the moment is:

$$M = -Wx^3/3l^2$$

For the conjugate beam, the intensity of loading is:

$$w' = -Wx^3/3EIl^2$$

and the rotation at 1 is given by:

$$\theta_1 = -W'$$
$$= \int_0^l Wx^3 \, dx/3EIl^2$$
$$= Wl^2/12EI \ \ldots \text{ anticlockwise}$$

and the deflection at 1 is given by:

$$\delta_1 = W'\bar{x}$$
$$= -\int_0^l w'x \, dx$$
$$= -\int_0^l Wx^4 \, dx/3EIl^2$$
$$= -Wl^3/15EI \ \ldots \text{ downward}$$

The slope at any section in the real beam is:

$$Q' = \theta_1 + \int w' \, dx$$
$$= Wl^2/12EI - \int Wx^3 \, dx/3EIl^2$$
$$= W(l^4 - x^4)/12EIl^2$$

The deflection at any section in the real beam is:

$$M' = \delta_1 + \int Q' \, dx$$
$$= -Wl^3/15EI + W \int (l^4 - x^4) \, dx/12EIl^2$$
$$= -W(4l^5 - 5l^4x + x^5)/60EIl^2$$

Example 4.2

A simply supported steel beam of 20 ft effective span carries a uniformly distributed load of 20 kips. Flange plates are added to the central 10 ft of the beam in order to limit the deflection to 1/480 of the span. Determine the second moment of area required for the central portion if the second moment of area of the plain beam is 200 in^4 and the modulus of elasticity is 29,000 kips/in^2.

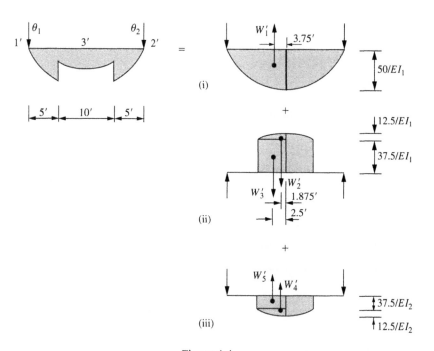

Figure 4.4

Solution

The loading on the conjugate beam is shown in Figure 4.4, which can be replaced by the three loading conditions (i), (ii), and (iii) where I_1 and I_2 are the second moment of areas of the plain beam and the plated beam, respectively, and the bending moment at the center of the real beam is 50 kip-ft.

Thus, the elastic loads are:

$$W'_1 = 0.67 \times 10 \times 50/EI_1$$
$$= 333/EI_1$$
$$W'_2 = -0.67 \times 5 \times 12.5/EI_1$$
$$= -42/EI_1$$
$$W'_3 = -5 \times 37.5/EI_1$$
$$= -187/EI_1$$
$$W'_4 = 0.67 \times 5 \times 12.5/EI_2$$
$$= 42/EI_2$$
$$W'_5 = 5 \times 37.5/EI_2$$
$$= 187/EI_2$$

and:

$$\theta_l = -W'_1 - W'_2 - W'_3 - W'_4 - W'_5$$
$$= -104/EI_1 - 229/EI_2$$

The central deflection of the real beam is:

$$\delta_3 = (10 \times \theta_l) + (3.75 \times W'_1) + (1.875 \times W'_2) + (2.5 \times W'_3)$$
$$+ (1.875 \times W'_4) + (2.5 \times W'_5)$$
$$= -339/EI_1 - 1745/EI_2$$
$$= -20/480$$

Thus:

$$1745/I_2 = 20 \times 29,000/(480 \times 144) - 339/200$$

and:

$$I_2 = 260 \text{ in}^4$$

Example 4.3

Determine the horizontal displacement of the roller 1 of the rigid frame shown in Figure 4.5. All the members of the frame have the same second moment of area and the same modulus of elasticity.

Solution

The bending moment diagram, drawn on the compression side of the members, is shown at (i), and the elastic load on the conjugate frame is shown at (ii). The elastic loads are:

$$W'_1 = 0.5 \times 15 \times 30W/EI$$
$$= 225W/EI$$
$$W'_2 = 9 \times 30W/EI$$
$$= 270W/EI$$

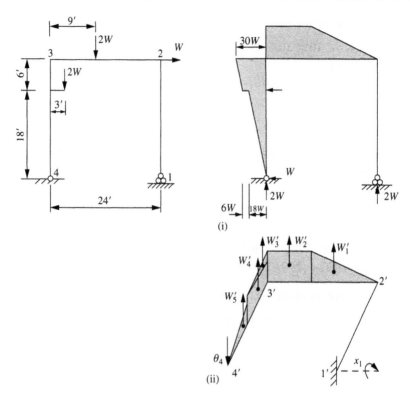

Figure 4.5

$$W'_3 = 0.5 \times 6 \times 6W/EI$$
$$= 18W/EI$$
$$W'_4 = 6 \times 24W/EI$$
$$= 144W/EI$$
$$W'_5 = 0.5 \times 18 \times 18W/EI$$
$$= 162W/EI$$

The horizontal displacement at 1 is:

$$x_1 = \sum W'y'$$
$$= 24W'_1 + 24W'_2 + 22W'_3 + 21W'_4 + 12W'_5$$
$$= 17{,}244W/EI$$

and is outward since the moment at $1'$ produces tension in the bottom fiber.

Example 4.4

Determine the bending moments in the continuous beam shown in Figure 4.6. Flange plates are added to the 24 ft portion of the beam between the applied loads so as to double the second moment of area.

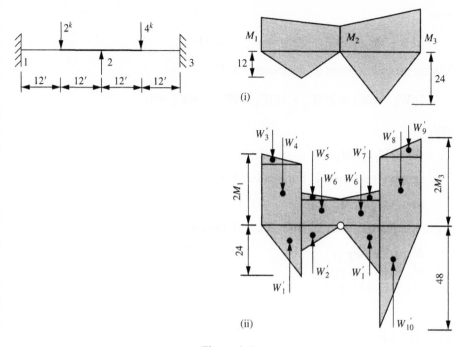

Figure 4.6

Solution

The bending moment diagram is shown at (i), where $-M_1$ and $-M_3$ are the fixed-end moments and $-M_2$ is the support moment.

The elastic loads on the conjugate beam are shown at (ii), where $EI = 1$ over the central portion of the beam and $EI = \frac{1}{2}$ over the end 12 ft portions. The elastic loads are:

$$W_1' = 144$$
$$W_2' = 72$$
$$W_3' = 6(M_1 - M_2)$$
$$W_4' = -12(M_1 + M_2)$$
$$W_5' = -3(M_1 - M_2)$$
$$W_6' = -12\,M_2$$
$$W_7' = -3(M_3 - M_2)$$
$$W_8' = -12(M_3 + M_2)$$
$$W_9' = -6(M_3 - M_2)$$
$$W_{10}' = 288$$

For equilibrium of the conjugate beam:

$$2W_1' + W_2' + W_{10}' + W_3' + W_4' + W_5' + 2W_6' + W_7' + W_8' + W_9' = 0$$

and:

$$216 - 7M_1 - 10M_2 - 7M_3 = 0 \quad \dots (1)$$

Taking moments about 2' for member 1'2':

$$16W'_1 + 8W'_2 + 20W'_3 + 18W'_4 + 8W'_5 + 6W'_6 = 0$$

and:

$$40 - 5M_1 - 2M_2 = 0 \quad \dots (2)$$

Taking moments about 2' for member 2'3':

$$80 - 5M_3 - 2M_2 = 0 \quad \dots (3)$$

Solving equations (1), (2), and (3) simultaneously:

$M_1 = 3.63$ kip-ft
$M_2 = 10.83$ kip-ft
$M_3 = 11.63$ kip-ft

Example 4.5

The grid shown in Figure 4.7 is simply supported at the four corners and consists of members of uniform section pinned together at their intersections. Determine the distribution of bending moment in each member.

Solution

The deflections produced at the interconnections are indicated at (i). The internal reactions at the interconnections may be considered as the redundants and are indicated at (ii). Along the diagonals, these reactions are zero due to the symmetry of the structure and applied loading. The applied loads and deflections of beams 11, 22, and 33 are indicated at (iii), (iv), and (v).

Let

δ_{ij} = deflection at i due to a unit load at j.

Then, the values of δ_{ib} may be obtained from (vi) and (vii) as:

$$\delta_{bb} = M_{b'}$$
$$= -7a^3/8 + 3a^3/24$$
$$= -3a^3/4$$
$$\delta_{cb} = M_{c'}$$
$$= -10a^3/8 + 2a^3/6$$
$$= -11a^3/12$$
$$= \delta_{bc}$$
$$= \delta_{dc}$$

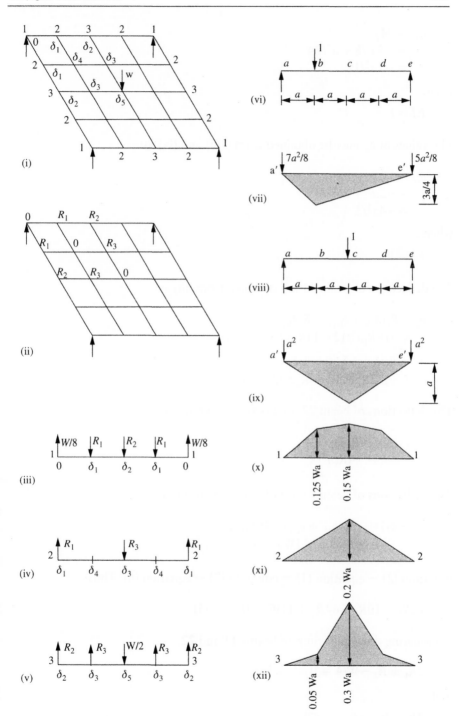

Figure 4.7

$$\delta_{db} = M_{d'}$$
$$= -5a^3/8 + a^3/24$$
$$= -7a^3/12$$

where:

$$EI = 1$$

The values of δ_{ic} may be obtained from (viii) and (ix) as:

$$\delta_{cc} = M_{c'}$$
$$= -2a^3 + 2a^3/3$$
$$= -4a^3/3$$

where:

$$EI = 1$$

The deflections of beam 11 are obtained from (iii) as:

$$\delta_1 = R_1(\delta_{bb} + \delta_{bd}) + R_2\delta_{bc}$$
$$= -16R_1a^3/12 - 11R_2a^3/12 \quad \dots (1)$$
$$\delta_2 = R_1(\delta_{cb} + \delta_{cd}) + R_2\delta_{cc}$$
$$= -22R_1a^3/12 - 4R_2a^3/3 \quad \dots (2)$$

The deflections of beam 22 are obtained from (iv) as:

$$(\delta_3 - \delta_1) = R_3\delta_{cc}$$
$$= -4R_3a^3/3 \quad \dots (3)$$

The deflections of beam 33 are obtained from (v) as:

$$(\delta_3 - \delta_2) = -R_3(\delta_{bb} + \delta_{bd}) + 0.5W\delta_{bc}$$
$$= 4R_3a^3/3 - 11Wa^3/24 \quad \dots (4)$$

Equation (2) − equation (1) = equation (3) − equation (4). Thus:

$$12R_1 + 10R_2 - 64R_3 + 11W = 0 \quad \dots (5)$$

Considering the equilibrium of beams 11 and 22:

$$2R_1 + R_2 - W/4 = 0 \quad \dots (6)$$

and:

$$2R_1 - R_3 = 0 \quad \dots (7)$$

Solving equations (5), (6) and (7) simultaneously, we obtain:

$$R_1 = 0.1W$$
$$R_2 = 0.05W$$
$$R_3 = 0.2W$$

and the bending moments in the members are shown at (x), (xi), and (xii).

When the number of internal reactions exceeds three, the matrix method given by Stanek[2] is a preferred method of solution.

4.6   Pin-jointed frames

The change in slope for a member of a rigid structure is given by:

$$d\delta = Mdx/EI$$

and constitutes the load on a small element dx of the conjugate beam.

The change in slope of a pin-jointed frame is concentrated at the pins and is known as the angle change. Thus, the deflections at the panel points of a pin-jointed frame are given by the bending moments at the corresponding points in a conjugate beam loaded with the total angle change at the panel points.

In the pin-jointed frame shown in Figure 4.8, the applied load produces a force in each member and a change in the length of each member. Due to these changes in length, all angles in the frame change. The total angle changes at the bottom chord panel points are $\Sigma\Delta_2$, $\Sigma\Delta_3$, $\Sigma\Delta_4$, $\Sigma\Delta_5$, and $\Sigma\Delta_6$ and are the applied loads on the conjugate beam at (i), with a decrease of angle giving a positive load. The deflections of panel points 2, 3, 4, 5, and 6 are given by the bending moments at 2′, 3′, 4′, 5′, and 6′ and are positive when the corresponding conjugate beam moment produces tension in the bottom fiber of the conjugate beam.

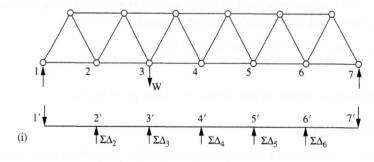

Figure 4.8

The member forces, P, due to the applied loads in a basic triangle of a pin-jointed frame shown in Figure 4.9(i) produce extensions in the members of δ_{12}, δ_{23}, δ_{31}. The angle change at 1 is given by:

$$\Delta_1 = \sum Pul/AE$$

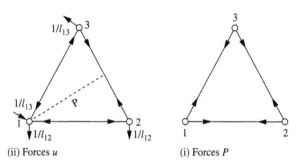

(ii) Forces u (i) Forces P

Figure 4.9

where u is the force in a member due to a unit couple applied to 1 and 3, as shown at (ii).

Thus:

$$\Delta_1 = u_{12}\delta_{12} + u_{23}\delta_{23} + u_{31}\delta_{31}$$
$$u_{12} = -\cot \angle 2/l_{12}$$
$$u_{31} = -\cot \angle 3/l_{31}$$
$$u_{23} = 1/p$$
$$= (\cot \angle 3 + \cot \angle 2)/l_{23}$$

Hence:

$$\Delta_1 = \cot \angle 2(\delta_{23}/l_{23} - \delta_{12}/l_{12}) + \cot \angle 3(\delta_{23}/l_{23} - \delta_{13}/l_{13})$$
$$= \cot \angle 2(\varepsilon_{23} - \varepsilon_{12}) + \cot \angle 3(\varepsilon_{23} - \varepsilon_{13})$$

where ε is the strain produced in a member by the applied loads on the pin-jointed frame with tensile strain positive. Similarly:

$$\Delta_2 = \cot \angle 3(\varepsilon_{13} - \varepsilon_{23}) + \cot \angle 1(\varepsilon_{13} - \varepsilon_{12})$$
$$\Delta_3 = \cot \angle 1(\varepsilon_{12} - \varepsilon_{13}) + \cot \angle 2(\varepsilon_{12} - \varepsilon_{23})$$

A comprehensive review of the method has been given by Lee[3].

Example 4.6

Determine the deflections at the panel points of the pin-jointed frame shown in Figure 4.10. All members of the frame have the same length, area, and modulus of elasticity.

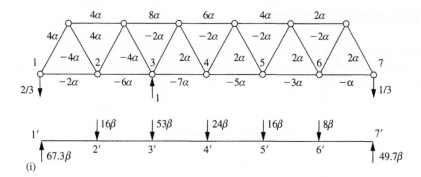

Figure 4.10

Solution

The member forces due to the applied load are indicated on the figure, where $\alpha = 1/3\sqrt{3}$. The total angle changes at the panel points are:

$$\sum \Delta_2 = \{(4 + 4 + 4 + 2) + (4 + 4 + 4 - 4) + (-4 + 6 - 4 - 4)\}\alpha/AE\sqrt{3}$$
$$= 16\beta$$

$$\sum \Delta_3 = \{(4 + 6 + 4 + 4) + (8 + 4 + 8 + 2) + (2 + 2 + 2 + 7)\}\alpha/AE\sqrt{3}$$
$$= 53\beta$$

$$\sum \Delta_4 = \{(-2 + 7 - 2 - 2) + (6 - 2 + 6 + 2) + (2 + 2 + 2 + 5)\}\alpha/AE\sqrt{3}$$
$$= 24\beta$$

$$\sum \Delta_5 = \{(-2 - 2 - 2 + 5) + (4 - 2 + 4 + 2) + (2 + 2 + 2 + 3)\}\alpha/AE\sqrt{3}$$
$$= 16\beta$$

$$\sum \Delta_6 = \{(-2 - 2 - 2 + 3) + (2 - 2 + 2 + 2) + (2 + 2 + 2 + 1)\}\alpha/AE\sqrt{3}$$
$$= 8\beta$$

where $\beta = 1/9AE$. The loads on the conjugate beam are shown at (i).

The panel point deflections are given by the bending moments at the corresponding points in the conjugate beam:

$$\delta_2 = M_{2'} = 67.3l\beta$$
$$\delta_3 = M_{3'} = 118.7l\beta$$
$$\delta_4 = M_{4'}$$
$$= 117.0l\beta$$
$$\delta_5 = M_{5'}$$
$$= 91.3l\beta$$
$$\delta_6 = M_{6'}$$
$$= 49.7l\beta$$

Supplementary problems

Use the conjugate beam method to solve the following problems.

S4.1 Determine the support moment at the end 1 of the fixed-ended beam shown in Figure S4.1.

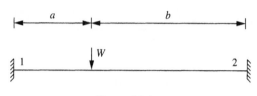

Figure S4.1

S4.2 For the non-uniform beam shown in Figure S4.2, determine the rotation produced at support 1 and the deflection produced at point 3 by the concentrated load W. The relative EI values are shown ringed.

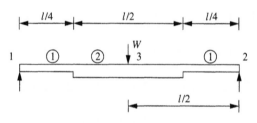

Figure S4.2

S4.3 Determine the deflection at the free end of the cantilever shown in Figure S4.3. A uniformly distributed load w is applied over a length a from the support.

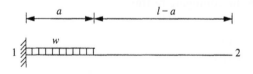

Figure S4.3

S4.4 Determine the fixed-end moments produced by the application of the moment M at point 3 in the non-prismatic beam shown in Figure S4.4. The relative EI values are shown ringed.

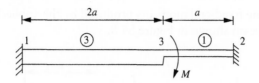

Figure S4.4

S4.5 Determine the fixed-end moments produced by the applied load indicated in the non-prismatic beam shown in Figure S4.5. The relative *EI* values are shown ringed.

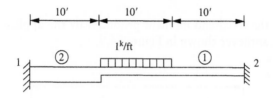

Figure S4.5

S4.6 Determine the fixed-end moments produced by the applied loads indicated in the non-prismatic beam shown in Figure S4.6. The relative *EI* values are shown ringed.

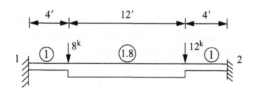

Figure S4.6

S4.7 Determine the fixed-end moments produced by the applied loads indicated in the non-prismatic beam shown in Figure S4.7. The relative *EI* values are shown ringed.

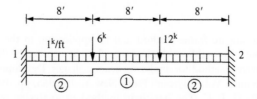

Figure S4.7

S4.8 Determine the fixed-end moments produced by the applied loads indicated in the prismatic beam shown in Figure S4.8.

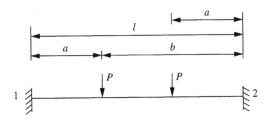

Figure S4.8

S4.9 Determine the fixed-end moment produced by the applied load indicated in the propped cantilever shown in Figure S4.9.

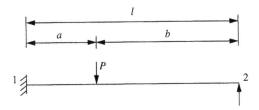

Figure S4.9

S4.10 Determine the fixed-end moments produced by the load of 10 kips applied at point 3 as indicated in the non-prismatic beam shown in Figure S4.10. The relative *EI* values are shown ringed.

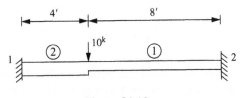

Figure S4.10

References

1. Lee, S. L. The conjugate frame method and its application in the elastic and plastic theory of structures. Journal Franklin Institute. 266. September 1958. pp. 207–222.
2. Stanek, F. J. A matrix method for analyzing beam grillages. Developments in theoretical and applied mechanics. Vol. 1. Plenum Press. New York. 1963. pp. 488–503.
3. Lee, S. L. and Patel, P. C. The bar-chain method of analysing truss deformations. Proc. Am. Soc. Civil Eng. 86 (ST3). May 1960. pp. 69–83.

5 Influence lines

Notation

A	cross–sectional area of a member
E	modulus of elasticity
I	second moment of area of a member
l	length of a member
M	bending moment in a member due to the applied loads
M'	bending moment in the conjugate beam
P	axial force in a member due to the applied loads
Q	shear force in a member due to the applied loads
Q'	shear force in the conjugate beam
R	redundant force in a member due to the applied loads
V	vertical reaction
w'	intensity of loading on the conjugate beam
W	applied load
W'	elastic load on the conjugate beam, $M\,dx/EI$
δ_{ij}	deflection at i due to a unit load at j
δx	element of length of a member
$\delta\theta$	relative rotation between two sections in a member due to the applied loads
θ	rotation

5.1 Introduction

An influence line for a structure is a curve showing the variation in shear, moment, member force, or external reaction due to a load traversing the structure. Influence lines for statically indeterminate structures may be obtained by the applications of Müller-Breslau's principle and Maxwell's reciprocal theorem.

Influence lines for arches and multibay frames may be obtained by the methods given in Sections 6.4 and 7.14. A comprehensive treatment of the determination of influence lines for indeterminate structures has been given by Larnach[1].

5.2   General principles

(a) Müller–Breslau's principle

The influence line for any restraint in a structure is the elastic curve produced by the corresponding unit virtual displacement applied at the point of application of the restraint. The term "displacement" is used in its general sense, the displacement corresponding to a moment is a rotation and to a force, a linear deflection. The displacement is applied in the same direction as the restraint.

To obtain the influence line for reaction in the prop of the propped cantilever shown in Figure 5.1, a unit virtual displacement is applied in the line of action of V. The displacement produced under a unit load at any point i is δ_i. Then, applying the virtual work principle:

$$V \times (\delta = 1) = (W = 1) \times \delta_i$$

that is: $V = \delta_i$

and the elastic curve is the influence line for V.

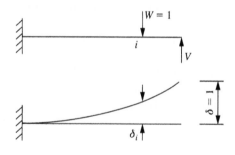

Figure 5.1

To obtain the influence line for moment at the fixed end of the propped cantilever shown in Figure 5.2, the cantilever is cut at the fixed end and a unit

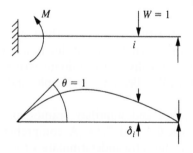

Figure 5.2

virtual rotation is imposed in the line of action of M. The deflection produced under a unit load at any point i is δ_i. Then, applying the virtual work principle:

$$M \times (\theta = 1) = (W = 1) \times \delta_i$$

that is: $M = \delta_i$

and the elastic curve is the influence line for M.

Example 5.1

Obtain the influence lines for V_1 and M_2 for the propped cantilever shown in Figure 5.3.

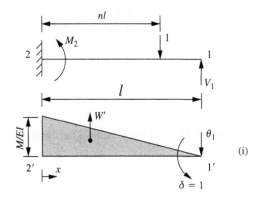

Figure 5.3

Solution

A unit upward displacement is applied to 1 and produces a moment M at the fixed end 2. The elastic load on the conjugate beam is shown at (i); taking moments about $1'$:

$$2W'l/3 = 1$$

Thus:

$$M = 3EI/l^2$$

The intensity of loading on the conjugate beam is:

$$w' = 3/l^2 - 3x/l^3$$

The shear is:

$$Q' = \int w' \, dx$$
$$= 3x/l^2 - 3x^2/2l^3$$

The moment is:

$$M' = \int Q' \, dx$$
$$= 3x^2/2l^2 - x^3/2l^3$$

and this is the equation of the elastic curve of the real beam.
Substituting $x = nl$, the expression for the influence line for V_1 is:

$$V_1 = 3n^2/2 - n^3/2$$

The expression for the influence line for M_2 is:

$$M_2 = nl - V_1 l$$
$$= l(n - 3n^2/2 + n^3/2)$$

Example 5.2

Obtain the influence line for M_{12} for the fixed-ended beam shown in Figure 5.4.

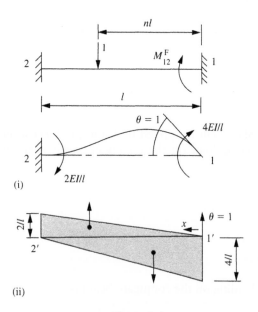

Figure 5.4

Solution

The beam is cut at 1 and a unit clockwise rotation imposed. This requires a clockwise moment of $M_{12} = 4EI/l$ and induces a clockwise moment of $M_{21} = 2EI/l$, as shown at (i).

The intensity of loading on the conjugate beam at (ii) is:

$$w' = 2x/l^2 - 4/l + 4x/l^2$$
$$= 6x/l^2 - 4/l$$

The shear is:

$$Q' = 3x^2/l^2 - 4x/l + 1$$

The moment is:

$$M' = x^3/l^2 - 2x^2/l + x$$

and this is the equation of the elastic curve of the real beam.

Substituting $x = nl$, the expression for the influence line for M_{12}^F:

$$M_{12}^F = nl(n - 1)^2$$

The values of the influence line ordinates, at intervals of $0.2 \times l$, are given in Table 5.1.

Table 5.1 Influence line ordinates for Example 5.2

n	0.2	0.4	0.6	0.8
$M_{12}^F \times 1/l$	0.128	0.144	0.096	0.032
$M_{21}^F \times 1/l$	−0.032	−0.096	−0.144	−0.128

Example 5.3

A non-prismatic beam deflects in the form of a sine wave when it is simply supported and displaced vertically at its center. The beam is then supported at its center to form a two-span continuous beam. Obtain the influence line for bending moment for a point midway between one end support and the center support.

Solution

The influence line for V_2 for the two-span beam shown in Figure 5.5 is:

$$V_2 = \sin(\pi x/l)$$

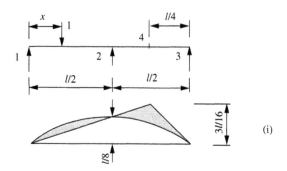

Figure 5.5

For unit load to the left of section 4 the influence line for M_4 is:

$$
\begin{aligned}
M_4 &= V_3 l/4 \\
&= \{x/l - (\tfrac{1}{2}) \times \sin(\pi x/l)\} \times l/4 \\
&= x/4 - (l/8) \times \sin(\pi x/l)
\end{aligned}
$$

For unit load to the right of section 4 the influence line for M_4 is:

$$
\begin{aligned}
M_4 &= 3V_1 l/4 + V_2 l/4 \\
&= 3(l - x)/4 - l/8 \times \sin(\pi x/l)
\end{aligned}
$$

The influence line may be plotted as shown at (i).

(b) Maxwell's reciprocal theorem

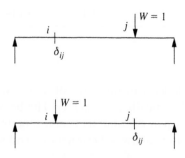

Figure 5.6

The displacement produced at any point i in a linear structure due to a force applied at another point j equals the displacement produced at j due to the same force applied at i. The displacement is measured in the line of action of the applied force; the displacement corresponding to an applied moment is a rotation and to a force, a linear deflection.

Referring to Figure 5.6, the deflection at i due to a unit load at j is:

$$\delta_{ij} = \int Mm \ dx/EI$$
$$= \int m_j m_i \ dx/EI$$

where m_i and m_j are the bending moments produced at any section due to a unit load applied at i and j, respectively.

Similarly, the deflection at j due to a unit load at i is:

$$\delta_{ji} = \int Mm \ dx/EI$$
$$= \int m_i m_j \ dx/EI$$
$$= \delta_{ij}$$

Thus, the influence line for deflection at i is the elastic curve produced by a unit load applied at i.

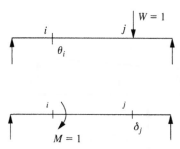

Figure 5.7

Referring to Figure 5.7, the rotation at i due to a unit load at j is:

$$\theta_i = \int Mm \ dx/EI$$
$$= \int m_j m_i \ dx/EI$$

where m_j is the bending moment at any section due to a unit load at j and m_i is the bending moment at any section due to a unit moment at i.

Similarly, the deflection at j due to a unit moment at i is:

$$\delta_j = \int Mm \ dx/EI$$
$$= \int m_i m_j \ dx/EI$$
$$= \theta_i$$

Thus, the influence line for rotation at i is the elastic curve produced by a unit moment applied at i.

A general procedure for obtaining the influence lines for any restraint in a structure is to apply a unit force to the structure in place of and corresponding to the restraint. The elastic curve produced is the influence line for displacement corresponding to the restraint. Dividing the ordinates of this elastic curve by the displacement occurring at the point of application of the unit force gives the influence line for the required restraint.

The influence line for relative rotation, $\delta\theta$, between two sections in a structure a distance δx apart is the elastic curve produced by equal and opposite unit moments applied to the two sections. From Figure 2.1:

$$\delta\theta = \delta x/R$$
$$= M\,\delta x/EI$$

Example 5.4

Obtain the influence line for V_1 for the propped cantilever shown in Figure 5.8. The second moment of area varies linearly from a value of I at end 1 to $2I$ at end 2.

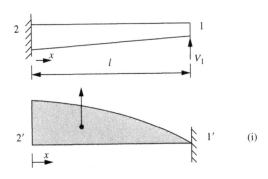

Figure 5.8

Solution

The second moment of area at any section is:

$$I_x = I(2 - x/l)$$

The reaction V_1 is replaced by a unit load acting vertically upward, and the elastic load on the conjugate beam is shown at (i). The intensity of loading on the conjugate beam is:

$$w' = M_x/EI_x$$
$$= l(l - x)/EI(2l - x)$$

The shear is:

$$Q' = \int w' dx$$
$$= \{lx + l_2 \log(2l - x)\}/EI - (l^2 \log 2l)/EI$$
$$= \{lx + l^2 \log(1 - x/2l)\}/EI$$

The moment is:

$$M' = \int Q' dx$$
$$= \{lx^2/2 + l^2(x - 2l) \times \log(1 - x/2l) - l^2 x\}/EI$$

and this is the equation of the elastic curve of the real beam for unit load at 1.
   At $x = l$ the deflection of the real beam is:

$$\delta_1 = M'_1$$
$$= -l^3(1 + 2 \log \tfrac{1}{2})/2EI$$
$$= 0.1932 l^3/EI$$

The expression for the influence line for V_1 is:

$$V_1 = M'/\delta_1$$
$$= \{lx^2/2 + l^2(x - 2l) \times \log(1 - x/2l) - l^2 x\}/0.1932 l^3$$

Example 5.5

Obtain the influence line for V_1 as unit load crosses the beam of the rectangular frame shown in Figure 5.9. The beam has a second moment of area three
times that of each column.

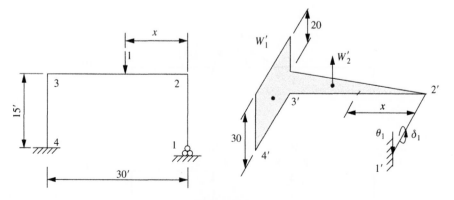

Figure 5.9

Solution

For each column $EI = 1$, and for the beam $EI = 3$.

The reaction V_1 is replaced by a unit load acting vertically upward; the elastic loads on the conjugate frame are shown at (i) and are given by:

$$W'_1 = 450$$
$$W'_2 = 150$$

The rotation and vertical deflection at 1 are:

$$\theta_1 = -600$$
$$\delta_1 = 30W'_1 + 20W'_2$$
$$= 16,500$$

The deflection of the real frame, due to unit load at 1 at any point a distance x from 2 is:

$$\delta_x = M'_x$$
$$= \delta_1 + x\theta_1 + (10x^2/60) \times (x/3)$$
$$= 16,500 - 600x + x^3/18$$

The expression for the influence line for V_1 is:

$$V_1 = \delta_x/\delta_1$$
$$= 1 - 6x/165 + x^3/297,000$$

5.3 Moment distribution applications

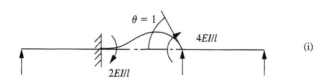

(i)

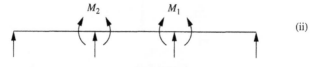

(ii)

Figure 5.10

To obtain the influence line for moment M_1 at the support 1 of the continuous beam shown in Figure 5.10, the beam is cut at 1 and a unit virtual rotation imposed at 1 with end 2 clamped as shown at (i). The moment required to produce the unit rotation is derived in Section 6.7 as:

$$s_{12} = 4EI/l$$

where l is the span length 12 and the moment induced at 2 is:

$$c_{12}s_{12} = 2EI/l$$

The cut ends at 1 are clamped together to maintain the unit rotation between the members meeting at 1, and these initial moments are distributed throughout the beam to produce the final moments shown at (ii). The elastic curve of the structure, due to these final moments, is, from Müller-Breslau's principle, the influence line for M_1.

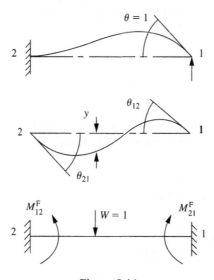

Figure 5.11

The elastic curve of the structure may be obtained from the values of fixed-end moments tabulated in Table 5.1. Thus, referring to Figure 5.11, the elastic curve of member 12 due to a unit rotation at 1 is the influence line for the fixed-end moment at 1. For any other rotation, the elastic curve is obtained as the product of the rotation and the influence line ordinate for M_{12}^F at the corresponding section. When a rotation also occurs at 2, the elastic curve is obtained as the algebraic sum of the two products. Thus, the elastic curve ordinate, y, at any section as shown is:

$$y = \theta_{12}M_{12}^F + \theta_{21}M_{21}^F$$

where clockwise rotations and clockwise fixed-end moments are positive.

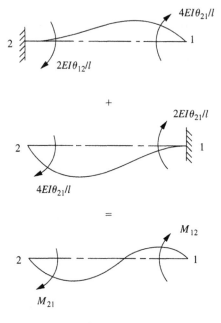

Figure 5.12

The rotations of the structure at the supports may be obtained from the final support moments as shown in Figure 5.12. Thus:

$$M_{12} = 4EI\theta_{12}/l + 2EI\theta_{21}/l$$
$$M_{21} = 4EI\theta_{21}/l + 2EI\theta_{12}/l$$

And: $\theta_{12} = l(2M_{12} - M_{21})/6EI$
$$\theta_{21} = l(2M_{21} - M_{12})/6EI$$

where clockwise rotations and moments are positive.

The method may be readily extended to structures with non–prismatic members using stiffness factors, carry-over factors and fixed-end moments that have been tabulated[4] for a large variety of non-prismatic members.

Example 5.6

Determine the influence line ordinates for M_2 over the central span of the continuous beam shown in Figure 5.13. The second moment of area of span 34 is twice that of spans 12 and 23.

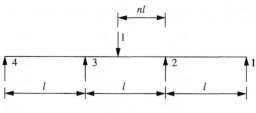

Figure 5.13

Solution

The beam is cut at 2 and clamped at 3. A rotation θ_{23} is produced by applying a clockwise moment of $M_{23} = 400$ units; this induces a clockwise moment of $M_{32} = 200$ units. These initial moments are distributed in Table 5.2.

The final rotations at 2 and 3 are:

$\theta_{23} = (330 - 55)/600$
    $= 0.46$ radians clockwise

$\theta_{32} = (110 - 165)/600$
    $= -0.09$ radians anticlockwise

Table 5.2 Distribution of moments in Example 5.6

Member	21	23	32	34
Relative EI/l	4	4	4	8
Modified stiffness	3	4	4	6
Distribution factor	3/7	4/7	2/5	3/5
		←	1/2	
Carry-over factor		1/2	→	
Initial moments		400	200	
Distribution		−228	−80	
Carry-over		−40	−114	
Distribution		23	46	
Carry-over		23	12	
Distribution		−13	−5	
Carry-over		−2	−6	
Distribution		1	2	
Carry-over		1	0	
Final moments		165	55	

The influence line ordinates, at intervals of $0.2l$ are obtained in Table 5.3 from:

$$M_2 = 0.46M_{23}^F - 0.09M_{32}^F$$

Table 5.3 Influence line ordinates for Example 5.6

n	0.2	0.4	0.6	0.8
$M_{23}^F \times 1/l$	0.128	0.144	0.096	0.032
$M_{32}^F \times 1/l$	−0.032	−0.096	−0.144	−0.128
$M_{23}^F \times 0.46 \times 1/l$	0.059	0.066	0.044	0.014
$M_{32}^F \times -0.09 \times 1/l$	0.003	0.009	0.013	0.012
Ordinates $\times 1/l$	0.062	0.075	0.057	0.026

Example 5.7

Determine the influence line ordinates for M_1, over the beam 12, for the frame shown in Figure 5.14. The second moment of area and modulus of elasticity is constant for all members.

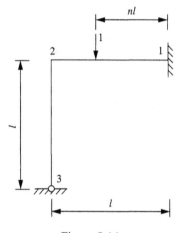

Figure 5.14

Solution

The frame is cut at 1 and clamped at 2. A rotation θ_{12} is produced by applying a clockwise moment of $M_{12} = 400$ units; this induces a clockwise moment of $M_{21} = 200$ units. These initial moments are distributed in Table 5.4.

The final rotations at 1 and 2 are:

$$\theta_{12} = (686 - 86)/600$$
$$= 1.0 \text{ radians clockwise}$$
$$\theta_{21} = (172 - 343)/600$$
$$= -0.285 \text{ radians anticlockwise}$$

The influence line ordinates, at intervals of $0.2l$, are obtained in Table 5.5 from:

$$M_1 = M_{12}^F - 0.285 M_{21}^F$$

Table 5.4 Distribution of moments in Example 5.7

Member	12	21	23	32
Relative EI/l	4	4	4	4
Modified stiffness		4	3	
Distribution factor		4/7	3/7	
Carry-over factor	←	1/2	0	→
Initial moments	400	200		
Distribution and carry-over	−57	−114	−86	
Final moments	343	86	−86	0

Table 5.5 Influence line ordinates for Example 5.7

n	0.2	0.4	0.6	0.8
$M_{12}^F \times 1/l$	0.128	0.144	0.096	0.032
$M_{21}^F \times 1/l$	−0.032	−0.096	−0.144	−0.128
$M_{21}^F \times -0.285 \times 1/l$	0.009	0.027	0.041	0.037
Ordinates $\times 1/l$	0.137	0.171	0.137	0.069

5.4 Non-prismatic members

When tabulated values for stiffness factors, carry-over factors, and fixed-end moments for non-prismatic members are not available, it is necessary to use the general procedure for obtaining influence lines for the structure.

To determine the influence line for V_1 for the two-span continuous beam shown in Figure 5.15, the beam is divided into short segments of length s. The second moment of area is regarded as constant over the length of each segment, as indicated at (i). The reaction V_1 is replaced by a unit load acting vertically upward, and the bending moment diagram on the cut-back structure is shown at (ii). The bending moment is regarded as constant over the length of each segment and has the values indicated. The elastic loads on the conjugate beam are given by $W' = xs/EI$ and are considered to be concentrated at the center of each segment, as shown at (iii). The bending moment in the conjugate beam at any section is the ordinate of the elastic curve of the real beam

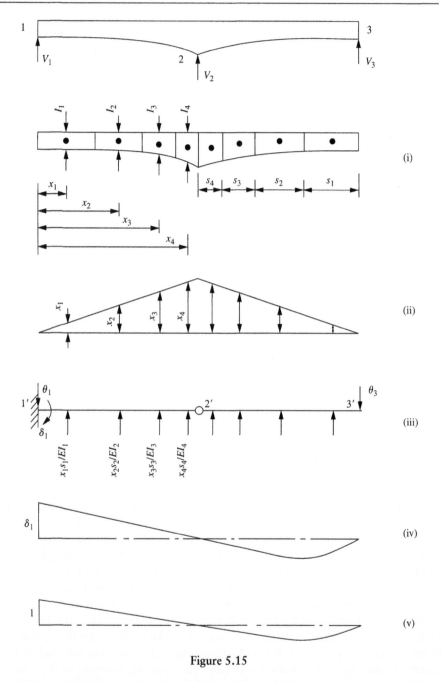

Figure 5.15

at the corresponding section, as shown at (iv), and these values, when divided by δ_1, are the required influence line ordinates, as shown at (v). The influence line for any other restraint may now be obtained by the direct application of statics.

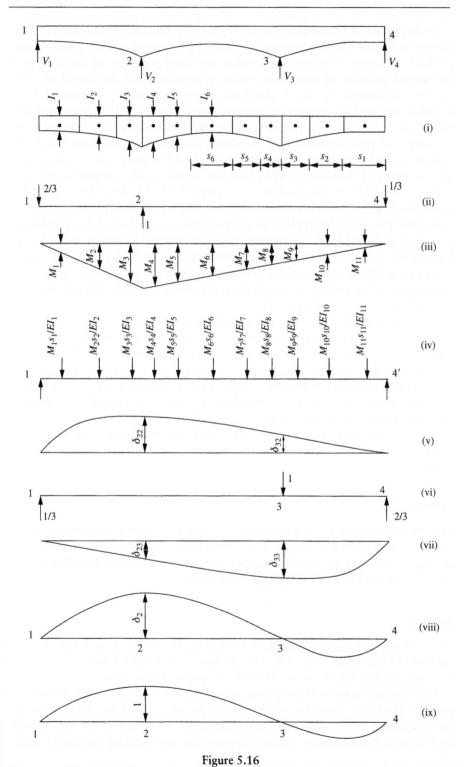

Figure 5.16

To determine the influence line for V_2 for the three-span continuous beam shown in Figure 5.16, reactions V_2 and V_3 are removed to produce the cut-back structure. The beam is divided into short segments of length s, and the second moment of area is regarded as constant over the length of each segment, as indicated at (i). A unit load is applied vertically upward at 2, as shown at (ii), and the bending moment at the center of each segment is determined, as shown at (iii). The bending moment is regarded as constant over the length of each segment, and the elastic loads on the conjugate beam are considered to be concentrated at the center of each segment, as shown at (iv). The ordinates of the elastic curve of the real beam, simply supported at 1 and 4 and with a unit load applied at 2, are given by the bending moments at the corresponding section in the conjugate beam and are shown at (v).

A correction must be applied to these ordinates to reduce the deflection at 3 to zero since, with V_3 in position, there can be no deflection at 3. A unit load acting vertically downward is applied at 3, as shown at (vi). The ordinates of the elastic curve for this loading condition are obtained by the above procedure and are shown at (vii). For a symmetrical structure, curve (vii) is the inverted mirror image of curve (v).

The ordinates of curve (vii) are multiplied by the factor δ_{32}/δ_{33} and added to curve (v) to give curve (viii), which is the elastic curve produced by unit load at 2 with supports 1, 3, and 4 in position. These ordinates, when divided throughout by δ_2, are the influence line ordinates for V_2 as shown at (ix).

Similarly, the influence line for V_3 may be obtained and, for a symmetrical structure, is the mirror image of the influence line for V_2. The influence line for any other restraint may now be obtained by the direct application of statics.

When the number of redundants exceeds 2, the flexibility matrix method, given in Section 10.3, is a preferred method of solution and has been used by Jenkins[5].

Example 5.8

Determine the influence line ordinates for V_1 and M_2 for the three-span symmetrical beam shown in Figure 5.17. The relative EI values are indicated on the figure and may be assumed to be constant over the lengths indicated, as may also the elastic loads on the conjugate beam.

Solution

The cut-back structure is produced by removing V_1 and V_4; a vertically upward load of 0.4 kip is applied at 1. This produces the bending moment diagram shown at (ii) and the elastic loads on the conjugate beam at (iii).

The bending moment in the conjugate beam at the center of each segment is given in line (2) of Table 5.6 and represents the ordinate of the elastic curve of the real beam, simply supported at 2 and 3 and with a load of 0.4 kip applied at 1. Due to the symmetry of the structure, the ordinates of the elastic curve

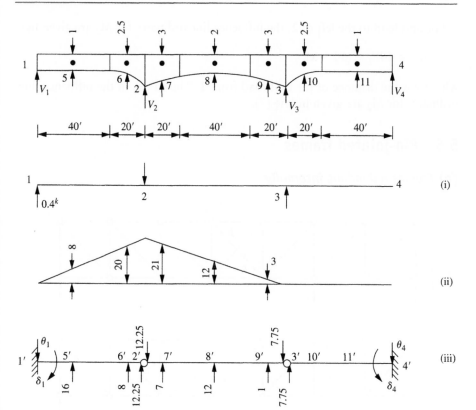

Figure 5.17

Table 5.6 Influence line ordinates for Example 5.8

1 Section	1	5	6 2	7	8	9 3	10	11	4
2 0.4 kip at 1	1455	730	122.5 0	−122.5	−280	−77.5 0	77.5	310	465
3 0.4 kip at 4	−465	−310	−77.5 0	77.5	280	122.5 0	−122.5	−730	−1455
4 (3) × 465/1455	−148	−99	−24.5 0	24.5	90	39.5 0	−39.5	−233	−465
5 (2) + (4)	1307	631	98 0	−98	−190	−38 0	38	77	0
6 (5) × 1/1307	1	0.483	0.075 0	−0.075	−0.146	−0.029 0	0.029	0.059	0
7 M_2	0	−11	−5.5 0	−4.5	−8.75	−1.75 0	1.75	3.55	0

due to a vertically downward load of 0.4 kip applied at 4 are as indicated in line (3). Line (4) gives the values of line (3) multiplied by the factor 465/1455; it is added to line (2) to give line (5). This represents the ordinates of the elastic curve produced by a load of 0.4 kip at 1 and with supports 2, 3, and 4 in position. Line (5) multiplied by the factor 1/1307 gives the influence line ordinates for V_1 which are given in line (6).

For unit load to the right of 2, the influence line ordinates for M_2 are given by:

$$M_2 = 60V_1$$

For unit load to the left of 2, the influence line ordinates for M_2 are given by:

$$M_2 = 60V_1 - (60 - x)$$

where x is the distance of the unit load from 1. The values of the influence line ordinates for M_2 are given in line (7).

5.5 Pin-jointed frames

(a) Frames redundant internally

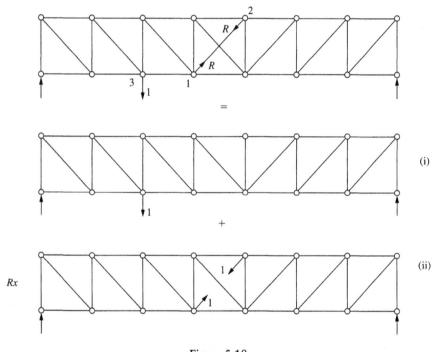

Figure 5.18

The pin-jointed frame shown in Figure 5.18 contains one redundant member 12, with its unknown force R assumed tensile when unit load is at panel point 3. The indeterminate frame can be replaced by system (i) plus $R \times$ system (ii). The relative inward movement of the points of application of R in system (i) is δ'_{12} and in system (ii) is δ''_{12}. The relative movement of the points of application of R in the actual structure is outward and consists of the extension δ_{12} in member 12, given by:

$$\delta_{12} = -Rl_{12}/A_{12}E_{12}$$
$$= \delta'_{12} + R\delta''_{12}$$

Thus: $R = -\delta'_{12}/(\delta''_{12} + l_{12}/A_{12}E_{12})$

From Maxwell's reciprocal theorem, the displacement δ'_{12}, in system (i) due to a unit load at 3 equals the vertical displacement of 3 in system (ii) due to unit load replacing R, and 3 may be any panel point. Thus, all the information required to obtain the influence line ordinates for R may be obtained from one Williot-Mohr diagram constructed for system (ii). The vertical deflection of each lower panel point is measured on the diagram and divided by $(\delta''_{12} + l_{12}/A_{12}E_{12})$ to give the influence line ordinate at that panel point.

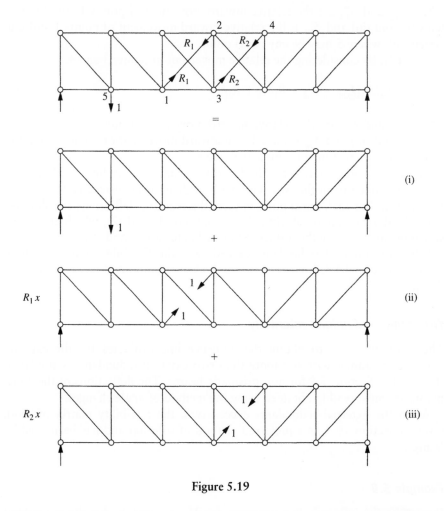

Figure 5.19

The pin-jointed frame shown in Figure 5.19 contains the two redundant members 12 and 23 with unknown forces R_1 and R_2, assumed tensile when unit load is applied at panel point 5. The indeterminate frame can be replaced by system (i) plus $R_1 \times$ system (ii) plus $R_2 \times$ system (iii). The relative movement of points 1 and 2 in the actual structure consists of the extension in member 12 and is:

$$\delta_{12} = -R_1 l_{12}/A_{12}E_{12}$$

and:

$$-R_1 l_{12}/A_{12}E_{12} = \delta'_{12} + R_1\delta''_{12} + R_2\delta'''_{12} \tag{1}$$

where δ''_{12} and δ'''_{12} are the relative inward movement of points 1 and 2 in systems (ii) and (iii) and δ'_{12} is the relative inward movement of points 1 and 2 in system (ii), where 5 may be any panel point.

Similarly, by considering the relative movement of points 3 and 4:

$$-R_2 l_{34}/A_{34}E_{34} = \delta'_{34} + R_1\delta''_{34} + R_2\delta'''_{34} \tag{2}$$

where δ''_{34} and δ'''_{34} are the relative inward movement of points 3 and 4 in systems (ii) and (iii) and δ'_{34} is the relative inward movement of points 3 and 4 in system (iii), where 5 may be any panel point.

Thus, all the information required to obtain the influence lines for R_1 and R_2 may be obtained from the two Williot-Mohr diagrams constructed for systems (ii) and (iii). For each panel point in turn, the required deflections are obtained from the diagrams and substituted in equations (1) and (2), which are solved simultaneously to give the influence line ordinates for R_1 and R_2.

When the internal redundants exceed two, the flexibility matrix method, given in Section 10.3, is a preferred method of solution and has been used by Wang[6].

(b) Frames redundant externally

The technique used to obtain the influence line ordinates for the external reactions in frames with not more than two external redundants is similar to that used in Section 5.4 for non-prismatic beams. The deflections of the panel points are most readily obtained using the method of angle changes.

When the external redundants exceed two, the flexibility matrix method, given in Section 10.3, is a preferred method of solution and has been used by Wang[6].

Example 5.9

Determine the influence line ordinates for V_1, V_3, and V_6 for the pin-jointed frame shown in Figure 5.20 and the influence lines for member force in members 27, 23, and 39. All members of the frame have the same cross-sectional area, modulus of elasticity, and length.

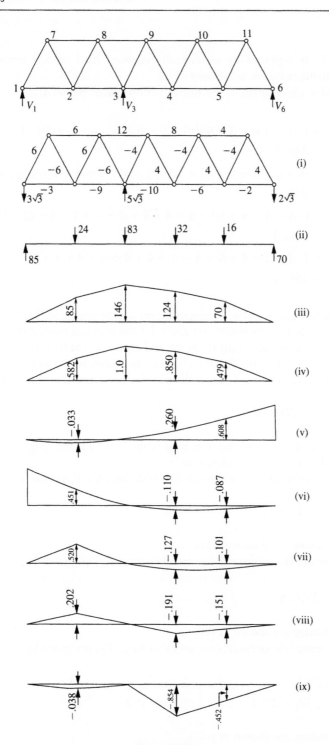

Figure 5.20

Solution

Reaction V_3 is replaced by a vertically upward load of $5\sqrt{3}$ units, and the resulting member forces are shown at (i).

The total angle changes at the bottom chord panel points are:

$$\sum \Delta_2 = (6 + 6 + 6 + 3 + 6 + 6 + 6 - 6 - 6 - 6 - 6 + 9)/\sqrt{3}$$
$$= 24/\sqrt{3}$$
$$\sum \Delta_3 = (6 + 6 + 6 + 9 + 12 + 6 + 12 + 4 + 4 + 4 + 4 + 10)/\sqrt{3}$$
$$= 83/\sqrt{3}$$
$$\sum \Delta_4 = (-4 - 4 - 4 + 10 + 8 - 4 + 8 + 4 + 4 + 4 + 4 + 6)/\sqrt{3}$$
$$= 32/\sqrt{3}$$
$$\sum \Delta_5 = (-4 - 4 - 4 + 6 + 4 - 4 + 4 + 4 + 4 + 4 + 4 + 2)/\sqrt{3}$$
$$= 16/\sqrt{3}$$

and the relative loads on the conjugate beam are shown at (ii). The deflections at the bottom chord panel points are given by the bending moments at the corresponding section in the conjugate beam and are shown at (iii). These deflections, when divided by 146, give the influence line ordinates for V_3 as shown at (iv).

Taking moments about 1, the influence line ordinates for V_6 are given by:

$$V_6 = (x - 2lR_3)/5l$$

where x is the distance of the unit load from 1 and l is the length of the frame members. These values are shown at (v).

Resolving vertically, the influence line ordinates for V_1 are given by:

$$V_1 = 1 - V_3 - V_6$$

and these values are shown at (vi).

The influence line ordinates for member force P_{27} are given by:

$$P_{27} = 2V_1/\sqrt{3}$$

and these values are shown at (vii).

The influence line ordinates for member force P_{23} are given by:

$$P_{23} = \sqrt{3}V_1 \qquad \text{... for unit load at 3, 4, 5 and 6}$$
$$P_{23} = \sqrt{3}V_1 - 1/\sqrt{3} \quad \text{... for unit load at 2}$$

and these values are shown at (viii).

The influence line ordinates for member force P_{39} are given by:

$$P_{39} = 2V_6/\sqrt{3} \qquad \text{... for unit load at 1, 2 and 3}$$
$$P_{39} = 2V_6/\sqrt{3} - 2/\sqrt{3} \qquad \text{... for unit load at 4, 5 and 6}$$

and these values are shown at (ix).

Example 5.10

Determine the influence line ordinates for V_3 for the three-span pin-jointed frame shown in Figure 5.21. All members have the same cross-sectional area, modulus of elasticity, and length.

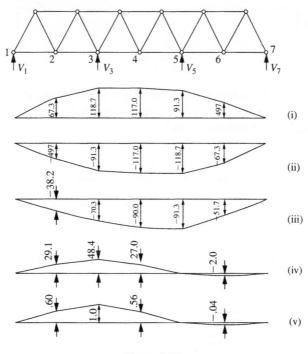

Figure 5.21

Solution

Supports V_3 and V_5 are removed, and a unit load is applied vertically upward at 3. The deflections of the lower panel points were obtained in Example 4.6, and the relative values are shown at (i). When unit load is applied vertically downward at 5, the relative deflections shown at (ii) are produced, and (ii) is the inverted mirror image of (i) due to the symmetry of the frame.

The ordinates at (ii) are multiplied by 91.3/118.7 to give (iii), which is added to (i) to give (iv). The ordinates at (iv) are multiplied by 1/48.4 to give (v), which is the influence line for V_3.

Supplementary problems

S5.1 The two-span continuous beam shown in Figure S5.1 has a second moment of area that varies linearly from a maximum value of I at the center to zero at the ends. Determine the influence line ordinates, at intervals of $0.2l$, for the central vertical reaction.

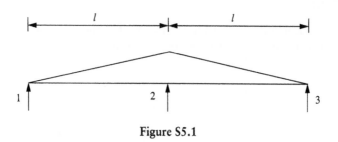

Figure S5.1

S5.2 The relative EI/l values are shown ringed for each member of the structure shown in Figure S5.2. Determine the influence line ordinates for M_2 at intervals of $0.2l$ over members 12 and 23.

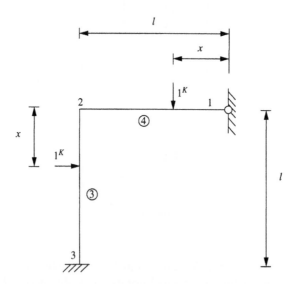

Figure S5.2

S5.3 The two-hinged symmetrical polygonal arch shown in Figure S5.3 has a second moment of area for the beam three times that of the inclined columns. Determine the influence line ordinates, at intervals of 20 ft, for horizontal thrust at the hinges as unit load moves from 5 to 7.

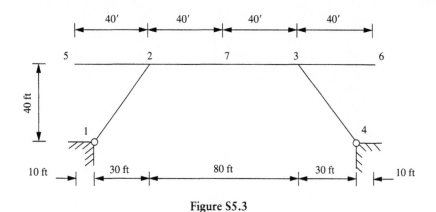

Figure S5.3

S5.4 Determine the influence line ordinates for R_1 for the two-span pin-jointed truss shown in Figure S5.4 as unit load crosses the bottom chord. All members have the same cross-sectional area, modulus of elasticity, and length.

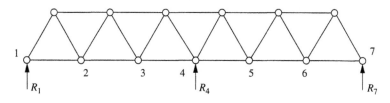

Figure S5.4

References

1. Larnach, W. J. Influence lines for statically indeterminate plane structures. London, Macmillan & Co. Ltd. 1964.
2. Portland Cement Association. Influence lines drawn as deflection curves. Chicago. 1948.
3. Thadani, B. N. Distribution of deformation method for the construction of influence lines. Civ. Eng. and Pub. Works. Rev. 51. June 1956. pp. 645–649.
4. Portland Cement Association. Handbook of frame constants. Chicago. 1948.
5. Jenkins, W. M. Influence line computations for structures with members of varying flexural rigidity using the electronic digital computer. Struct. Engineer 39. Sept. 1961. pp. 269–276.
6. Wang, C. K. Matrix analysis of statically indeterminate trusses. Proc. Am. Soc. Civil Eng. 85 (ST4). April 1959. pp. 23–36.

6 Elastic center and column analogy methods

Notation

A	elastic area
c_{12}	carry-over factor for a member 12 from the end 1 to the end 2
c'_{12}	modified carry-over factor
e_x	eccentricity of the elastic load with respect to the x-axis through O
e_y	eccentricity of the elastic load with respect to the y-axis through O
E	Modulus of elasticity
H	horizontal reaction
H_O	horizontal reaction acting at the elastic center
H^F	horizontal thrust in a fixed-ended arch due to the applied loads
$H^{\delta=1}$	translational stiffness of an arch
$H^{\theta=1}$	horizontal thrust induced by a unit rotation at one end of an arch
I	second moment of area
I_o	second moment of area of an arch at its crown
$\mathbf{I_x}$	second moment of the elastic area about the x-axis through O
$\mathbf{I_y}$	second moment of the elastic area about the y-axis through O
l	length of a member, span of an arch
L_{12}	equivalent length of an elastically restrained member 12, $l_{12} + 3EI\eta_{12}$
M	bending moment in the cut-back structure
M_1	actual bending moment at section 1
M_c	actual bending moment at the crown of an arch
M_O	bending moment acting at the elastic center
$\mathbf{M_x}$	moment of the elastic load about the x-axis through O
$\mathbf{M_y}$	moment of the elastic load about the y-axis through O
M^F	fixed-end moment
M^{FE}	fixed-end moment in an elastically restrained member
$M^{\delta=1}$	fixed-end moment induced by a unit horizontal translation at one end of an arch
O	elastic center, centroid of the elastic area
s	length of segment
s_{12}	restrained stiffness at the end 1 of a member 12, moment required to produce a unit rotation at the end 1, end 2 being fixed
s'_{12}	modified stiffness
t	change in temperature
V	vertical reaction

V_O vertical reaction acting at the elastic center
W elastic load
x_O horizontal displacement of elastic center
y_O vertical displacement of elastic center
α temperature coefficient of expansion
δ_x spread of arch abutments
δ_y settlement of arch abutment
η_{12} elastic connection factor at the end 1 of an elastically restrained member 12
θ angle of rotation
ϕ'_{12} relative rotation between an elastically connected beam and column at the end 1 of beam 12

6.1   Introduction

The elastic center method provides a rapid means for the solution of the two-hinged polygonal arch and for the determination of influence lines for the fixed-ended arch. The method is applied here to symmetrical structures only but may be readily extended to unsymmetrical frames and arches[1].

The column analogy method provides the most useful means for the determination of fixed-end moments, stiffness, and carry-over factors for non-prismatic members. The method is derived here from the elastic center method, but may be derived independently[2] if required.

6.2   Elastic center method

The symmetrical fixed-ended frame shown in Figure 6.1 (i) is three degrees redundant, and these redundants may be regarded as the restraints M_1, H_1, V_1

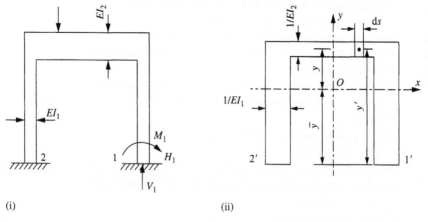

(i)   (ii)

Figure 6.1

at end 1. An analogous frame, known as the elastic frame, is shown in Figure 6.1 (ii) and consists of members with a width at any section of $1/EI$, where E and I are the modulus of elasticity and the second moment of area at the corresponding section in the real frame. The area of the elastic frame is:

$$A = \int ds/EI$$

The centroid O of the elastic area is known as the elastic center and must lie on the vertical axis of symmetry at a distance $\bar{y}$ from the base, given by:

$$A\bar{y} = \int y' \, ds/EI$$

The second moment of the elastic area about the x-axis through O is:

$$I_x = \int y^2 \, ds/EI$$

The second moment of the elastic area about the y-axis through O is:

$$I_y = \int x^2 \, ds/EI$$

The symmetrical fixed-ended frame shown in Figure 6.2 may be replaced by systems (i) and (ii). The cut-back structure is produced by releasing end 1. An infinitely rigid arm, with its free end at the elastic center, is attached to 1. Since the arm is infinitely rigid, it has no effect on the flexural properties of the real frame or on the elastic area, as its $\int ds/EI$ value is zero. To the cut-back structure is applied the external load in system (i) and the redundants M_O, H_O, V_O in system (ii). Displacement of the elastic center in systems (i) and (ii) is transferred through the rigid arm and causes a corresponding displacement at 1. The bending moment at any point in the frame in system (i) is M, with moment producing tension on the inside of the frame regarded as positive.

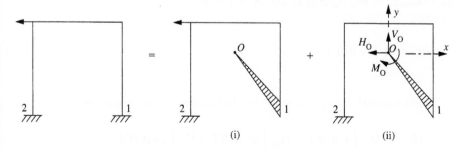

Figure 6.2

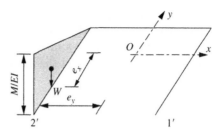

Figure 6.3

The elastic frame is placed in a horizontal plane as shown in Figure 6.3 and loaded with an elastic load of:

$$W = \int M \, ds/EI$$

where M is the bending moment in the cut-back structure at a given section and E and I are the modulus of elasticity and the second moment of area at the same section. A positive moment in the cut-back structure produces a vertically downward load on the elastic frame. The moment of W about the x-axis through O is:

$$M_x = My \, ds/EI$$
$$= We_x$$

The moment of W about the y-axis through O is:

$$M_y = \int Mx \, ds/EI$$
$$= We_y$$

The horizontal displacement of the elastic center in system (i) is:

$$x'_o = \int Mm \, ds/EI$$

where m is the bending moment at any section due to a unit virtual load acting horizontally to the right at O and $m = 1 \times y$.
Thus:

$$x'_O = \int My \, ds/EI$$
$$= M_x$$

The horizontal displacement of the elastic center in system (ii) is:

$$x''_O = -M_O \int y \, ds/EI - H_O \int y^2 \, ds/EI - V_O \int xy \, ds/EI$$
$$= 0 - H_O I_x - 0$$

In the original structure the displacement of end 1 is zero and thus:

$$0 = x'_O + x''_O$$

Hence:

$$H_O = \mathbf{M_x}/\mathbf{I_x}$$

The vertical displacement of the elastic center in system (i) is:

$$y'_O = \int Mm \; ds/EI$$

where m is the bending moment at any section due to a unit virtual load acting vertically downward at O and $m = 1 \times x$.

Thus:

$$\begin{aligned} y'_O &= \int Mx \; ds/EI \\ &= \mathbf{M_y} \end{aligned}$$

The vertical displacement of the elastic center in system (ii) is:

$$\begin{aligned} y''_O &= -M_O \int x \; ds/EI - H_O \int xy \; ds/EI - V_O \int x^2 \; ds/EI \\ &= -0 - 0 - V_O \mathbf{I_y} \end{aligned}$$

In the original structure the displacement of end 1 is zero and thus:

$$0 = y'_O + y''_O$$

Hence:

$$V_O = \mathbf{M_y}/\mathbf{I_y}$$

The rotation of the elastic center in system (i) is:

$$\theta'_O = \int Mm \; ds/EI$$

where m is the bending moment at any section due to a unit virtual anticlockwise moment acting at O and $m = 1$.

Thus:

$$\begin{aligned} \theta'_O &= \int M \; ds/EI \\ &= \mathbf{W} \end{aligned}$$

The rotation of the elastic center in system (ii) is:

$$\theta_O'' = -M_O \int ds/EI - H_O \int y \, ds/EI - V_O \int x \, ds/EI$$
$$= -M_O \mathbf{A} - 0 - 0$$

In the original structure the displacement of end 1 is zero and thus:

$$0 = \theta_O' + \theta_O''$$

Hence:

$$M_O = \mathbf{W/A}$$

The bending moment at any point in the original structure with coordinates x and y is given by:

$$M - M_O - H_O y - V_O x = M - \mathbf{W}/\mathbf{A} - \mathbf{M_x} y/\mathbf{I_x} - \mathbf{M_y} x/\mathbf{I_y}$$

Allowing for rib-shortening effects, spread of the abutments, and temperature rise, the horizontal displacements of the elastic center in the original structure in system (i) and system (ii) are:

$$x_O = \delta_x$$

where δ_x is the spread of the abutments:

$$x_O' = \mathbf{M_x} + \alpha t l$$

where α is the temperature coefficient of expansion, t is the rise in temperature and l is the span of the arch.
And:

$$x_O'' = -H_O \mathbf{I_x} - H_O l/AE$$

where $H_O l/AE$ is the approximate allowance for rib shortening derived in Section 3.6:

$$x_O = x_O' + x_O''$$

and:

$$H_O = (\mathbf{M_x} + \alpha t l - \delta_x)/(\mathbf{I_x} + l/AE)$$

Allowing for vertical settlement of one abutment, the vertical displacements of the elastic center in the original structure in system (i) and in system (ii) are:

$$y_O = \delta_y$$

where δ_y is the settlement of the abutment:

$$y'_O = M_y$$

and:

$$y''_O = -V_O I_y$$

Thus:

$$V_O = (M_y - \delta_y)/I_y$$

Allowing for a clockwise rotation of θ at end 1 of the frame shown in Figure 6.4, the displacements of the elastic center are:

$$x_O = \theta \bar{y}$$
$$y_O = -\theta \bar{x}$$
$$\theta_O = -\theta$$

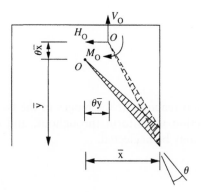

Figure 6.4

The values of the redundants at the elastic center are, then:

$$H_O = (M_x - \theta \bar{y})/I_x$$
$$V_O = (M_y + \theta \bar{x})/I_y$$
$$M_O = (W + \theta)/A$$

6.3   Two-hinged polygonal arch

The symmetrical, two-hinged polygonal arch shown in Figure 6.5 (i) is one degree redundant, the redundant being the horizontal restraint H. The two hinges have zero stiffness, and the elastic arch has two infinite areas concentrated at the foot of the columns, as shown at (ii). The elastic center lies centrally between the two feet, and the elastic area and the second moment of the elastic area about the y-axis are infinite.

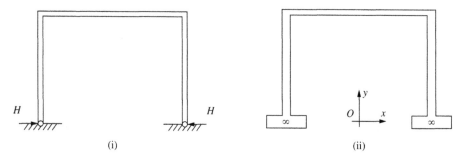

Figure 6.5

Hence:

$$V_O = M_O$$
$$= 0$$

and:

$$H_O = M_x/I_x$$
$$= H$$

Example 6.1

Determine the horizontal reaction at the hinges of the two-hinged arch shown in Figure 6.6. The section is uniform throughout, and deformations due to axial effects and shear may be neglected.

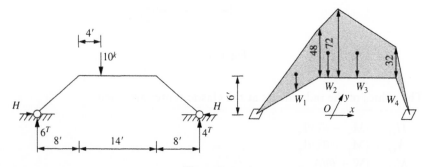

Figure 6.6

Solution

The cut-back structure is produced by replacing one hinge with a roller, and the elastic load on the elastic arch is shown at (i). The value of $1/EI$ may be taken as unity.

Then:

$$
\begin{aligned}
\mathbf{W_1} &= 10 \times 48/2 \\
&= 240 \\
\mathbf{W_2} &= 4 \times 120/2 \\
&= 240 \\
\mathbf{W_3} &= 10 \times 104/2 \\
&= 520 \\
\mathbf{W_4} &= 10 \times 32/2 \\
&= 160 \\
\mathbf{M_x} &= 4 \times 240 + 6 \times 760 + 4 \times 160 \\
&= 6160 \\
\mathbf{I_x} &= 14 \times 6^2 + 2 \times 10 \times 6^2 \times 1/3 \\
&= 744
\end{aligned}
$$

and:

$$
\begin{aligned}
H &= \mathbf{M_x}/\mathbf{I_x} \\
&= 8.28 \text{ kips}
\end{aligned}
$$

6.4   Influence lines for fixed-ended arches

The influence lines for the restraints at the end of 1 of the fixed-ended symmetrical arch shown in Figure 6.7 may be most readily obtained from the influence lines for M_O, H_O, and V_O. The arch is divided into short segments of length s, and the second moment of area is regarded as constant over the length of each

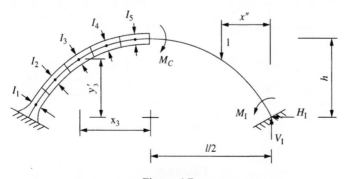

Figure 6.7

segment, as indicated in the figure. The elastic area and the height of the elastic center above the base of the arch are given by:

$$A = \Sigma s/I$$
$$\bar{y} = \Sigma y's/I \times 1/A$$

where I is the second moment of area at the center of a given segment and y' is its height above the base.

The cut-back structure is produced by introducing a cut in the crown of the arch to form two identical cantilevers, one of which is shown in Figure 6.8 (i). The reactions at the elastic center for the other cantilever are equal and opposite to those shown. To determine the influence line for H_O, a unit load is applied in place of H_O, as shown at (ii). The bending moment at the center of each segment is given by $M = -y$ and is regarded as constant over the

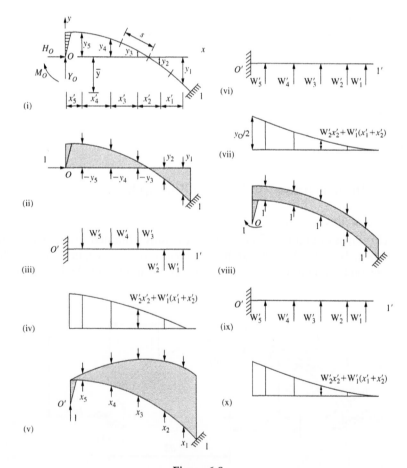

Figure 6.8

length of each segment. The elastic loads on the conjugate beam are given by $W' = -ys/EI$ and are considered as concentrated at the center of each segment, as shown at (iii). The ordinates of the elastic curve of the cut-back structure, with a unit load applied in place of H_O, are given by bending moments at the corresponding section in the conjugate beam and are shown at (iv). These ordinates, when divided throughout by the total horizontal displacement between the two cantilevers at O due to the unit load, are the influence line ordinates for H_O. The horizontal displacement of O due to the unit load is:

$$x_O = \int y^2 \, ds/EI$$
$$= I_x$$

To determine the influence line for V_O, a unit load is applied in place of V_O, as shown at (v). The bending moment at the center of each segment is given by $M = x$ and is regarded as constant over the length of each segment. The elastic loads on the conjugate beam are given by $W' = xs/EI$ and are considered as concentrated at the center of each segment, as shown at (vi). The ordinates of the elastic curve of the cut-back structure are given by the bending moments at the corresponding section in the conjugate beam and are shown at (vii). These ordinates, when divided throughout by y_O, the total vertical displacement between the two cantilevers at O, are the influence line ordinates for V_O. The influence line for V_O over the left half of the span is the inverted mirror image of the influence line over the right half.

To determine the influence line for M_O, a unit moment is applied in place of M_O, as shown at (viii). The bending moment at the center of each segment is given by $M = 1$. The elastic loads on the conjugate beam are given by $W' = s/EI$ and are considered as concentrated at the center of each segment, as shown at (ix). The ordinates of the elastic curve of the cut-back structure are given by the bending moments at the corresponding section in the conjugate beam and are shown at (x). These ordinates, when divided throughout by θ_O, the total rotation at O due to the unit load, are the influence line ordinates for M_O. The rotation of O is given by:

$$\theta_O = \int ds/EI$$
$$= A$$

The influence line ordinates for H_1 are given by:

$$H_1 = H_O$$

The influence line ordinates for V_1 are given by:

$$V_1 = -V_O \ \dots \ \text{left half}$$
$$V_1 = 1 - V_O \ \dots \ \text{right half}$$

The influence line ordinates for M_c are given by:

$M_c = M_O - H_O(h - \bar{y})$ with tension in the bottom fiber positive

The influence line ordinates for M_1 are given by:

$M_1 = M_O + H_O\bar{y} + V_O l/2$ ... left half
$M_1 = -x'' + M_O + H_O\bar{y} + V_O l/2$ ... right half

where x'' is the horizontal distance of the unit load from 1, and tension in the bottom fiber is positive.

Example 6.2

Determine the influence line ordinates for the elastic center reactions, the springing reactions, and the bending moment at the crown of the arch shown in Figure 6.9. The equation of the arch is:

$$y = x - 0.0133x^2$$

and the second moment of area varies as indicated in Table 6.1.

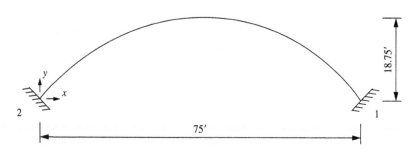

Figure 6.9

Table 6.1 Properties of elastic arch for Example 6.2

Segment	1	2	3	4	5	$2 \times \Sigma$
Length s	8.60	8.60	8.60	8.60	8.60	
Relative EI	5.10	3.07	2.03	1.45	1.11	
y'	2.95	8.45	13.05	16.50	18.50	
s/EI	1.69	2.80	4.24	5.94	7.75	44.84
$y's/EI$	5.00	23.70	55.30	98.00	139.50	643.00
y	−11.55	−6.05	−1.35	2.15	4.15	
y^2s/EI	225.00	102.50	7.70	27.40	133.60	992.40

Solution

The properties of the elastic arch are obtained in Table 6.1. The elastic area, the height of the elastic center, and the second moment of the elastic area about the x-axis are:

$$A = 44.84$$
$$\bar{y} = 643.0/44.84$$
$$= 14.4 \text{ ft}$$
$$I_x = 992.4$$

The influence line ordinates for the elastic center reactions are derived in Table 6.2 and plotted in Figure 6.10. The influence line ordinates for V_O over the left half of the span are $-1 \times$ the corresponding ordinate over the right half of the span.

Table 6.2 Influence line ordinates for Example 6.2

1	Segment	1	2	3	4	5	Crown
2	y	−11.55	−6.05	−1.35	2.15	4.15	4.35
3	$-ys/EI$	19.50	16.95	5.72	−12.75	−32.20	−
4	x'	6.60	7.25	7.80	8.40	4.35	−
5	M' for $H_O = 1$	0	129	393	723	970	957
6	$(5) \times 1/992.4 = H_O$	0	0.13	0.40	0.73	0.98	0.97
7	x	34.40	27.80	20.55	12.75	4.35	0
8	xs/EI	58.20	78.00	87.00	75.70	33.70	−
9	M' for $V_O = 1$	0	384	1372	3122	5632	7082
10	$(9) \times 1/(2 \times 7082) = V_O$	0	0.027	0.097	0.221	0.398	0.500
11	s/EI	1.69	2.80	4.24	5.94	7.75	−
12	M' for $M_O = 1$	0	11.20	43.70	112	235	333
13	$(12) \times 1/44.84 = M_O$	0	0.25	0.97	2.50	5.25	7.43
14	M_c	0	−0.32	−0.77	−0.67	0.99	3.08
15	V_1 left half	0	0.027	0.097	0.221	0.398	0.500
16	V_1 right half	1	0.973	0.903	0.779	0.602	0.500
17	M_1 left half	0	1.11	3.09	4.70	4.35	3.08
18	M_1 right half	−3.10	−6.57	−6.58	−3.45	1.20	3.08

The influence line ordinates for M_c are given by:

$$M_c = M_O - 4.35\, H_O \; \text{... tension in the bottom fiber positive}$$

The influence line ordinates for M_1 are given by:

$$M_1 = M_O + 14.4 H_O + 37.5 V_O \; \text{... left half}$$
$$M_1 = -x'' + M_O + 14.4\, H_O + 37.5 V_O \; \text{... right half}$$

where x'' is the horizontal distance of the unit load from 1 and bending moment producing tension in the bottom fiber is positive.

The influence line ordinates for M_c, M_1, and V_1 are derived in Table 6.2 and plotted in Figure 6.11.

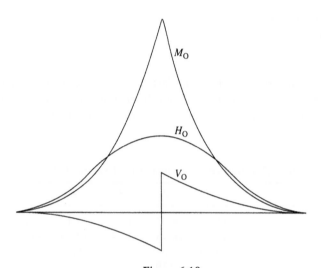

Figure 6.10

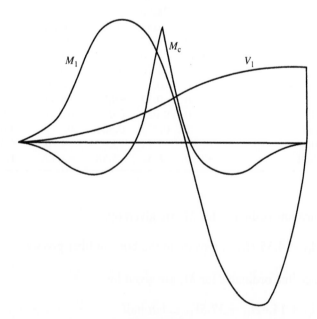

Figure 6.11

6.5 Column analogy method

The fixed-ended beam shown in Figure 6.12 may be replaced by systems (i) and (ii), the cut-back structure being produced by introducing a moment release at 1 and 2. For a beam with a straight axis, the elastic center O is on the beam axis and $M_x = H_O = 0$. Then, the bending moment at any section, with coordinates x in the original beam is given by:

$$M - (M_O + V_O x) = M - (W/A + W e_y x/I_y)$$

where M is the bending moment in the cut-back structure and the expression in brackets is the bending moment due to the elastic center reactions, with bending moment producing tension in the bottom fiber regarded as positive.

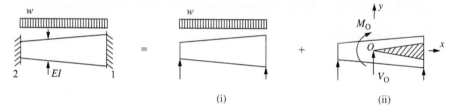

(i)   (ii)

Figure 6.12

The analogous column shown in Figure 6.13 consists of a short column with a width at any section of $1/EI$ subjected to an applied knife-edge load W with an intensity at that section of M/EI, where E and I are the modulus of elasticity and the second moment of area at the corresponding section in the real beam. That is, the cross-section of the column consists of the elastic area A, with its centroid at O, loaded with the elastic load W. A positive moment in the cut-back structure

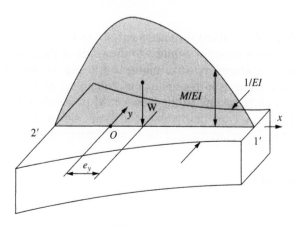

Figure 6.13

produces a vertically downward load on the analogous column. Then, the stress at any section, with coordinates x in the analogous column, is given by:

$$W/A + We_y x/I_y$$

with compressive stress regarded as positive.

This expression for stress in the analogous column is identical to that for moment in the cut-back structure due to the elastic center reactions. Thus, a value for the stress at any section in the analogous column is equivalent to the moment at the corresponding section in the cut-back structure due to the elastic center reactions.

While the column analogy method has been restricted here to straight beams, it may be readily extended, in the same manner as the elastic center method, to frames and arches, which may be symmetrical or unsymmetrical.

A vertical displacement δ_y of end 1 of the fixed-ended beam shown in Figure 6.14 produces an equal displacement of the elastic center, which is given by:

$$\delta_y = \int Mx \ ds/EI$$
$$= M_y$$

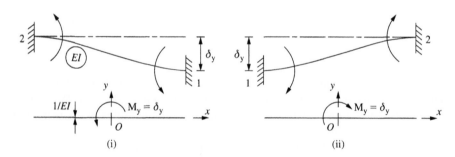

Figure 6.14

Thus, the effect of a vertical displacement on the real beam is equivalent to the effect of a bending moment applied to the y-axis of the analogous column. The displacement produces positive moment at 1 and negative moment at 2, and the equivalent stresses in the analogous column are tension at 1 and compression at 2. For cases (i) and (ii) the sense of M_y is as indicated.

A rotation θ at end 1 of the fixed-ended beam shown in Figure 6.15 produces a rotation and displacement of the elastic center, which are given by:

$$\theta_O = \theta$$
$$= \int M \ ds/EI$$
$$= W$$
$$y_O = \theta e_y$$
$$= We_y$$

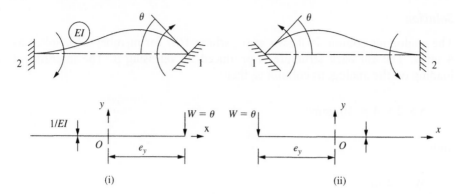

Figure 6.15

Thus, the effect of a rotation θ at 1 is equivalent to the effect of a concentrated load **W** applied to the analogous column at 1. The rotation produces a negative moment at 1 and a positive moment at 2, and the equivalent stresses in the analogous column are compression at 1 and tension at 2. For cases (i) and (ii) the sense of **W** is as indicated.

The column analogy method may be readily extended[3] to the solution of frames and continuous beams in which there are displacements and rotations of the supports.

6.6   Fixed-end moments

Example 6.3

Determine the bending moments in the column shown in Figure 6.16. The relative EI values are shown ringed.

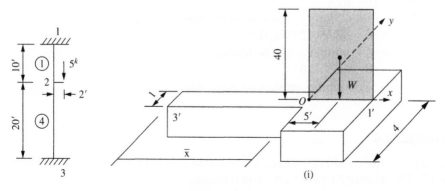

Figure 6.16

Solution

The analogous column is shown at (i), where $1/EI$ for member 12 is taken as 4 units. The cut-back structure is produced by releasing 3. The intensity of loading on the analogous column is, then:

$5 \times 2 \times 4 = 40$ units

and:

$W = 400$

The properties of the analogous column are:

$A = 10 \times 4 + 20 \times 1$
$\quad = 60$
$\bar{x} = (40 \times 25 + 20 \times 10)/60$
$\quad = 20 \text{ ft}$
$I_y = 40 \times 100/3 + 20 \times 400/3$
$\quad = 4000$
$e_y = 5 \text{ ft}$

The moments in the real column are:

$M_{12} = 10 - (W/A + 10We_y/I_y)$
$\quad\quad = 10 - (400/60 + 20{,}000/4000)$
$\quad\quad = 10 - 6.67 - 5$
$\quad\quad = -1.67 \text{ kip-ft} \dots \text{ tension on left side}$
$M_{21} = 10 - W/A$
$\quad\quad = 3.33 \text{ kip-ft} \dots \text{ tension on right side}$
$M_{23} = 0 - W/A$
$\quad\quad = -6.67 \text{ kip-ft} \dots \text{ tension on left side}$
$M_{32} = 0 - (W/A - 20We_y/I_y)$
$\quad\quad = 0 - (400/60 - 40{,}000/4000)$
$\quad\quad = 0 - 6.67 + 10$
$\quad\quad = 3.33 \text{ kip-ft} \dots \text{ tension on right side}$

Example 6.4

Determine the fixed-end moments in the non-prismatic beam shown in Figure 6.17. The relative EI values are shown ringed.

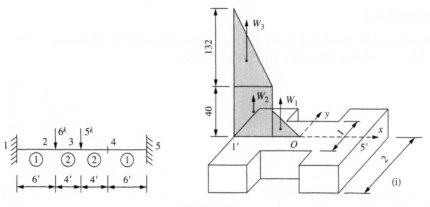

Figure 6.17

Solution

The analogous column is shown at (i), where $1/EI$ for member 12 is taken as 2 units. The cut-back structure is produced by releasing 5, and the column loads, eccentricities, and moments about the y-axis are:

$$W_1 = 40$$
$$e_1 = 2.67$$
$$M_1 = 107$$
$$W_2 = 240$$
$$e_2 = 7$$
$$M_2 = 1680$$
$$W_3 = 396$$
$$e_3 = 8$$
$$M_3 = 3168$$

The properties of the analogous column are:

$$A = 20 \times 2 - 8 \times 1$$
$$= 32$$

$$I_y = 40 \times 400/12 - 8 \times 64/12$$
$$= 1290$$

The fixed-end moments in the real beam are:

$$M_{12} = -86 - (-W/A - 10We_y/I_y)$$
$$= -86 - (-676/32 - 10 \times 4955/1290)$$
$$= -26.5 \text{ kip-ft ... tension in the top fiber}$$
$$M_{54} = 0 - (-W/A + 10We_y/I_y)$$
$$= 0 - (-676/32 + 10 \times 4955/1290)$$
$$= -17.3 \text{ kip-ft ... tension in the top fiber}$$

Example 6.5

Determine the fixed-end moments in the non-prismatic beam shown in Figure 6.18. The relative EI values are shown ringed.

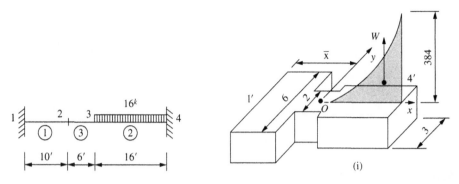

Figure 6.18

Solution

The analogous column is shown at (i), where $1/EI$ for member 12 is taken as 6 units. The cut-back structure is produced by releasing 1, and the column properties are:

$$A = 60 + 12 + 48$$
$$= 120$$
$$\bar{x} = (60 \times 5 + 12 \times 13 + 48 \times 24)/120$$
$$= 13.4 \text{ ft}$$
$$I_y = 60 \times 100/12 + 60 \times (8.4)^2 + 12 \times 36/12 + 12 \times (0.4)^2 + 48 \times 256/12 + 48 \times (10.6)^2$$
$$= 11190$$
$$W = 384 \times 16/3$$
$$= 2050$$
$$e_y = 18.6 - 4$$
$$= 14.6 \text{ ft}$$

The fixed-end moments in the real beam are:

$$M_{12} = 0 - (-2050/120 + 13.4 \times 2050 \times 14.6/11,190)$$
$$= -18.7 \text{ kip-ft ... tension in the top fiber}$$
$$M_{43} = -128 - (-2050/120 - 18.6 \times 2050 \times 14.6/11,190)$$
$$= -61.3 \text{ kip-ft ... tension in the top fiber}$$

Example 6.6

Determine the fixed-end moments in the non-prismatic beam shown in Figure 6.19 due to support 2 settling a distance δ. The EI values are shown ringed.

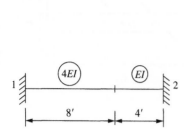
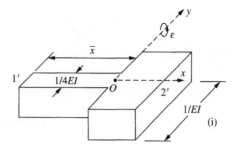

Figure 6.19

Solution

The analogous column is shown at (i), and the settlement of 2 is equivalent to a moment δ acting on the analogous column as shown. The column properties are:

$$A = 2/EI + 4/EI$$
$$= 6/EI$$
$$\bar{x} = (2 \times 4 + 4 \times 10)/6$$
$$= 8 \text{ ft}$$
$$I_y = 2/EI \times 64/3 + 4/EI \times 16/3$$
$$= 64/EI$$

The fixed-end moments in the real beam are:

$$M_{12} = 0 - 8\delta/I_y$$
$$= -EI\delta/8 \ ... \text{ tension in the top fiber}$$
$$M_{21} = 0 + 4\delta/I_y$$
$$= EI\delta/16 \ ... \text{ tension in the bottom fiber}$$

6.7 Stiffness and carry-over factors

(a) Prismatic members

The restrained stiffness s_{12} at the end 1 of a member 12, which is fixed at 2, as shown in Figure 6.20, is defined as the bending moment required at 1 to produce a rotation there of one radian. The carry-over factor is defined as the ratio of the moment induced at 2 to the moment required at 1.

The analogous column is shown at (i), and the unit rotation of 1 is equivalent to a unit load acting on the analogous column as shown. The column properties are:

$$A = l/EI$$
$$I_y = l^3/12EI$$

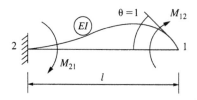 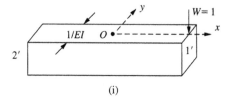

(i)

Figure 6.20

The moments in the real beam are:

$$M_{12} = 0 - (1/A + l^2/4I_y)$$
$$= -4EI/l \dots \text{tension in the top fiber}$$
$$M_{21} = 0 - (1/A - l^2/4I_y)$$
$$= 2EI/l \dots \text{tension in the bottom fiber}$$

Thus, the restrained stiffness and carry-over factors are:

$$s_{12} = M_{12}$$
$$= 4\,EI/l \dots \text{clockwise positive}$$
$$c_{12} = M_{21}/M_{12}$$
$$= \tfrac{1}{2}$$

For a beam hinged at 2, as shown in Figure 6.21, the analogous column is shown at (i) and has an infinite area concentrated at $2'$. The column properties are:

$$A = \infty$$
$$I_y = l^3/3EI$$

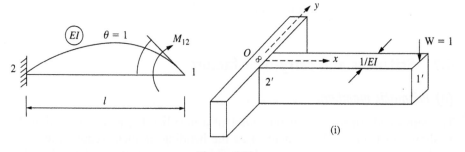

(i)

Figure 6.21

The moments in the real beam are:

$$M_{12} = 0 - (1/\infty - l^2/I_y)$$
$$= -3EI/l \dots \text{tension in the top fiber}$$
$$M_{21} = 0 - (1/\infty + 0)$$
$$= 0$$

Thus, the stiffness and carry-over factors are:

$$s'_{12} = M_{12}$$
$$= 3EI/l \text{ ... clockwise positive}$$
$$c'_{12} = M_{21}/M_{12}$$
$$= 0$$

(b) Non-prismatic members

Stiffness and carry-over factors have been tabulated for a large range of non-prismatic beams with parabolic and straight haunches[4,5], and for beams with discontinuous flanges[6]. The characteristics of members for which tabulated values are not available may be obtained by the methods detailed in Examples 6.7 and 6.8.

Example 6.7

Determine the stiffness s_{21} and the carry-over factor c_{21} for the non-prismatic beam shown in Figure 6.22. The EI values are shown ringed.

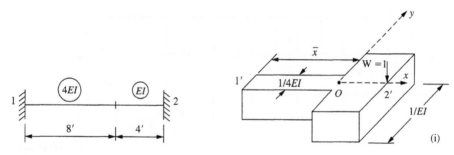

Figure 6.22

Solution

The analogous column is shown at (i), and the unit rotation of 2 is equivalent to a unit load acting on the analogous column as shown. The column properties are:

$$A = 2/EI + 4/EI$$
$$= 6/EI$$
$$\bar{x} = (2 \times 4 + 4 \times 10)/6$$
$$= 8 \text{ ft}$$
$$I_y = 2 \times 64/3EI + 4 \times 16/3EI$$
$$= 64/EI$$

The moments in the real beam are:

$$M_{21} = 0 - (1/A + 4 \times 4 \times 1/I_y)$$
$$= -5EI/12 \text{ ... tension in the top fiber}$$
$$M_{12} = 0 - (1/A - 8 \times 4 \times 1/I_y)$$
$$= EI/3 \text{ ... tension in the bottom fiber}$$

Thus the stiffness and carry-over factors are:

$$s'_{21} = M_{21}$$
$$= 5EI/12 \text{ ... clockwise positive}$$
$$c'_{21} = M_{12}/M_{21}$$
$$= 4/5$$

Example 6.8

Determine the stiffness, carry-over factor, and influence line for the fixed-end moment at the end 1 of the steel beam shown in Figure 6.23. The modulus of elasticity may be taken as 30,000 kips/in^2.

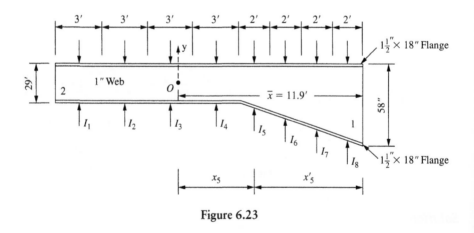

Figure 6.23

Solution

The properties of the analogous column are obtained in Table 6.3 with $E = 1$ and I in ft units:

$$A = 27.80$$
$$\bar{x} = 330.5/27.8$$
$$= 11.9 \text{ ft}$$
$$I_y = 676 + 18$$
$$= 694$$

Table 6.3 Properties of analogous column for Example 6.8

Segment	1	2	3	4	5	6	7	8	Σ
Length, s, ft	3	3	3	3	2	2	2	2	
I, ft^4	0.56	0.56	0.56	0.56	0.73	1.16	1.7	2.37	
x'	18.50	15.5	12.5	9.5	7	5	3	1	
s/I	5.33	5.33	5.33	5.33	2.73	1.73	1.18	0.84	27.8
$x's/I$	98.6	82.6	66.6	50.6	19.1	8.7	3.5	0.8	330.5
x	−6.60	−3.60	−0.60	2.4	4.9	6.9	8.9	10.9	
x^2s/I	232	69	2	31	65	82	94	101	676
$x^3/12I$	4	4	4	4	1	0.6	0.4	0	18

The influence line for M_1 is given by the elastic curve produced by a unit rotation at 1 and is equivalent to a unit load at $1'$ on the analogous column. The bending moment at the center of each segment is given by:

$$M = 0 - (1/\mathbf{A} + x\bar{x}/\mathbf{I_y})$$

and is regarded as constant over the length of each segment. The ordinates of the elastic curve are given by the bending moments at the corresponding sections in the conjugate beam, and the load on the conjugate beam is:

$$\mathbf{W'} = sM/I$$

where s is the length of the segment. The influence line ordinates for M_1 are obtained in Table 6.4 and shown in Figure 6.24.

Table 6.4 Influence line ordinates for Example 6.8

Segment	1	2	3	4	5	6	7	8
M	0.077	0.026	−0.026	−0.077	−0.120	−0.154	−0.189	−0.223
$\mathbf{W'}$	0.415	0.138	−0.138	−0.415	−0.322	−0.267	−0.223	−0.188
Influence ordinates	0	1.25	2.9	4.15	4.15	3.51	2.33	0.7

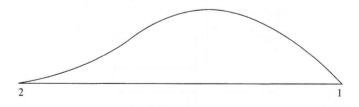

Figure 6.24

The moments in the real beam due to a unit rotation at 1 are:

$$M_{12} = 0 - (1/A + \bar{x}^2/I_y)$$
$$= -(0.036 + 0.204)$$
$$= -0.240$$
$$M_{21} = 0 - (1/A - \bar{x}(l - \bar{x})/I_y)$$
$$= -(0.036 - 0.139)$$
$$= 0.103$$

Thus, the stiffness and carry-over factors are:

$$s'_{12} = 0.240E$$
$$= 0.240 \times 30,000 \times 144$$
$$= 1360 \text{ kip-ft}$$
$$c'_{21} = 0.103/0.240$$
$$= 0.430$$

(c) Curved members

In addition to rotational stiffness and carry-over factors, the translational stiffness of curved members is also required. The translational stiffness is the horizontal force required to produce unit translation at one end of a curved member, all other displacements being prevented. The characteristics of a large range of symmetrical segmental, parabolic, and elliptical members are available[1,7]. The characteristics of unsymmetrical curved members may be determined in a similar manner to that detailed in Example 6.2.

Example 6.9

The parabolic arch shown in Figure 6.25 has a second moment of area that varies directly as the secant of the slope of the arch axis. Determine (i) the rotational stiffness and carry-over factor, (ii) the translational stiffness, (iii) the fixed-end moments and thrust due to a unit load at the crown.

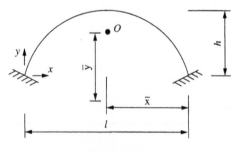

Figure 6.25

The second moment of area of the arch at any section is:

$$I = I_O \sec \alpha$$
$$= I_O \, ds/dx$$

where I_O is the second moment of area at the crown.
The equation of the arch axis is:

$$y = 4hx(l - x)/l^2$$

The properties of the analogous column are:

$$A = \int ds/EI$$
$$= \int_0^1 dx/EI_O$$
$$= l/EI_O$$
$$A\bar{y} = \int y \, ds/EI$$
$$= \int_0^1 y \, dx/EI_O$$
$$= 4h \int_0^1 (xl - x^2)dx/l^2 EI_O$$
$$= 2lh/3EI_O$$
$$\bar{y} = 2h/3$$
$$I_x = \int (y - \bar{y})^2 \, ds/EI$$
$$= 4h^2/EI_O \int_0^1 (4x^4/l^4 - 8x^3/l^3 + 16x^2/3l^2 - 4x/3l + 1/9)dx$$
$$= 4h^2 l/45EI_O$$
$$I_y = \int (x - l/2)^2 ds/EI$$
$$= \int_0^1 (x^2 - lx + l^2/4)dx/EI_O$$
$$= l^3/12EI_O$$

(i) The rotational stiffness s_{12} at the end 1 of the arch is the bending moment required at 1 to produce unit rotation there, as shown in Figure 6.26. This is equivalent to a unit load acting on the analogous column in the sense indicated for case (i) and (ii). Using the convention of clockwise moments positive, the moments in the arch are:

$$s_{12} = 1/A + \bar{x}^2/I_y + \bar{y}^2/I_x$$
$$= EI_O(1/l + 3/l + 5/l)$$
$$= 9EI_O/l$$
$$c_{12}s_{12} = -1/A + \bar{x}^2/I_y - \bar{y}^2/I_x$$
$$= -3EI_O/l$$

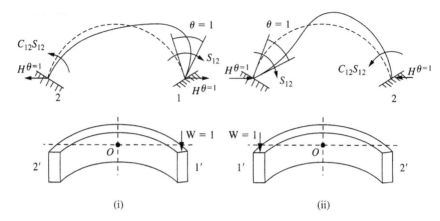

Figure 6.26

Thus:

$$c_{12} = -1/3$$

The horizontal thrust associated with the unit rotation is given by Section 6.2 as:

$$H^{\theta=1} = H_O$$
$$= \bar{y}/I_x$$
$$= 15EI_O/2hl$$

and has the sense indicated for case (i) and (ii).

(ii) The translational stiffness $H^{\delta=1}$ at the end 1 of the arch is the thrust required at 1 to produce unit translation there, as shown in Figure 6.27 and is given by Section 6.2 as:

$$H^{\delta=1} = H_o$$
$$= 1/I_x$$
$$= 45EI_O/4h^2l$$

and has the sense indicated for case (i) and (ii).

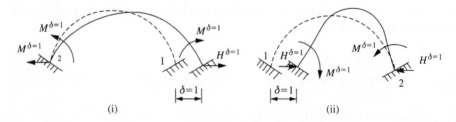

Figure 6.27

The fixed-end moment associated with the unit translation is given by Section 6.2 as:

$$M^{\delta=1} = \bar{y}H_O$$
$$= \bar{y}/I_x$$
$$= 15EI_O/2hl$$

and has the sense indicated for case (i) and (ii).

(iii) The cut-back structure is shown in Figure 6.28, and the elastic load is:

$$\mathbf{W} = 2\int_0^{l/2} M \, ds/EI$$
$$= \int_0^{l/2} x \, dx/EI_O$$
$$= l^2/8EI_O$$

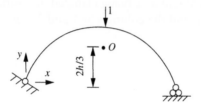

Figure 6.28

The moment of **W** about a vertical axis through O is:

$$\mathbf{M_y} = 0$$

The moment of **W** about a horizontal axis through O is:

$$\mathbf{M_x} = 2\int_0^{l/2} M(y - \bar{y})ds/EI$$
$$= 2h\int_0^{l/2}(2x^2l - 2x^3 - l^2x/3)dx/l^2EI_O$$
$$= hl^2/48EI_O$$

The moments in the arch are:

$$M_{12}^F = M_{21}^F = 0 - (\mathbf{W}/A - 2h\mathbf{M_x}/3\mathbf{I_x})$$
$$= -l/8 + 5l/32$$
$$= l/32 \dots \text{tension in the bottom fiber}$$

The horizontal thrust is given by Section 6.2 as:

$$H^F = H_O$$
$$= \mathbf{M_x}/\mathbf{I_x}$$
$$= 15l/64h \dots \text{inwards}$$

(d) Elastically restrained members

A bending moment transmitted through a riveted or bolted beam-column connection causes a relative angular displacement between the beam and the column. A linear relationship may be assumed between the transmitted moment and the relative rotation and is:

$$\phi' = \eta M$$

where η is the elastic connection factor. Expressions for determining the value of η have been derived[8] for a number of different types of connections. For a fully restrained connection $\eta = 0$ and for a hinge $\eta = \infty$.

The beam 12 shown in Figure 6.29 is attached to columns at 1 and 2 by means of elastic connections with factors of η_{12} and η_{21}. The stiffness of the beam at 1 is the bending moment, s'_{12} required to produce a unit rotation of the column there, the column at 2 being clamped in position. The relative rotations between the beam and the columns at 1 and 2 are:

$$\phi'_{12} = \eta_{12}s'_{12}$$
$$\phi'_{21} = \eta_{21}c'_{12}s'_{12}$$

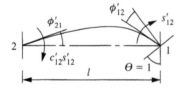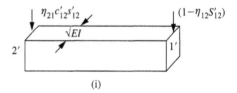

Figure 6.29

The analogous column and its loading is shown at (i), and the column properties are:

$$A = l/EI$$
$$I_y = l^3/12EI$$

Using the convention of clockwise moments positive, the moments in the real beam are:

$$s'_{12} = (1 - \eta_{12}s'_{12} + \eta_{21}c'_{12}s'_{12})EI/l + (1 - \eta_{12}s'_{12} - \eta_{21}c'_{12}s'_{12})3EI/l$$
$$c'_{12}s'_{12} = -(1 - \eta_{12}s'_{12} + \eta_{21}c'_{12}s'_{12})EI/l + (1 - \eta_{12}s'_{12} - \eta_{21}c'_{12}s'_{12})3EI/l$$

Thus: $s'_{12} = 12EIL_{21}/(4L_{12}L_{21} - l^2)$

and: $c'_{12} = l/2L_{21}$

where the equivalent lengths of the elastically restrained member are:

$$L_{12} = l + 3EI\eta_{12}$$

and: $L_{21} = l + 3EI\eta_{21}$

When the connection at end 2 is fully rigid:

$$s'_{12} = 4EI/(l + 4EI\eta_{12})$$

and: $c'_{12} = \dfrac{1}{2}$

When the connection at end 2 is hinged:

$$s'_{12} = 3EI/(l + 3EI\eta_{12})$$

and: $c'_{12} = 0$

When the connection at end 1 is fully rigid:

$$s'_{12} = 4EI(l + 3EI\eta_{21})/l(l + 4EI\eta_{21})$$

and: $c'_{12} = l/2(l + 3EI\eta_{21})$

and these two values also apply to a column 12 with an elastically restrained base at 2.

The fixed-end moments due to a unit lateral translation of the end 1 of column 12 shown in Figure 6.30 may be obtained from the analogous column shown at (i). Using the convention of clockwise moments positive, the moments in the real column are:

$$\begin{aligned}
M_{21}^{FE} &= -EI\eta_{21}M_{21}^{FE}/l - 3EI\eta_{21}M_{21}^{FE}/l + 6EI/l^2 \\
&= 6EI/l(l + 4EI\eta_{21}) \\
M_{12}^{FE} &= EI\eta_{21}M_{21}^{FE}/l + 3EI\eta_{21}M_{21}^{FE}/l - 6EI/l^2 \\
&= 6EI(l + 2EI\eta_{21}/l^2(l + 4EI\eta_{21})
\end{aligned}$$

When lateral loads are applied to a member that is elastically restrained at its ends, as shown in Figure 6.31, the ends of the member rotate, and this produces the modified fixed-end moments shown. These may be obtained in any particular instance from the analogous column shown at (i).

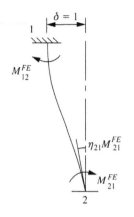
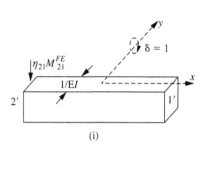

Figure 6.30

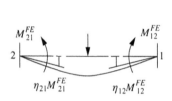
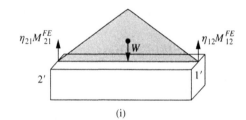

Figure 6.31

6.8 Closed rings

The distribution of bending moment in closed rings is readily obtained by the column analogy method.

Displacements in a closed ring are given by the moment of the elastic load about an axis through the elastic center parallel to the displacement. Since this requires the determination of the position of the elastic center, displacements are more conveniently determined by the conjugate structure method given in Examples 6.10 and 6.13.

Example 6.10

Determine the distribution of bending moment and the extension of the horizontal diameter for the uniform circular ring shown in Figure 6.32.

Solution

The cut-back structure is shown at (i), and the bending moment at any section defined by the angle θ is:

$$M = PR(1 - \cos \theta)/2$$

with moment producing tension on the inside of the ring regarded as positive.

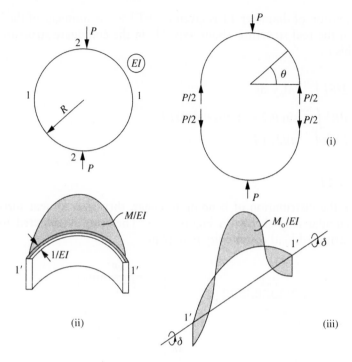

Figure 6.32

The elastic load on half the analogous column is shown at (ii) and is given by:

$$\mathbf{W} = 4/EI \int_0^{\pi/2} M \, ds$$
$$= 2PR^2 \int_0^{\pi/2} (1 - \cos\theta) d\theta / EI$$
$$= PR^2 (\pi - 2)/EI$$

The elastic load is symmetrical and $\mathbf{M_x} = \mathbf{M_y} = 0$.
The area of the analogous column is:

$$\mathbf{A} = 2\pi R/EI$$

The moments in the ring are:

$$M_1 = 0 - \mathbf{W/A}$$
$$= -PR(\pi - 2)/2\pi$$
$$M_\theta = PR(1 - \cos\theta)/2 - \mathbf{W/A}$$
$$= PR(2 - \pi \cos\theta)/2\pi$$
$$M_2 = PR/2 - \mathbf{W/A}$$
$$= PR/\pi$$

The extension of diameter 11 is given by $1/EI \times$ the moment of the bending moment in the real structure about axis $1'1'$ in the conjugate structure shown at (iii). This is:

$$
\begin{aligned}
\delta &= 2/EI \int_0^{\pi/2} M_\theta y \ ds \\
&= PR^3 \int_0^{\pi/2} \sin \theta (2 - \pi \cos \theta) d\theta / \pi EI \\
&= PR^3 (4 - \pi)/2\pi EI
\end{aligned}
$$

Example 6.11

Determine the distribution of bending moment, thrust, and shear force in the uniform circular ring shown in Figure 6.33. The ring is subjected to a uniformly distributed load of intensity w in plan.

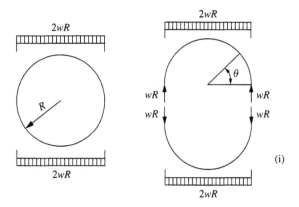

Figure 6.33

Solution

The cut-back structure is shown at (i), and the bending moment at any section defined by the angle θ is:

$$
\begin{aligned}
M &= wR^2(1 - \cos \theta) - wR^2(1 - \cos \theta)^2/2 \\
&= (wR^2/2)\sin^2 \theta
\end{aligned}
$$

The elastic load is given by:

$$
\begin{aligned}
\mathbf{W} &= 4\int_0^{\pi/2} M \ ds/EI \\
&= 2wR^3 \int_0^{\pi/2} \sin^2 \theta \ d\theta / EI \\
&= \pi wR^3/2EI
\end{aligned}
$$

The elastic load is symmetrical and $M_x = M_y = 0$.
The area of the analogous column is:

$$A = 2\pi R/EI$$

The moment in the ring is:

$$
\begin{aligned}
M_\theta &= (wR^2/2)\sin^2\theta - W/A \\
&= (wR^2/2)\sin^2\theta - wR^2/4 \\
&= -(wR^2/4)\cos(2\theta)
\end{aligned}
$$

The shear in the ring is:

$$
\begin{aligned}
Q_\theta &= \sin\theta\{wR - wR(1 - \cos\theta)\} \\
&= wR\sin\theta\cos\theta
\end{aligned}
$$

The thrust in the ring is:

$$
\begin{aligned}
P_\theta &= \cos\theta\{wR - wR(1 - \cos\theta)\} \\
&= wR\cos^2\theta
\end{aligned}
$$

Example 6.12

The uniform square culvert shown in Figure 6.34 is filled with water of density w and stands on a rigid foundation. Determine the terminal moments in the culvert.

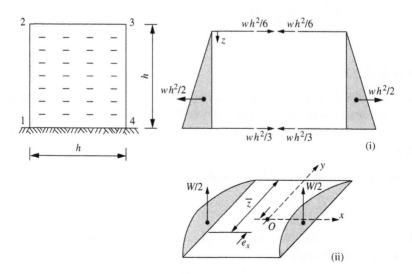

Figure 6.34

Solution

The properties of the analogous column, with $1/EI = 1$, are:

$$A = 4h$$
$$I_x = 2(h^3/12 + h^3/4)$$
$$= 2h^3/3$$

The pressure distribution on the cut-back structure is shown at (i), and the elastic load on the analogous column is:

$$W = 2\int_0^h Mdz$$
$$= 2\int_0^h (-wzh^2/6 + wz^3/6)dz$$
$$= -wh^4/12$$
$$W\bar{z} = 2\int_0^h Mz\, dz$$
$$= 2\int_0^h (-wz^2h^2/6 + wz^4/6)dz$$
$$= -2wh^5/45$$
$$\bar{z} = 8h/15$$
$$e_x = 8h/15 - h/2 = h/30$$

The elastic load is symmetrical about the y-axis as shown at (ii). The moments in the culvert are:

$$M_1 = 0 - W/A - We_x h/2I_x$$
$$= wh^3/48 + wh^3/480$$
$$= 11wh^3/480$$
$$M_2 = 0 - W/A + We_x h/2I_x$$
$$= 9wh^3/480$$

Example 6.13

Determine the distribution of bending moment and the extension of the horizontal diameter for the uniform link shown in Figure 6.35.

Solution

The area of the analogous column, with $1/EI = 1$, is:

$$A = 2\pi R + 4h$$

The cut-back structure is shown at (i), and the bending moment at any section defined by the angle θ is:

$$M = -PR(1 - \cos \theta)/2$$

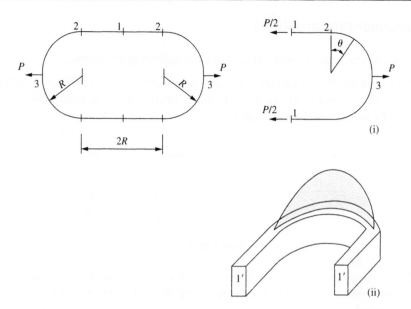

Figure 6.35

The elastic load on half the analogous column is shown at (ii) and is given by:

$$W = 4\int_0^{\pi/2} M\,ds$$
$$= -PR^2 \int_0^{\pi/2} (1 - \cos\theta)d\theta$$
$$= -PR^2(\pi - 2)$$

The elastic load is symmetrical, and $M_x = M_y = 0$.
The moments in the link are:

$$M_1 = M_2$$
$$= 0 - W/A$$
$$= PR(\pi - 2)/2(\pi + 2)$$
$$M_\theta = -PR(1 - \cos\theta)/2 - W/A$$
$$= PR/2\{\cos\theta - 4/(\pi + 2)\}$$
$$M_3 = -2PR/(\pi + 2)$$

The extension of the horizontal axis is:

$$\delta = 2RM_1/EI + 2\int_0^{\pi/2} M_\theta y\,ds/EI$$
$$= PR^3(\pi - 2)/(\pi + 2)EI + PR^3 \int_0^{\pi/2} \cos\theta\{\cos\theta - 4/(\pi + 2)\}d\theta/EI$$
$$= PR^3(\pi - 2)/(\pi + 2)EI + PR^3(\pi^2 + 2\pi - 16)/4(\pi + 2)EI$$
$$= PR^3(\pi^2 + 6\pi - 24)/4(\pi + 2)EI$$

Supplementary problems

Use the column analogy method to solve the following problems.

S6.1 Determine the fixed-end moments produced by the application of the moment M at point 3 in the non-prismatic beam shown in Figure S6.1. The relative EI values are shown ringed.

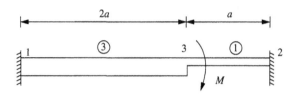

Figure S6.1

S6.2 Determine the fixed-end moments produced by the applied load indicated in the non-prismatic beam shown in Figure S6.2. The relative EI values are shown ringed.

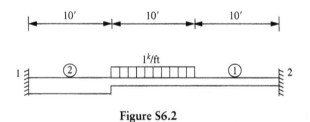

Figure S6.2

S6.3 Determine the bending moment at the base of the column shown in Figure S6.3. The relative EI values are shown ringed.

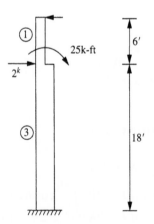

Figure S6.3

S6.4 Determine the fixed-end moments produced by the applied loads indicated in the non-prismatic beam shown in Figure S6.4. The relative *EI* values are shown ringed.

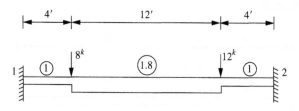

Figure S6.4

S6.5 Determine the fixed-end moments produced by the applied loads indicated in the non-prismatic beam shown in Figure S6.5. The relative *EI* values are shown ringed.

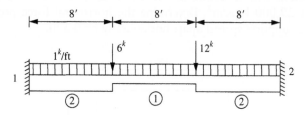

Figure S6.5

S6.6 Determine the fixed-end moments produced by the applied load indicated in the prismatic beam shown in Figure S6.6.

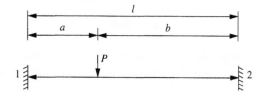

Figure S6.6

S6.7 Determine the fixed-end moments produced by the applied load indicated in the propped cantilever shown in Figure S6.7.

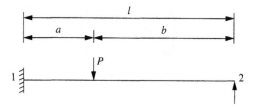

Figure S6.7

S6.8 Determine the stiffness s_{12} and the carry-over factor c_{12} for the non-prismatic beam shown in Figure S6.8. The relative EI values are shown ringed. Determine the fixed-end moments produced by the load of 10 kips applied at point 3 as indicated

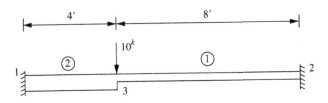

Figure S6.8

S6.9 The two-hinged portal frame shown in Figure S6.9 is fabricated from a uniform section with a second moment of area of 280 in^4 and a modulus of elasticity of 29,000 kips/in^2. Determine the horizontal force produced at the hinges if the supports spread apart horizontally by 0.5 in.

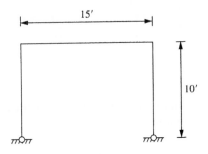

Figure S6.9

S6.10 The circular pipe shown in Figure S6.10 has two concentrated loads applied as shown. Determine the bending moments produced at the location of the loads.

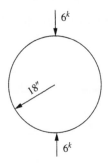

Figure S6.10

S6.11 Figure S6.11 shows a monolithic reinforced concrete culvert of constant section. Determine the bending moments produced at sections 1 and 2 by the lateral earth pressure, assuming that this is equivalent to the pressure of a fluid having a density of 30 lb/ft³.

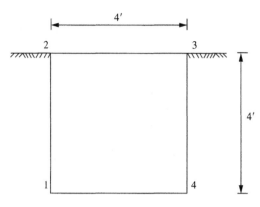

Figure S6.11

References

1. Lightfoot, E. Moment distribution. London, E. and F. N. Spon Ltd. 1961. pp. 255–272.
2. Fox, D. Y. and Au, T. Solution of rigid frames by the column analogy. Proc. Am. Soc. Civil Eng. 85 (ST1). Jan. 1959. pp. 103–112.
3. Chin, F. K. Some extensions of the column analogy. Civil Eng. and Pub. Works Rev. 58. Feb. 1963. pp. 209–210.
4. Cross, H. and Morgan, N. D. Continuous frames of reinforced concrete. New York. J. Wiley and Sons. 1932. pp. 137–155.
5. Portland Cement Association. Handbook of frame constants. Chicago. 1948.
6. Turner, F. H. Moment distribution factors for flanged beams. Conc. and Constr. Eng. 58. Aug. 1963. pp. 315–319.
7. Bannister, A. Characteristics of symmetrical segmental members. Conc. and Constr. Eng. 55. July 1960. pp. 257–262.
8. Lothers, J. E. Advanced design in structural steel. Englewood Cliffs, NJ. Prentice–Hall Inc. 1960. pp. 367–405.

7 Moment distribution methods

Notation

a_{12} — flexibility factor at the end 1 of a member 12 = restrained stiffness of the member/(ditto + adjacent actual stiffnesses) = $s_{12}/(s_{12} + \sum s^a_{1n} - s^a_{12})$

a'_{12} — $s_{12}/(s_{21} + \sum s^a_{2n} - s^a_{21})$

b_{12} — moment transmission coefficient, the proportion of the out-of-balance moment at the joint 1, which is transmitted to the fixed end 2 of a member when joint 1 is balanced = $-a_{12}c_{12}$

b'_{12} — rotation transmission coefficient, equals the proportion of the angle of rotation imposed at the end 1 of a member 12, which is transmitted to the end 2 = $-a'_1 2c_{12}$

c_{12} — carry-over factor for a member 12 from the end 1 to the end 2, = ½ for a straight prismatic member

c' — modified carry-over factor

C^S — left-hand side of sway equation after substituting M^S values

C^W — left-hand side of sway equation after substituting M^W values

d_{12} — restrained distribution factor at the end 1 of a member 12 = $s_{12}/\sum s_{1n}$

d^a_{12} — actual distribution factor at the end 1 of member 12 = $s^a_{12}/\sum s^a_{1n}$

E — Young's modulus

G — modulus of torsional rigidity

h — height of a column

H — horizontal reaction

H_{12} — outward thrust exerted at the end 1 by a curved member 12

I — second moment of area of a straight prismatic member

I_o — second moment of area of a curved member at its crown

J — torsional inertia

l_{12} — length of a member 12

M — clockwise moment of the applied loads on a structure above a particular story about the top of the story

M_{12} — moment acting at the end 1 of a member 12

M^F_{12} — fixed-end moment at the end 1 of a member 12

M^S — final moments in a structure due to an arbitrary sway displacement

M^W — final moments in a structure due to the applied loads after a non-sway distribution

Q — shear force, acting from left to right in a structure, due to all the loads applied above the base of a particular story

r — ratio of distances between bottom of columns and top of columns for a particular story in a frame with inclined columns

s_{12} restrained stiffness at the end 1 of a member 12, moment required to produce a unit rotation at the end 1, end 2 being fixed, $= 4EI/l_{12}$ for a straight prismatic member 12

s' modified stiffness

s_{12}^{a} actual stiffness at the end 1 of a member 12 = moment required to produce a unit rotation at the end 1, end 2 having its actual stiffness = $s_{12}(1 - a_{21}c_{12}c_{21})$

V vertical reaction

W applied load

x horizontal sway displacement

y vertical sway displacement

δ deflection

θ_{12} angle of rotation at the end 1 of a member 12

ϕ_{12} angle of rotation at the end 1 of a member 12 due to sway deformation

ψ_1 rotation at the end 1 of a curved member 12

η elastic connection factor

7.1 Introduction

Since its introduction in 1930[1] the moment distribution method has become established as the most useful hand-computational method in the analysis of rigid frames and continuous beams. Numerous developments have been made to the original system, and these have been recorded in comprehensive textbooks[2,3,4].

Moment distribution is essentially a relaxation technique where the analysis proceeds by a series of approximations until the desired degree of accuracy has been obtained. The method may be considered as an iterative form of the more recently developed stiffness matrix method. The moment distribution method provides a quantitative solution to a problem but may also be used to provide a qualitative understanding of structural behavior. In this way, it may be used to develop a clearer perception of the relationship between load and deformation[5]. Alternative relaxation techniques have been developed on the continent of Europe[6,7,8] for the analysis of rigid frames. These will not be considered here, as moment distribution is the customary method employed in English-speaking countries and has a more extensive literature.

7.2 Sign convention and basic concepts

Bending moments at the ends of a member are shown acting from the support to the member: i.e., the support reactions are considered. These are positive when acting in a clockwise direction, as also are clockwise rotations at the ends of a member. The directions and sense of the terminal moments in a continuous beam are indicated in Figure 7.1, and it will be noticed that the arrow heads point towards the face of the member, which is in tension.

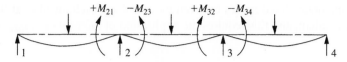

Figure 7.1

The restrained stiffness at the end 1 of a member 12 that is fixed at 2, as shown in Figure 7.2, is defined as the bending moment required at 1 to produce a rotation there of one radian.

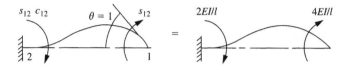

Figure 7.2

The carry-over factor is defined as the ratio of the moment induced at 2 to the moment required at 1.

In Section 6.7 the restrained stiffness and carry-over factors for a straight, prismatic member were obtained as:

$$s_{12} = 4EI/l_{12}$$
$$c_{12} = \frac{1}{2}$$

Consider the clockwise moment s_0 applied to the rigid joint 0 of Figure 7.3, causing the joint to rotate one radian. Then, s_0 is the stiffness of the joint. Since the

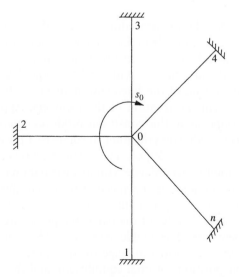

Figure 7.3

rigid joint maintains fixed angles between the members, each member at 0 rotates through one radian, and this requires the application of moments s_{01}, s_{02}, etc. Hence:

$$s_0 = s_{01} + s_{02} + \cdots + s_{0n}$$
$$= \Sigma s_{0n}$$

The distribution factor for a member is defined as the ratio of the stiffness of the member to the stiffness of the joint:

$$d_{01} = s_{01}/s_0$$
$$= s_{01}/\Sigma s_{0n}$$

and is that proportion of the moment applied at a joint that is absorbed by the member. It is clear that at any joint:

$$\Sigma d_{0n} = 1$$

The fixed-end moments for a number of common cases of loading applied to straight, prismatic members are summarized in Table 7.1. These are readily obtained by the column analogy method given in Section 6.6. A number of handbooks[9,10,11] provide a comprehensive list of fixed-end moments.

7.3 Distribution procedure for structures with joint rotations and specified translations

The first stage in the distribution procedure consists of considering the structure as a number of fixed-ended members. Each member may then be analyzed individually and the fixed-end moments derived as shown in stage (i) of Figure 7.4. In general, these fixed-end moments will not be in equilibrium at a particular joint, and the algebraic sum of the fixed-end moments is the support reaction required there to prevent the rotation of the members. Applying a balancing moment equal and opposite to this constraint produces equilibrium at the support and is equivalent to allowing the joint to rotate to a position of temporary equilibrium, where it is again clamped. The balancing moment is distributed to each member at the joint in accordance with its distribution factor.

The constraints at each joint are balanced in turn, and, due to the carry-over of moments to adjacent clamped joints, previously balanced joints become unbalanced. Hence, several cycles of the balancing procedure are required until the carry-over moments are of negligible magnitude. The sum of all the required balancing cycles is then as shown in stage (ii) of Figure 7.4, where the joints have rotated to their positions of final equilibrium, and the terminal moments are obtained by summing (i) and (ii).

Table 7.1 Fixed-end moments

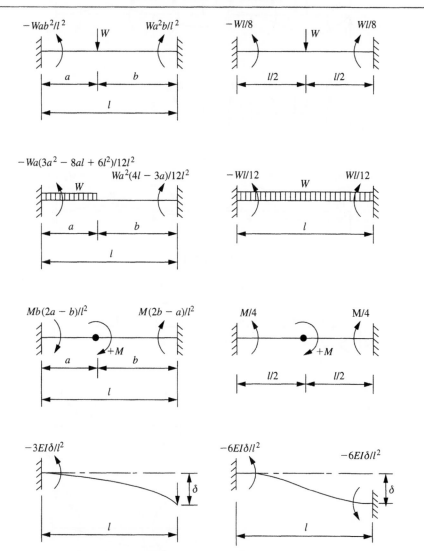

The magnitudes of the joint rotations may be readily obtained if required once distribution is completed but are not an essential part of the procedure. Figure 7.5 shows prismatic member 12, with a stiffness s, which forms part of a framework. The final terminal moments are given by:

$$M_{12} = M_{12}^F + s\theta_1 + s\theta_2/2$$
$$M_{21} = M_{21}^F + s\theta_2 + s\theta_1/2$$

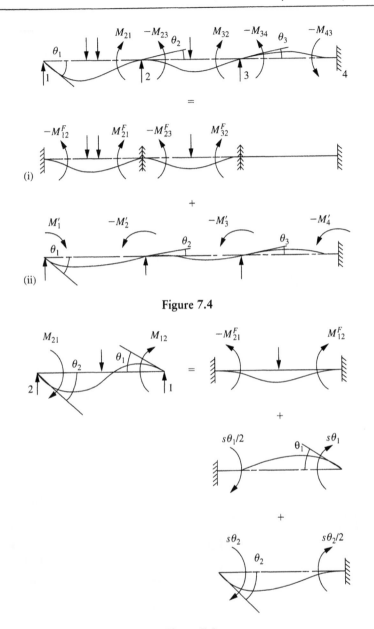

Figure 7.4

Figure 7.5

and hence, θ_1 and θ_2 may be obtained. Due regard must be paid to the signs of the moments and rotations.

Example 7.1

Determine the bending moments in each member of the structure shown in Figure 7.6. The relative EI/l values are shown ringed alongside the members.

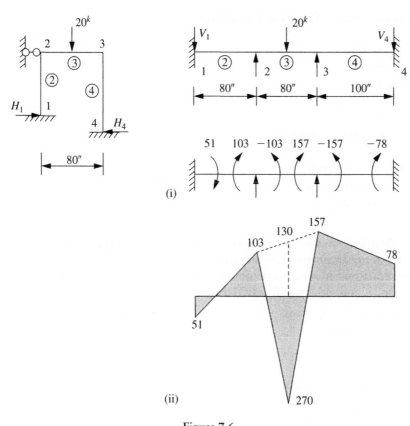

Figure 7.6

Solution

Both the propped frame and the continuous beam shown in Figure 7.6 are analyzed in an identical manner. The relative EI/l values are also the relative stiffness values, and the distribution factors at 2 are:

$$d_{21} = 2/(2 + 3)$$
$$= 2/5$$
$$d_{23} = 3/(2 + 3)$$
$$= 3/5$$

and the distribution factors at 3 are:

$$d_{32} = 3/(3 + 4)$$
$$= 3/7$$
$$d_{34} = 4/(3 + 4)$$
$$= 4/7$$

and the distribution factors at 1 and 4 are:

$$d_{12} = 2/\infty$$
$$= 0$$
$$d_{43} = 4/\infty$$
$$= 0$$

The fixed-end moments are:

$$M_{32}^F = 20 \times 80/8$$
$$= 200 \text{ kip-in}$$
$$= -M_{23}^F$$

The distribution procedure is shown in Table 7.2, and the final moments for the continuous beam are shown at (i) on Figure 7.6.

The reactions at 1 and 4 are obtained by taking moments about 2 and 3 for members 12 and 43, respectively. Then:

$$H_1 = V_1 = (M_{12} + M_{21})/l_{12}$$
$$= 154/80$$
$$= 1.93 \text{ kips}$$
$$H_4 = V_4$$
$$= (M_{43} + M_{34})/l_{34}$$
$$= 235/100$$
$$= 2.35 \text{ kips}$$

and act in the direction shown.

Table 7.2 Distribution of moments in Example 7.1

Joint	1		2		3		4
Member	12	21	23	32	34	43
Relative EI/l	2	2	3	3	4	4
Distribution factor	0	$\frac{2}{5}$	$\frac{3}{5}$	$\frac{3}{7}$	$\frac{4}{7}$	0
Carry-over factor			$\leftarrow$	$\frac{1}{2}$		
	$\leftarrow$	$\frac{1}{2}$	$\frac{1}{2}$	$\rightarrow$	$\frac{1}{2}$	$\rightarrow$
Fixed-end moments			-200	200		
Distribution		80	120	-86	-114	
Carry-over	40		-43	60		-57
Distribution		17	26	-26	-34	
Carry-over	8		-13	13		-17
Distribution		5	8	-6	-7	
Carry-over	3		-3	4		-3
Distribution and carry-over		1	2	-2	-2	-1
Final moments, kip-in	51	103	-103	157	-157	-78

The bending moment diagram for the continuous beam, drawn on the tension side of the structure, is shown at (ii) on Figure 7.6 and is obtained by combining the diagram for the fixing moments and the free moments.

7.4 Abbreviated methods

(a) Hinged ends

When the end 2 of a prismatic member 12 is hinged, it is possible to derive a modified value for the stiffness at 1 that eliminates the carry-over of moment to end 2. Referring to Figure 7.7, where the restrained stiffness of 12 is denoted by s, it is seen that the modified stiffness at 1 is given by:

$$s'_{12} = s(1 - c^2)$$
$$= 4EI/l \times 3/4$$
$$= 3EI/l$$

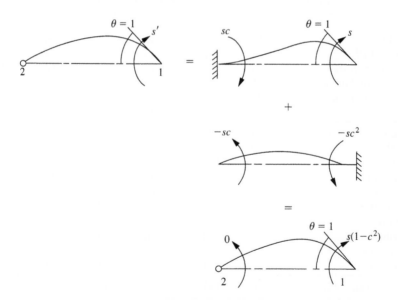

Figure 7.7

Example 7.2

Determine the bending moments in each member of the structure shown in Figure 7.8. The relative EI/l values are shown ringed alongside the members.

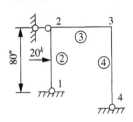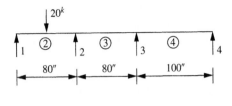

Figure 7.8

Solution

The distribution procedure is shown in Table 7.3 and is identical for both the propped frame and the continuous beam.

Table 7.3 Distribution of moments in Example 7.2

Joint	1		2		3	4
Member	12	21	23	32	34	43
Relative EI/l	2	2	3	3	4	4
Modified stiffness	8	6	12	12	12	16
Distribution factor	1	⅓	⅔	½	½	1
Carry-over factor	½	→	←	½	←	½
	←	0	½	→	0	→
Fixed-end moments	−200	200				
Distribution	200	−67	−133			
Carry-over		100		−66		
Distribution		−33	−67	33	33	
Carry-over			16	−34		
Distribution		−5	−11	17	17	
Carry-over			9	−6		
Distribution		−3	−6	3	3	
Carry-over			2	−3		
Distribution		−1	−1	2	1	
Final moments, kip-in	0	191	−191	−54	54	0

(b) Cantilevers

The stiffness s'_{12}, of a cantilever 12 with the free end 2 not restrained in position or direction is clearly zero. Hence, the distribution factor, a_{12}, is also zero.

Example 7.3

Determine the bending moments in each member of the propped frame shown in Figure 7.9. The relative EI/l values are shown ringed alongside the members.

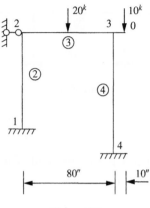

Figure 7.9

Solution

The distribution procedure is shown in Table 7.4.

Table 7.4 Distribution of moments in Example 7.3

Joint	1	2			3		4
Member	12	21	23	32	30	34	43
Relative stiffness	2	2	3	3	0	4	4
Distribution factor	0	$\frac{2}{5}$	$\frac{3}{5}$	$\frac{3}{7}$	0	$\frac{4}{7}$	0
Carry-over factor				$\leftarrow$	$\frac{1}{2}$		
	$\leftarrow$	$\frac{1}{2}$	$\frac{1}{2}$	$\rightarrow$		$\frac{1}{2}$	$\rightarrow$
Fixed-end moments			-200	200	-100		
Distribution		80	120	-43		-57	
Carry-over	40		-22	60			-29
Distribution		9	13	-26		-34	
Carry-over	4		-13	6			-17
Distribution		5	8	-2		-4	
Carry-over	3		-1	4			-2
Distribution and carry-over		1	0	-2		-2	-1
Final moments, kip-in	47	95	-95	197	-100	-97	-49

Example 7.4

Determine the bending moments in each member of the continuous beam shown in Figure 7.10. The relative EI/l values are shown ringed alongside the members.

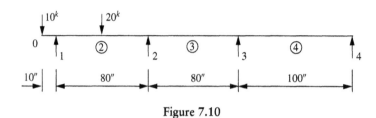

Figure 7.10

Solution

The stiffness s'_{21} is $3EI/l$ since 1 is a hinged end. The distribution procedure is shown in Table 7.5.

Table 7.5 Distribution of moments in Example 7.4

Joint		1		2		3	4
Member	10	12	21	23	32	34	43
EI/l		2	2	3	3	4	4
Modified stiffness	0	8	6	12	12	12	16
Distribution factor	0	1	$\frac{1}{3}$	$\frac{2}{3}$	$\frac{1}{2}$	$\frac{1}{2}$	1
Carry-over factor		$\frac{1}{2}$	$\rightarrow$	$\leftarrow$	$\frac{1}{2}$	$\leftarrow$	$\frac{1}{2}$
		$\leftarrow$	0	$\frac{1}{2}$	$\rightarrow$	0	$\rightarrow$
Fixed-end moments	100	−200	200				
Distribution		100	−67	−133			
Carry-over			50		−66		
Distribution			−17	−33	33	33	
Carry-over			16	−16			
Distribution			−5	−11	8	8	
Carry-over				4	−6		
Distribution			−1	−3	3	3	
Carry-over				2	−2		
Distribution			−1	−1	1	1	
Final moments, kip-in	100	−100	159	−159	−45	45	0

(c) Symmetry in structure and applied loading

In a symmetrical structure subjected to symmetrical loading, the bending moments and rotations at corresponding points in the structure are equal and of opposite sense.

For the continuous beam shown at (i) in Figure 7.11, which has an even number of spans, there is zero rotation at the center support and the members meeting there can be considered fixed-ended. Hence, distribution is required in only half the structure, there is no carry-over between the two halves, and the modified stiffnesses are as shown on the figure.

(i)

$3EI/l$ $4EI/l$ $4EI/l$ $3EI/l$

(ii)

$3EI/l$ $4EI/l$ $2EI/l$ $4EI/l$ $3EI/l$

Figure 7.11

The modified stiffnesses for a continuous beam with an odd number of spans are shown at (ii) in Figure 7.11. The modified stiffness for the central span is derived as shown in Figure 7.12 and is given by:

$$s'_{12} = s(1 - c)$$
$$= 4EI/l \times 1/2$$
$$= 2EI/l$$

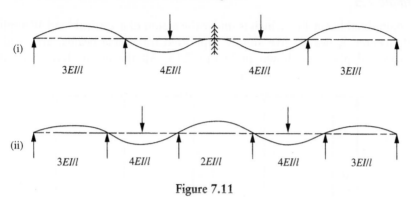

Figure 7.12

Again, distribution is required in only half the structure, there is no carry-over between the two halves, and the usual values of the initial fixed-end moments apply in all members.

Example 7.5

A longitudinal beam 11′, which is under the action of a vertical load W applied at the center 3, lies on five simply supported transverse beams, arranged in a horizontal plane as shown in Figure 7.13. All the beams have the same

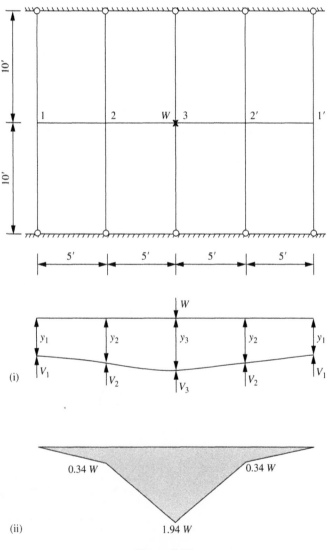

Figure 7.13

second moment of area. Draw the bending moment diagram for the longitudinal beam.

Solution

The deflection of the longitudinal beam is shown at (i) in Figure 7.13, and the relationship between deflection of the transverse beams and support reaction is:

$$y = Vl^3/48EI$$
$$= 500V/3EI$$

Due to these deflections, the fixed-end moments for the longitudinal beam are:

$$M^F_{21} = -3EI(y_2 - y_1)/25$$
$$= -20(V_2 - V_1)$$
$$M^F_{23} = M^F_{32}$$
$$= -6EI(y_3 - y_2)/25$$
$$= -40(V_3 - V_2)$$

and the distribution procedure is shown in Table 7.6. Considering clockwise moments at 2 for member 12:

$$35V_1 + 80V_1 - 200V_2 + 120V_3 = 0$$
$$115V_1 - 200V_2 + 120V_3 = 0$$

Table 7.6 Distribution of moments in Example 7.5

Joint	2		3
Member	21	23	32
Modified stiffness	3	4	4
Distribution factor	$3/7$	$4/7$	0
Carry-over factor		$\frac{1}{2}$	$\rightarrow$
Fixed-end moments	$-20(V_2 - V_1)$	$-40(V_3 - V_2)$	$-40(V_3 - V_2)$
Distribution and carry-over	$60(2V_3 - V_2 - V_1)/7$	$80(2V_3 - V_2 - V_1)/7$	$40(2V_3 - V_2 - V_1)/7$
Final moments	$40(2V_1 - 5V_2 + 3V_3)/7$	$-40(2V_1 - 5V_2 - 3V_3)/7$	$40(-V_1 + 6V_2 - 5V_3)/7$

Considering clockwise moments at 3 for member 13:

$$70V_1 + 35V_2 - 40V_1 + 240V_2 - 200V_3 = 0$$
$$30V_1 + 275V_2 - 200V_3 = 0$$

Considering the equilibrium of the longitudinal beam:

$$2V_1 + 2V_2 + V_3 = W$$

Solving these three equations simultaneously, we obtain:

$$V_1 = 0.067W$$
$$V_2 = 0.253W$$
$$V_3 = 0.36W$$

and the bending moment diagram may be drawn as shown at (ii) in Figure 7.13.

Example 7.6

Determine the bending moments for all the members of the two-story frame shown in Figure 7.14. The second moments of area of the beams are $160\,\text{in}^4$ and of the columns $100\,\text{in}^4$.

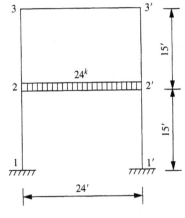

Figure 7.14

Solution

The fixed-end moments are:

$$M_{22'}^F = 24 \times 24/12$$
$$= 48 \text{ kip-ft}$$

The distribution procedure for the left half of the frame is shown in Table 7.7, and the final moments in the right half of the frame are equal and of opposite sense to those obtained for the left half. Because of the symmetry of the structure and the loading, no relative displacements occur between the ends of the members and only joint rotations need be considered.

Table 7.7 Distribution of moments in Example 7.6

Joint	1		2		3	
Member	12	21	22'	23	32	33'
I/l	100/15	100/15	160/24	100/15	100/15	160/24
Relative EI/l	1	1	1	1	1	1
Modified stiffness	4	4	2	4	4	2
Distribution factor	0	$\frac{2}{5}$	$\frac{1}{5}$	$\frac{2}{5}$	$\frac{2}{3}$	$\frac{1}{3}$
Carry-over factor	0	$\rightarrow$		$\leftarrow$	$\frac{1}{2}$	
	$\leftarrow$	$\frac{1}{2}$			$\frac{1}{2}$	$\rightarrow$
Fixed-end moments			-48			
Distribution and carry-over	9	19	10	19	9	
Distribution and carry-over				-3	-6	-3
Distribution and carry-over	1	1	1	1	1	-1
Final moments, kip-ft	10	20	-37	17	4	-4

(d) Skew symmetry

A symmetrical structure subjected to loading that is of opposite sense at corresponding points undergoes skew symmetrical deformation. The bending moments and rotations at corresponding points in the structure are equal and of the same sense, and a point of contraflexure, which is equivalent to a hinge, occurs at the center of the structure.

For a continuous beam with an even number of spans, both members at the central support can be considered hinged, and the modified stiffnesses are as shown at (i) in Figure 7.15. For a continuous beam with an odd number of

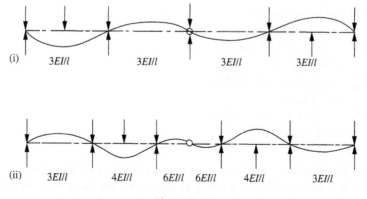

(i) $3EI/l$ $3EI/l$ $3EI/l$ $3EI/l$

(ii) $3EI/l$ $4EI/l$ $6EI/l$ $6EI/l$ $4EI/l$ $3EI/l$

Figure 7.15

spans, the central member can be considered as two hinged members each of length $l/2$, which results in the modified stiffness shown at (ii) in Figure 7.15.

As in the case of symmetrical loading, distribution is required in only half the structure, there is no carry-over between halves, and the usual values of the fixed-end moments apply in all members.

7.5 Illustrative examples

Example 7.7

Determine the reactions at the support 3 of the rigid frame shown in Figure 7.16. The relative EI values are shown ringed.

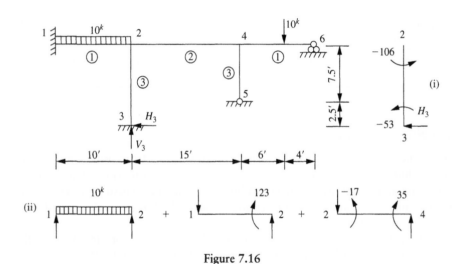

Figure 7.16

Solution

The fixed-end moments are:

$$M_{21}^F = -M_{12}^F = 10 \times 120/12$$
$$= 100 \text{ kip-in}$$
$$M_{64}^F = 10 \times 4 \times 36 \times 12/100$$
$$= 173 \text{ kip-in}$$
$$M_{46}^F = -10 \times 6 \times 16 \times 12/100$$
$$= -115 \text{ kip-in}$$

and the distribution procedure is shown in Table 7.8. Distribution to the tops of the columns is unnecessary, as the moments there may be obtained after completion of the distribution by considering the algebraic sum of the moments at joints 2 and 4. The moment at the foot of column 23 is half the

Table 7.8 Distribution of moments in Example 7.7

Joint	1		2			4		6
Member	12	21	23	24	42	45	46	64
Relative EI/l	1/10	1/10	3/10	2/15	2/15	6/15	1/10	1/10
Modified stiffness	4/10	4/10	12/10	8/15	8/15	18/15	3/10	4/10
Distribution factor	1	9/61	36/61	16/61	16/61	36/61	9/61	1
Carry-over factor	½	→		½	→		←	½
	←	0		←	½		0	→
Fixed-end moments	−100	100					−115	173
Distribution	100	−15		−26	30		17	−173
Carry-over		50		15	−13		−86	
Distribution		−10		−17	26		15	
Carry-over				13	−9			
Distribution		−2		−3	2		1	
Carry-over and distribution				1	−1			
Final moments, kip-in	0	123	−106	−17	35	133	−168	0

moment at the top since the carry-over factor is ½ and there are no sway or initial fixed-end moments.

The reaction H_3 is obtained by taking moments about 2 for member 23, as shown at (i) in Figure 7.16. Then:

$$H_3 = (106 + 53)/120$$
$$= 1.33 \text{ kips}$$

The reaction V_3 is obtained by considering the vertical reaction at 2 due to the fixing moments on spans 12 and 24, due to the distributed load on span 12 treated as a simply supported beam. Then, referring to (ii) in Figure 7.16, the vertical reaction is given by:

$$V_3 = 5 + 123/120 - 18/180$$
$$= 5.92 \text{ kips}$$

Example 7.8

Determine the slope of the beam at 2 and the final moments in the continuous beam shown in Figure 7.17 when the support 2 sinks by ½ in and the support 3 sinks by 1 in. The second moment of area of the beam is 120 in^4, and the modulus of elasticity is 29,000 kips/in^2.

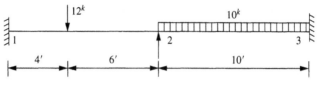

Figure 7.17

Solution

Due to the applied loading:

$$M_{21}^F = 12 \times 6 \times 16 \times 12/100$$
$$= 138 \text{ kip-in}$$
$$M_{12}^F = -12 \times 4 \times 36 \times 12/100$$
$$= -207 \text{ kip-in}$$
$$M_{32}^F = -M_{23}^F$$
$$= 10 \times 10 \times 12/12$$
$$= 100 \text{ kip-in}$$

Due to the sinking of the supports:

$$M_{12}^F = M_{21}^F$$
$$= M_{23}^F$$
$$= M_{32}^F$$
$$= -6 \times 29{,}000 \times 120 \times 1/2 \times 1/14{,}400$$
$$= -725 \text{ kip-in}$$

The distribution procedure is shown in Table 7.9.

Table 7.9 Distribution of moments in Example 7.8

Joint	1	2		3
Member	12	21	23	32
Relative EI/l	1	1	1	1
Distribution factor	0	½	½	0
Carry-over factor	←	½	½	→
Fixed-end moments	−932	−587	−825	−625
Distribution and carry-over	353	706	706	353
Final moments, kip-in	−579	119	−119	−272

Since there is no rotation at the fixed ends, we have:

$$M_{21} = M_{21}^F + s\theta_2$$
$$119 = -587 + \theta_2 \times 4 \times 29{,}000 \times 120/120$$

and $\theta_2 = 0.00609$ radians, clockwise

7.6   Secondary effects

Triangulated trusses are normally analyzed as if they are pin-jointed, although the members are invariably rigidly connected. The direct stresses obtained in this analysis are referred to as the primary stresses. These direct stresses cause axial deformations in the members, which produce relative lateral displacements of the ends of each member. The bending moments caused by these lateral displacements produce additional or secondary stresses in the members. It is customary to neglect the secondary stresses in light trusses with flexible members, but in heavy trusses the secondary stresses may be considerable.

The procedure of allowing for secondary effects consists of determining the primary stresses and, from a Williot–Mohr diagram, the lateral displacements of the ends of each member. The fixed-end moments due to these displacements are then distributed in the normal way, and the secondary stresses are obtained from the final moments.

The reactions due to the final moments may be obtained at the ends of each member. In general, these reactions will not be in equilibrium at a particular joint, and the algebraic sum of the reactions is equivalent to a constraint required at the joint to prevent its displacement. Applying a force equal and opposite to the constraint produces equilibrium. This causes additional axial forces and deformations in the members and additional secondary moments. The magnitude of the constraints is usually small compared with the applied loads and may be neglected.

Example 7.9

Determine the secondary bending moments in the members of the rigidly jointed truss shown in Figure 7.18. The second moment of area of members 14, 34, and 24 is $30\,\text{in}^4$, and that of members 12 and 23 is $40\,\text{in}^4$. The cross-sectional area of members 14, 34, and 24 is $4\,\text{in}^2$, and that of members 12 and 23 is $5\,\text{in}^2$.

Solution

The primary structure is shown at (i) in Figure 7.18, and the axial forces and deformations are listed in Table 7.10. The Williot–Mohr diagram is shown at (iii) in Figure 7.18, and the relative lateral displacement of the ends of members 12 and 14 is indicated; no lateral displacement is produced in member 24 due to the symmetry of the structure.

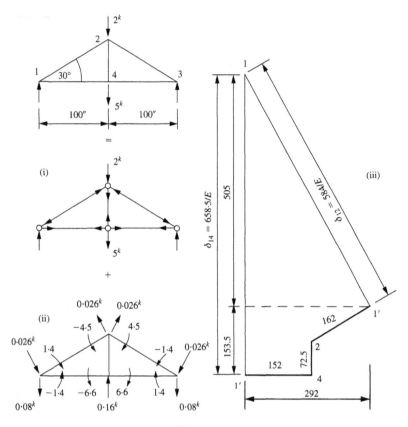

Figure 7.18

Table 7.10 Determination of forces and displacements in Example 7.9

Member	I	A	l	P	Pl/A	$\delta \times E$
12	40	5	116	7	162	584
14	30	4	100	6.07	152	658.5
24	30	4	58	5	72.5	0

Due to these displacements, the fixed-end moments are:

$$M_{12}^F = M_{21}^F$$
$$= -6EI\delta/l^2$$
$$= -6 \times 40 \times 584/13,300$$
$$= -10.5 \text{ kip-in}$$
$$M_{14}^F = M_{41}^F$$
$$= -6 \times 30 \times 658.5/10,000$$
$$= -11.8 \text{ kip-in}$$

Due to symmetry, member pairs 21 and 23 and 41 and 43 can be considered fixed-ended at 2 and 4, respectively; member 24 carries no moment; and the distribution procedure is shown in Table 7.11.

Table 7.11 Distribution of moments in Example 7.9

Joint	2		1		4
Member	21	12		14	41
Relative EI/l		346		300	
Distribution factor	0	0.535		0.465	0
Carry-over factor	←	½		½	→
Fixed-end moments	−10.5	−10.5		−11.8	−11.8
Distribution and carry-over	6	12.1		10.4	5.2
Final moments, kip-in	−4.5	1.4		−1.4	−6.6

The final moments, together with the forces required to maintain equilibrium at the joints, are shown at (ii) in Figure 7.18. These forces are approximately 3% of the applied loads and may be neglected.

7.7 Non-prismatic members

The methods of obtaining the stiffness, carry-over factors, and fixed-end moments for non-prismatic members were given in Sections 6.6 and 6.7. In addition, tabulated functions are available for a large range of non-prismatic members[2,3,4,11,12].

Example 7.10

Determine the bending moments in the frame shown in Figure 7.19. The second moments of area of the members are shown ringed.

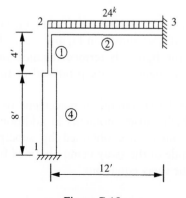

Figure 7.19

Solution

The stiffness and carry-over factors for member 21 have been obtained in Example 6.7 and are:

$$s'_{21} = 5EI/12$$
$$c'_{21} = 4/5$$

The fixed-end moments are:

$$M^F_{32} = -M^F_{23}$$
$$= 24 \times 12/12$$
$$= 24 \text{ kip-ft}$$

The distribution procedure is shown in Table 7.12.

Table 7.12 Distribution of moments in Example 7.10

Joint	1		2		3
Member	12	21	23		32
Stiffness		$5EI/12$	$8EI/12$		
Distribution factor	0	5/13	8/13		0
Carry-over factor	←	⁴⁄₅	½		→
Fixed-end moments			−24		24
Distribution and carry-over	7.4	9.25	14.75		7.38
Final moments, kip-ft	7.4	9.25	−9.25		31.38

7.8 Distribution procedure for structures subjected to unspecified joint translation

(a) Introduction

All frames subjected to lateral loads and frames unsymmetrical in shape or loading will deflect laterally, as shown in Figure 7.20. The horizontal displacement, denoted by x in the figure, is termed the side sway. In a similar manner, vertical sway displacements, y, are produced in the structures shown in Figure 7.21.

The sway produces relative lateral displacement of the ends of some of the frame members, which causes moments in addition to those due to joint rotations. The final moments are obtained by superposition, as shown in Figure 7.22, the magnitude of the sway being obtained from equations of static equilibrium known as the sway equations.

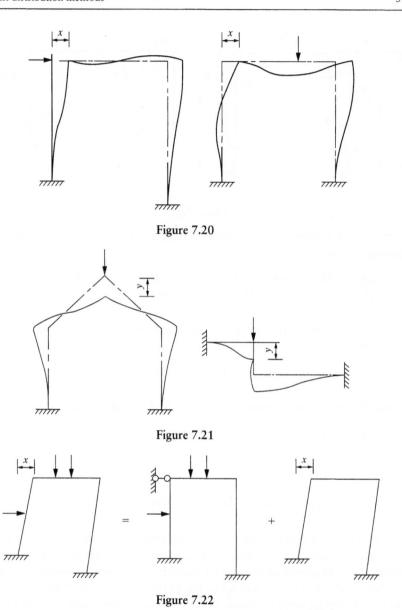

Figure 7.20

Figure 7.21

Figure 7.22

(b) The sway equations

The frame shown at (i) in Figure 7.23 may be considered to deform as the mechanism indicated under the action of the applied loads. Applying the equation of virtual work to the small displacements involved, the external work done by the applied loads equals the internal work done by the terminal moments in the columns rotating through an angle ϕ. Due to the sign convention adopted (clockwise moments positive) these terminal moments are all

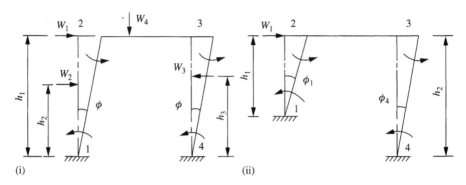

Figure 7.23

negative, and a change of sign must be introduced to satisfy the virtual work equation. Then:

$$-(M_{12} + M_{21})\phi - (M_{34} + M_{43})\phi = W_1 h_1 \phi + W_2 h_2 \phi - W_3 h_3 \phi$$
$$-M_{12} - M_{21} - M_{34} - M_{43} = W_1 h_1 + W_2 h_2 - W_3 h_3$$

and the final terminal moments in the frame must satisfy this sway equation. Similarly, for the frame shown at (ii) in Figure 7.23, the sway equation is:

$$-(M_{12} + M_{21})\phi_1 - (M_{34} + M_{43})\phi_4 = W_1 h_1 \phi_1$$

Since the sway displacements at the tops of the columns are equal:

$$h_1 \times \phi_1 = h_2 \times \phi_4$$

and:

$$-(M_{12} + M_{21}) - h_1(M_{34} + M_{43})/h_2 = W_1 h_1$$

In the case of a frame with inclined columns, as shown in Figure 7.24, the beam also rotates, and it is necessary to construct the displacement diagram, shown at (i), to establish the sway equation. Since axial deformations in the members of the frame are negligible, the points 1, 2, 3, and 4 coincide. A unit horizontal displacement is imposed on 2, and, as 2 must move perpendicularly to the original direction of 12, the point 2′ is obtained. Similarly, 3 must move perpendicularly to 23 and 34, and the point 3′ is obtained. The member rotations are obtained from the diagram as:

$$\phi_1 = 22'/l_{12}$$
$$\phi_2 = 2'3'/l_{23}$$
$$\phi_4 = 33'/l_{34}$$

where ϕ_2 is anticlockwise and produces clockwise moments M_{23} and M_{32}.

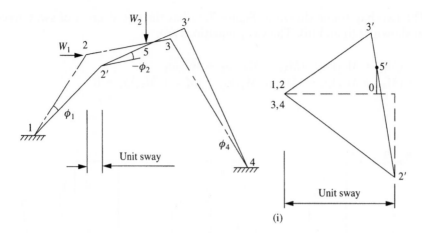

Figure 7.24

The vertical displacement of 5 is $05'$ where $5'$ divides $2'3'$ in the ratio 5 divides 23. The sway equation is:

$$-(M_{12} + M_{21})\phi_1 + (M_{23} + M_{32})\phi_2 - (M_{34} + M_{43})\phi_4 = W_1 \times 1 - W_2 \times (05')$$

and the rotations may be eliminated by using the expressions obtained from the displacement diagram.

The two-story frame shown in Figure 7.25 has two degrees of sway freedom, as shown at (i) and (ii), corresponding to the different translations possible at the beam levels. Considering sway (1) and (2) in turn, the sway equations are:

$$-(M_{32} + M_{23}) - (M_{45} + M_{54}) = W_2 h_2$$
$$-(M_{21} + M_{12}) - (M_{56} + M_{65}) = (W_1 + W_2) h_1$$

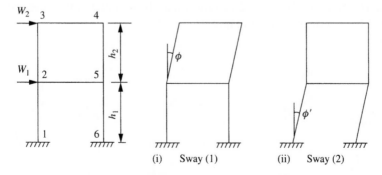

(i)   Sway (1)   (ii)   Sway (2)

Figure 7.25

and the terminal moments in the frame must satisfy both these equations.

The two-bay frame shown in Figure 7.26 has the two degrees of sway freedom shown at (i) and (ii). The sway equations are:

$$-(M_{12} + M_{21})\phi_1 - (M_{34} + M_{43})\phi_2 = Wl_{12}\phi_1$$
$$-(M_{45} + M_{54})\phi_4 - (M_{34} + M_{43})\phi_5 - (M_{67} + M_{76})\phi_3 = 0$$

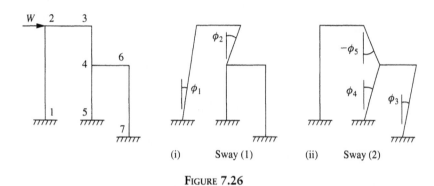

(i)   Sway (1)   (ii)   Sway (2)

FIGURE 7.26

and the rotations may be eliminated by considering the geometry of the structure.

The ridged portal shown in Figure 7.27 has the two degrees of sway freedom shown at (i) and (ii). Considering sway (1) and (2) in turn, the sway equations are:

$$-(M_{12} + M_{21}) - (M_{54} + M_{45}) = W_1 l_{12}$$
$$-(M_{12} + M_{21})\phi_1 - (M_{54} + M_{45})\phi_1 - (M_{23} + M_{32})\phi_2$$
$$-(M_{34} + M_{43})\phi_2 - (M_{43} + M_{34})\phi_2 = W_2 \times 1 - W_1 l_{12}\phi_1$$

and the rotations in the second equation may be eliminated using expressions obtained from a displacement diagram.

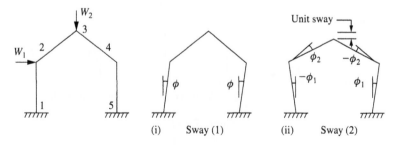

(i)   Sway (1)   (ii)   Sway (2)

Figure 7.27

In all cases, the sway equations can be considered as consisting of a left-hand side involving moments in the structure and a right-hand side involving the loads applied to the structure.

(c) Sway procedure

The analysis of structures subjected to sway proceeds in two stages as shown at (i) and (ii) in Figure 7.28. The first stage consists of a non-sway distribution with the external loads applied to the structure, which is prevented from displacing laterally. The fixed-end moments are obtained, and the distribution proceeds normally to give the final moments M^W. The second stage consists of determining the moments produced by a unit sway displacement. The fixed-end moments due to the unit displacement are obtained, and the distribution proceeds normally to give the final moments M^S. The actual moments in the structure are:

$$M = M^W + xM^S$$

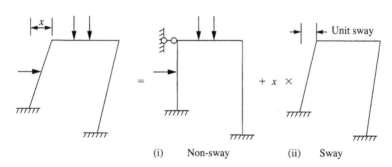

(i)   Non-sway   (ii)   Sway

Figure 7.28

and these moments satisfy the sway equation. Hence, if substituting the moments M^W and M^S in the left-hand side of the sway equation produces the values C^W and C^S, respectively, then:

$$C^W + xC^S = \text{right-hand side of sway equation}$$

and the value of x may be obtained.

The analysis of a structure with two degrees of sway freedom proceeds in three stages as shown at (i), (ii), and (iii) of Figure 7.29. The moments produced by the non-sway, sway (1), and sway (2) stages are M^W, M^{S1}, and M^{S2}, and these, when substituted in turn in the left-hand side of sway equation (1) and sway equation (2), give the values $C_1^W, C_1^{S1}, C_1^{S2}$ and $C_2^W, C_2^{S1}, C_2^{S2}$ respectively.

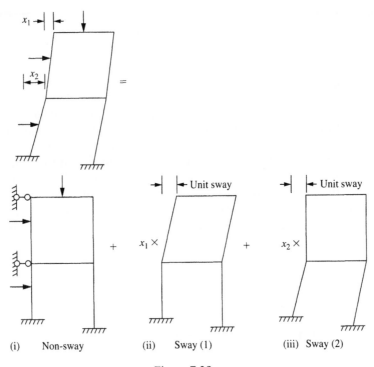

Figure 7.29

Then $C_1^W + x_1 C_1^{S1} + x_2 C_1^{S2} = $ right-hand side of sway equation (1)

and $C_2^W + x_1 C_2^{S1} + x_2 C_2^{S2} = $ right-hand side of sway equation (2)

The values of x_1 and x_2 are obtained by solving these two equations simultaneously, and the actual moments in the structure are:

$$M = M^W + x_1 M^{S1} + x_2 M^{S2}$$

In a similar fashion, structures having more than two degrees of freedom may be analyzed.

In practice, it is not necessary to impose unit displacement on the structure; any arbitrary displacement may be imposed that produces convenient values for the initial fixed-end moments. The value obtained for x will thus not be the actual displacement of the structure, but this in any case is generally not required.

When the external loads are applied to the joints of the structure, the non-sway distribution is not required, since the loading produces no fixed-end moments.

(d) Illustrative examples

Example 7.11

Determine the bending moments in the frame shown in Figure 7.30. The second moments of area of the members are shown ringed.

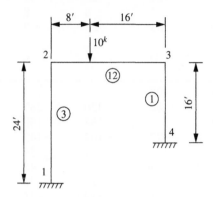

Figure 7.30

Solution

The sway equation is derived in a manner similar to that used for the frame shown at (ii) in Figure 7.23 and is:

$$M_{12} + M_{21} + 3(M_{34} + M_{43})/2 = 0$$

The fixed-end moments due to the applied loads are:

$$M_{23}^F = -10 \times 8 \times 16 \times 16 \times 12 /(24 \times 24)$$
$$= -427 \text{ kip-in}$$
$$M_{32}^F = 10 \times 16 \times 8 \times 8 \times 12 /(24 \times 24)$$
$$= 213 \text{ kip-in}$$

The fixed-end moments due to an arbitrary sway displacement are:

$$M_{12}^F = M_{21}^F$$
$$= -6E\delta \times 48/24^2$$
$$= -6E\delta /12$$
$$= -400x$$
$$M_{34}^F = M_{43}^F$$
$$= -6E\delta \times 16/16^2$$
$$= -6E\delta/16$$
$$= -300x$$

The distribution procedure for the sway and non-sway stages is shown in Table 7.13.

Table 7.13 Distribution of moments in Example 7.11

Joint	1	2		3		4
Member	12	21	23	32	34	43
Relative EI/l	2	2	8	8	1	1
Distribution factor	0	1/5	4/5	8/9	1/9	0
Carry-over factor	←	½	←	½		
			½	→	½	→
M^F, sway	−400	−400			−300	−300
Distribution		80	320	267	33	
Carry-over	40		134	160		16
Distribution		−27	−107	−142	−18	
Carry-over	−14		−71	−54		−9
Distribution		14	57	48	6	
Carry-over	7		24	28		3
Distribution		−5	−19	−25	−3	
Carry-over	−2		−12	−9		−1
Distribution		2	10	8	1	
Carry-over	1		4	5		
Distribution		−1	−3	−4	−1	
M^S	−368	−337	337	282	−282	−291
M^F, non-sway			−427	213		
Distribution		85	342	−189	−24	
Carry-over	42		−95	171		−12
Distribution		19	76	−152	−19	
Carry-over	10		−76	38		−10
Distribution		15	61	−34	−4	
Carry-over	8		−17	30		−2
Distribution		3	14	−27	−3	
Carry-over	2		−13	7		−2
Distribution		3	10	−6	−1	
Carry-over	1		−3	5		
Distribution		1	2	−4	−1	
M^W	63	126	−126	52	−52	−26
$0.046 \times M^S$	−17	−16	16	13	−13	−13
M kip-in	46	110	−110	65	−65	−39

Substituting the final non-sway moments in the left-hand side of the sway equation gives:

$$189 - 3 \times 78/2 = 72$$

Substituting the final sway moments in the left-hand side of the sway equation gives:

$$-705 - 3 \times 573/2 = -1565$$

Then: $72 - 1565x = 0$
And: $x = 0.046$

The actual moments are obtained by adding the final non-sway moments to $x \times$ the final sway moments.

Example 7.12

Determine the position on the beam, for the frame shown in Figure 7.31, at which a unit load may be applied without causing sway. All the members are of uniform cross-section.

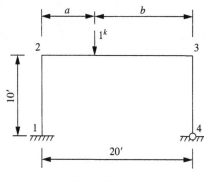

Figure 7.31

Solution

The sway equation is:

$$M_{12} + M_{21} + M_{34} = 0$$

and a sway distribution is not required since there is no sway.
The fixed-end moments due to the applied loads are:

$$M_{23}^F = -ab^2/400$$
$$M_{32}^F = ba^2/400$$

and these are in the ratio $-300b:300a$.

The distribution procedure is shown in Table 7.14; substituting the final moments in the sway equation gives:

$$279b - 125a = 0$$

Table 7.14 Distribution of moments in Example 7.12

Joint	1	2		3		4
Member	12	21	23	32	34	43
Relative EI/l	2	2	1	1	2	2
Modified stiffness		4	2	2	3	
Distribution factor	0	2/3	1/3	2/5	3/5	1
Carry-over factor	←	½	←	½		
			½	→	0	→
M^F, non-sway			−300b	300a		
Distribution		200b	100b	−120b	−180a	
Carry-over	100b		−60a	50b		
Distribution		40a	20a	−20b	−30b	
Carry-over	20a		−10b	10a		
Distribution		7b	3b	−4a	−6a	
Carry-over	3b		−2a	2b		
Distribution		a	a	−b	−b	
M^W	20a	41a	−41a	186a	−186a	0
	+103b	+207b	−207b	+31b	−31b	

Also:

$$b + a = 20$$

Solving these equations simultaneously, we obtain:

$$a = 13.8\,\text{ft}$$

Example 7.13

Determine the bending moments in the frame shown in Figure 7.32. All the members of the frame have the same second moment of area.

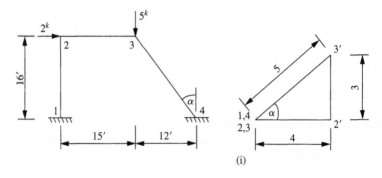

(i)

Figure 7.32

Solution

The displacement diagram, for a horizontal displacement of 4 ft units imposed at 2 is shown at (i), and the sway equation is:

$$-(M_{12} + M_{21}) \times 22'/16 + (M_{23} + M_{32})$$
$$\times 2'3'/15 - (M_{34} + M_{43}) \times 33'/20 = 2 \times 22' - 5 \times 2'3'$$
$$- (M_{12} + M_{21}) \times 4/16 + (M_{23} + M_{32}) \times 3/15$$
$$- (M_{34} + M_{43}) \times 5/20 = 2 \times 4 - 5 \times 3$$
$$- 5(M_{12} + M_{21}) - 4(M_{23} + M_{32}) + 5(M_{34} + M_{43}) = 140$$

The fixed-end moments due to the sway displacements are:

$$M_{12} = M_{21}$$
$$= -6EI \times (22')/16^2$$
$$= -6EI \times 4/16^2 = -750x$$
$$M_{23} = M_{32}$$
$$= 6EI \times (2'3')/15^2$$
$$= 6EI \times 3/15^2$$
$$= 640x$$
$$M_{34} = M_{43}$$
$$= -6EI \times (32')/20^2$$
$$= -6EI \times 5/20^2$$
$$= -600x$$

Table 7.15 Distribution of moments in Example 7.13

Joint	1		2	3		4
Member	12	21	23	32	34	43
Relative EI/l	1/16	1/16	1/15	1/15	1/20	1/20
Distribution factor	0	15/31	16/31	4/7	3/7	0
Carry-over factor	←	½	←	½		
			½	→	½	→
M^S, sway	−750	−750	640	640	−600	−600
Distribution		53	57	−24	−16	
Carry-over	26		−12	28		−8
Distribution		6	6	−16	−12	
Carry-over	3		−8	3		−6
Distribution and carry-over	2	4	4	−2	−1	0
M^S	−719	−687	687	629	−629	−614
$-0.00757 \times M^S$	5.4	5.2	−5.2	−4.8	4.8	4.7

The distribution procedure is shown in Table 7.15, and substituting the final sway moments in the sway equation gives:

$$(-7030 - 5264 - 6215)x = 140$$
$$x = -0.00757$$

The final moments are shown in the table.

Example 7.14

Determine the bending moments in the symmetrical ridged portal frame shown in Figure 7.33, which carries a uniformly distributed load of 1 kip/ft on plan. All the members have the same cross-section.

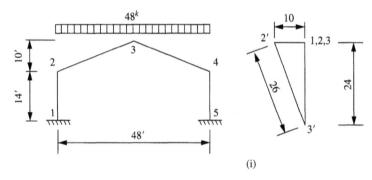

(i)

Figure 7.33

Solution

Due to symmetry of the structure and the loading only one mode of sway, shown at (i) in Figure 7.27, is possible. The displacement diagram for a vertical displacement of 24 ft units imposed at 3 is shown at (i) in Figure 7.33, and the sway equation is:

$$(M_{12} + M_{21}) \times 10/14 - (M_{23} + M_{32}) \times 26/26 + (M_{34} + M_{43}) \times 26/26$$
$$- (M_{45} + M_{54}) \times 10/14 = 48 \times 12 \times 24/2$$

Since the bending moments at corresponding points of the symmetrical frame are equal and of opposite sense, this reduces to:

$$-5(M_{12} + M_{21}) - 7(M_{23} + M_{32}) = 7 \times 24 \times 12 \times 24/2 = 24,190 \text{ kip-in}$$

The fixed-end moments due to the applied loads are:

$$M_{32}^F = -M_{23}^F$$
$$= 24 \times 24 \times 12/12$$
$$= 576 \text{ kip-in}$$

The fixed-end moments due to the sway displacement are:

$$M_{12}^F = M_{21}^F$$
$$= 6EI \times 10/14^2$$
$$= 6EI \times 10/196$$
$$= 260y$$
$$M_{32}^F = M_{23}^F$$
$$= -6EI \times 26/26^2$$
$$= -6EI \times 1/26$$
$$= -196y$$

The distribution procedure for the sway and non-sway stages is shown in Table 7.16, where joint 3 is considered fixed and there is no carry-over between the two halves of the frame. Substituting the final moments for these two stages in the left-hand side of the sway equation gives:

$$5 \times 457 + 7 \times 425 = 5260$$

Table 7.16 Distribution of moments in Example 7.14

Joint	1		2	3
Member	12	21	23	32
Relative EI/l	1/14	1/14	1/26	1/26
Distribution factor	0	13/20	7/20	0
Carry-over factor	←	½	½	→
M^F, sway	260	260	−196	−196
Distribution and carry-over	−21	−42	−22	−11
M^S	239	218	−218	−207
M^F, non-sway			−576	576
Distribution and carry-over	187	374	202	101
M^W	187	374	−374	677
$4.48 \times M^S$	1065	975	−975	−930
M kip-in	1252	1349	−1349	−253

and: $5 \times 561 - 7 \times 303 = 684$

Then: $684 + 5260y = 24{,}190$

$y = 4.48$

and the final moments are shown in the table.

Example 7.15

Determine the bending moments in the two bay frame shown in Figure 7.34. The relative EI/l values are shown ringed.

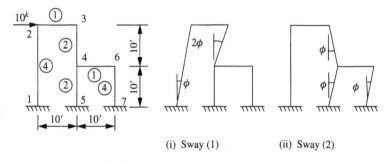

(i) Sway (1)   (ii) Sway (2)

Figure 7.34

Solution

The two sway equations are:

sway (1), $-(M_{12} + M_{21}) - 2(M_{34} + M_{43}) = 200$ kip-ft
sway (2), $-(M_{34} + M_{43}) - (M_{45} + M_{54}) - (M_{67} + M_{76}) = 0$

The fixed-end moments due to sway (1) are:

$$M_{12}^F = M_{21}^F$$
$$= 6\delta \times 4/20$$
$$= -100x_1$$
$$M_{43}^F = M_{34}^F$$
$$= -6\delta \times 2/10$$
$$= -100x_1$$

The fixed-end moments due to sway (2) are:

$$M_{43}^F = M_{34}^F$$
$$= -M_{45}^F$$
$$= -M_{54}^F$$
$$= -6\delta \times 2/10$$
$$= -100x_2$$

Table 7.17 Distribution of moments in Example 7.15

Joint	1	2		3		4				6	7	5
Member	12	21	23	32	34	43	45	46	64	67	76	54
Relative EI/l	4	4	1	1	2	2	2	1	1	4	4	2
Distribution factor	0	4/5	1/5	1/3	2/3	2/5	2/5	1/5	1/5	4/5	0	0
Carry-over factor	↓	½	↓ ½	↓ ½	← ½	½	½	↓ ½	↑	½ ↑	↑	↑
M^F, sway (1)	−100	−100			−100	−100						
Distribution		80	20	33	67	40	40	20				
Carry-over	40		17	10	20	34			10			20
Distribution		−13	−4	−10	−20	−13	−13	−8	−2	−8		
Carry-over	−7		−5	−2	−7	−10			−4		−4	−7
Distribution		4	1	3	6	4	4	2	1	3		
Carry-over	2		2		2	3			2		2	2
Distribution		−1	−1	−1	−1	−1	−1	−1	−1	−1		
M^{S1}	−65	−30	30	33	−33	−43	30	13	6	−6	−2	15
M^F, sway (2)					100	100	−100	0	0	−200	−200	−100
Distribution				−33	−67	0	0	0	40	160		
Carry-over			−17		0	−34	0	20	0		80	0
Distribution		14	13			5	5	4				
Carry-over	7			2	3				2			3
Distribution				−2	−3				−2	−2		
M^{S2}	7	14	−14	−33	33	71	−95	24	42	−42	−120	−97
$1.05 \times M^{S1}$	−68	−32	32	35	−45	31	14	6	11	−11	−25	16
$0.26 \times M^{S2}$	2	4	−4	−9	9	18	−24	6	11	−11	−31	−25
M kip-ft	−66	−28	28	26	−26	−27	7	20	17	−17	−33	−9

$$M^F_{67} = M^F_{76}$$
$$= -6\delta \times 4/10$$
$$= -200x_2$$

The distribution procedure for the two sway stages is shown in Table 7.17. Substituting the final sway (1) and sway (2) moments in the left-hand side of sway equation (1) gives 247 and − 229, respectively. Substituting the final sway (1) and sway (2) moments in the left-hand side of sway equation (2) gives −113 and 458, respectively. Then:

$$247x_1 - 229x_2 = 200$$
and: $$-113x_1 + 458x_2 = 0$$

Solving these two equations simultaneously, we obtain:

$$x_1 = 1.05$$
$$x_2 = 0.26$$

and the actual moments are obtained as shown in the table.

7.9   Symmetrical multi-story frames with vertical columns

The analysis of a single bay frame with lateral loads at the joints and vertical loads on the beams proceeds in two stages, as shown at (i) and (ii) in Figure 7.35.

The first stage consists of a symmetrically loaded frame, which is readily analyzed as there is no sway. Only the left half of the frame need be considered, with a modified stiffness of $2EI/l$ applied to the beams and no carry-over between the two halves. The final moments in the right half are equal and of opposite sense to the corresponding moments in the left half.

The second stage consists of a skew symmetrical distribution. Again, only the left half of the frame need be considered, with a modified stiffness of $6EI/l$ applied to the beams and no carry-over between the two halves. In addition, it is possible to impose initial lateral displacements causing fixed-end moments in the columns, which satisfy the sway equations for each story, and to use modified stiffness and carry-over factors for the columns, which will ensure that the sway equations remain balanced throughout the distribution. Since the structure is symmetrical, the sway equation for the second story is:

$$-2(M_{12} + M_{21}) = (2W_3 + 2W_2)h_2$$

For symmetrical columns, the fixed-end moments produced by a sway displacement are equal at the top and bottom of each column. Then, the initial fixed-end moments required to satisfy the sway equation are:

$$M^F_{12} = M^F_{21}$$
$$= -(2W_3 + 2W_2)h_2/4$$

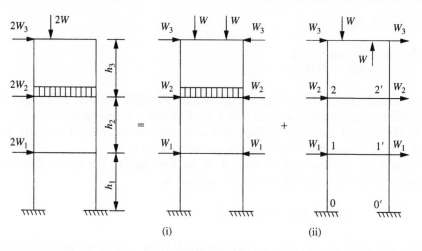

Figure 7.35

The out of balance moment at joint 1 is distributed while allowing joint 2 to translate laterally without rotation so that the sway equation remains satisfied. Then, the distribution and carry-over moments in column 12 must sum to zero.

From Figure 7.36, where s' and c' refer to the modified stiffness of 12 and the modified carry-over factor from 1 to 2:

$$s' + s'c' = 0$$

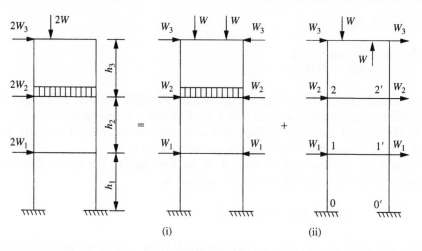

Figure 7.36

and

$$c' = -1$$
$$s' = s - xs(1 + c)/l$$
$$s'c' = -s'$$
$$\quad\; = sc - xs(1 + c)/l$$

Hence: $x = l/2$

and: $s' = s - s(1 + c)/2$
$= s/4 = EI/1$

for a straight prismatic column. Similarly, the modified stiffness of 21 and the modified carry-over factor from 2 to 1 are EI/l and -1, respectively.

In the bottom story the stiffness and carry-over factor from 1 to 0 are EI/l and -1, as before, but there is no carry-over from 0 to 1.

Should the column feet be hinged at the foundation, the initial fixed-end moment required to satisfy the sway equation is:

$$M_{10}^F = -(2W_3 + 2W_2 + 2W_1)h_1/2$$

For the sway equation to remain satisfied, there can be no distribution to member 10, and the modified stiffness of member 10 is zero.

Any lateral loads applied to the columns in the skew symmetrical case, as shown in Figure 7.37, will produce fixed-end moments in the columns. The initial fixed-end moments due to sway must now be adjusted so that, when added to the fixed-end moments due to the applied loading, the combined fixed-end moments in the columns satisfy the sway equations for each story.

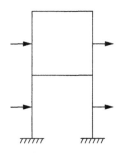

Figure 7.37

The Vierendeel girder, shown in Figure 7.38, in which the top and bottom chords in each panel are parallel and of equal stiffness, may be analyzed in a similar manner. For the skew symmetrical distribution, the stiffness of the vertical posts is $6EI/l$, and the stiffnesses and carry-over factors for the chords are EI/l and -1, respectively. The sway equations are derived as shown at (i), (ii), and (iii) and are:

$$2(M_{12} + M_{21}) = -2Wl$$
$$-2(M_{23} + M_{32}) = Wl$$
$$2(M_{34} + M_{43}) = Wl$$

for panels 1, 2, and 3, respectively.

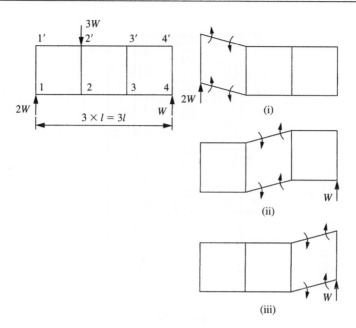

3W

1'   2'   3'   4'

1   2   3   4

2W   W   2W

3 × l = 3l

(i)

(ii)

(iii)

Figure 7.38

Multi-bay frames of the type shown in Figure 7.39 can be replaced by the two substitute frames at (i) and (ii). The ratio of applied load to member stiffness is the same for frame (i) and frame (ii), and the joint rotations and sway displacements are identical in frame (i), frame (ii), and the original frame. Such a frame is said to satisfy the principle of multiples, and the moments obtained in the analysis of the two substitute frames sum to give the moments in the original structure.

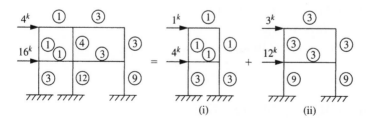

Figure 7.39

Example 7.16

Determine the bending moments in the symmetrical frame shown in Figure 7.40. The relative EI/l values are shown ringed.

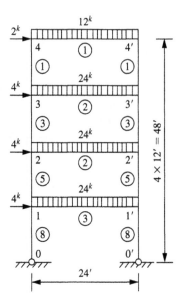

Figure 7.40

Solution

The sway equations and the initial fixed-end moments due to sway are:
top story:

$$2(M_{43} + M_{34}) = -2 \times 12 \times 12$$
$$M_{43}^F = M_{34}^F$$
$$= -72 \text{ kip-in}$$

third story:

$$2(M_{23} + M_{32}) = -6 \times 12 \times 12$$
$$M_{23}^F = M_{32}^F$$
$$= -216 \text{ kip-in}$$

second story:

$$2(M_{12} + M_{21}) = -10 \times 12 \times 12$$
$$M_{12}^F = M_{21}^F$$
$$= -360 \text{ kip-in}$$

bottom story:

$$2 \times M_{10} = -14 \times 12 \times 12$$
$$M_{10}^F = -1008 \text{ kip-in}$$

Table 7.18 Distribution of moments in Example 7.16

	0	1			2			3		4	
Joint	0	1			2			3		4	
Member	10	11'	12	21	22'	23	32	33'	34	43	44'
Relative EI/l	8	3	5	5	2	3	3	2	1	1	1
Modified stiffness	24	6	20	10	2	6	6	2	2	2	1
Distribution factor	12/25	3/25	2/5	5/9	1/9	1/3	3/5	1/5	1/5	2/3	1/3
Carry-over factor	0	1	½	½	½	½	½		½	½	
M^F, non-sway		−576			−576			−576			−288
Distribution	276	69	231	320	64	192	346	115	115	192	96
Carry-over	−77		160	115		173	96		96	57	
Distribution	38	−19	−64	−160	−32	−96	−116	−38	−38	−38	−19
Carry-over	−12		−80	−32		−58	−48		−19	−19	
Distribution	5	10	32	50	10	30	40	13	14	12	7
Carry-over	−1		25	16		20	15		−6	7	
Distribution		−3	−10	−20	−4	−12	−12	−5	−4	−4	−3
Carry-over			−10	−5		−6	−6		−2	−2	
Distribution		1	4	6	1	4	4	2	2	2	1
Carry-over			3	2		2	2		1		
Distribution		−1	−1	−2		−2	−2	−1		−1	
M^W	229	−519	290	290	−537	247	319	−489	170	206	−206
Modified stiffness	0	18	5	5	12	3	3	12	1	1	6
Distribution factor	0	18/23	5/23	1/4	3/5	3/20	3/16	3/4	1/16	1/7	6/7

(Continued)

Table 7.18 (Continued)

	0'1'	1'1	1'2'	2'1'	2'2	2'3'	3'2'	3'3	3'4'	4'3'	4'4
Joint	0	1		2			3			4	
Carry-over factor	→	→	−1 ↑	−1 ↑	→	−1 ↑	−1 ↑	→	−1 ↑	−1 ↑	→
M^F, sway		−1008	−360	−360		−216	−216		−72	−72	
Distribution		1070	298	144	346	86	54	216	18	10	62
Carry-over			−144	−298		−54	−86		−10	−18	
Distribution		113	31	88	211	53	18	72	6	3	15
Carry-over			−88	−31		−18	−53		−3	−6	
Distribution		69	19	12	30	7	11	42	3	1	5
Carry-over			−12	−19		−11	−7		−1	−3	
Distribution		9	3	7	18	5	1	6	1	0	3
Carry-over			−7	−3		−1	−5		−1	−1	
Distribution		6	1	1	2	1	1	4	0	0	1
M^S	0	259	−259	−459	607	−148	−282	340	−58	−86	86
M kip-in	0	1267	−31	169	−70	−99	−37	149	−112	−120	120
Member	0'1'	1'1	1'2'	2'1'	2'2	2'3'	3'2'	3'3	3'4'	4'3'	4'4
M kip-in	0	1786	−549	−749	1144	−395	−601	829	−228	−292	292

The fixed-end moments due to the applied loads are:

$$M^F_{44'} = -12 \times 24 \times 12/12$$
$$= -288 \text{ kip-in}$$

$$M^F_{11'} = M^F_{22'}$$
$$= M^F_{33'}$$
$$= -24 \times 24 \times 12/12$$
$$= -576 \text{ kip-in}$$

The distribution procedure for the non-sway and sway cases is tabulated in Table 7.18, and the final moments are obtained as shown.

Example 7.17

Determine the bending moments in the Vierendeel girder shown in Figure 7.41. All the members are of uniform cross-section.

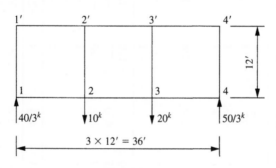

Figure 7.41

Solution

The sway equations and the initial fixed-end moments due to sway are: panel 1:

$$2(M_{12} + M_{21}) = -40 \times 12 \times 12/3$$
$$M^F_{12} = M^F_{21}$$
$$= -480 \text{ kip-in}$$

panel 2:

$$2(M_{23} + M_{32}) = -10 \times 12 \times 12/3$$
$$M^F_{23} = M^F_{32}$$
$$= -120 \text{ kip-in}$$

Table 7.19 Distribution of moments in Example 7.17

Joint	1		2			3			4	
Member	11'	12	21	22'	23	32	33'	34	43	44'
Relative EI/l	1	1	1	1	1	1	1	1	1	1
Modified stiffness	6	1	1	6	1	1	6	1	1	6
Distribution factor	6/7	1/7	1/8	3/4	1/8	1/8	3/4	1/8	1/7	6/7
Carry-over factor	↓	−1 ↑	−1 ↑		↓ −1	−1 ↑		↓ −1	−1 ↑	
M^F, sway		−480	−480		−120	−120		600	600	
Distribution	411	69	75	450	75	−60	−360	−60	−86	−514
Carry-over		−75	−69		60	−75		86	60	
Distribution	64	11	1	7	1	−1	−9	−1	−9	−51
Carry-over		−1	−11		1	−1		9	1	
Distribution	1	0	1	8	1	−1	−6	−1	0	−1
M kip-in	476	−476	−483	465	18	−258	−375	633	566	−566
Member	1'1	1'2'	2'1'	2'2	2'3'	3'2'	3'3	3'4'	4'3'	4'4'
M kip-in	476	−476	−483	465	18	−258	−375	633	566	−566

panel 3:

$$2(M_{34} + M_{43}) = 50 \times 12 \times 12 / 3$$
$$M_{43}^F = M_{34}^F$$
$$= 600 \text{ kip-in}$$

There are no fixed-end moments due to applied loading, and the sway distribution procedure is shown in Table 7.19.

Example 7.18

Determine the bending moments in the symmetrical two-bay frame shown in Figure 7.42. The relative EI/l values are shown ringed.

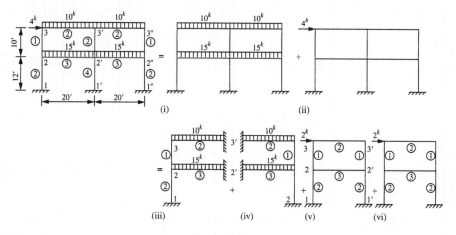

Figure 7.42

Solution

The original structure and loading are equivalent to the two cases shown at (i) and (ii). Case (i) consists of a symmetrical structure subjected to symmetrical loads. The moments at corresponding points are equal and of opposite sign, and there are no moments in the interior column. This can be replaced by the two identical substitute frames shown at (iii) and (iv). The fixed-end moments due to the applied loading are:

$$M_{3'3}^F = -M_{33'}^F$$
$$= 10 \times 20 \times 12 / 12$$
$$= 200 \text{ kip-in}$$

$$M_{2'2}^F = -M_{22'}^F$$
$$= 15 \times 20 \times 12 / 12$$
$$= 300 \text{ kip-in}$$

Table 7.20 Distribution of moments in Example 7.18

Joint	1	2			3		3′		2′			1′
Member	12	21	22′	23	32	33′	3′3	3′2′	2′2	2′3′	2′1′	1′2′
Relative EI/l	2	2	3	1	1	2	2					
Distribution factor	0	1/3	1/2	1/6	1/3	2/3	0	0				
Carry-over factor	←	½	½	←½	½→	½	→		↑			
M^F, non-sway			−300			−200	200		300			
Distribution		100	150	50	66	134						
Carry-over	50			33	25		67		75			
Distribution		−11	−17	−5	−8	−17						
Carry-over	−5			−4	−2		−8		−8			
Distribution and carry-over	1	1	2	1	1	1	1		1			
M^W	45	90	−165	75	82	−82	259	0	368	0	0	0
EI/l, substitute frame	2	2	3	1	1	2						
Modified stiffness	0	2	18	1	1	12						
Distribution factor	0	2/21	6/7	1/21	1/13	12/13						
Carry-over factor	←	−1		←−1	←−1							
M^F, sway	−72	−72	114	−60	−60	55						
Distribution	12	12			5							
Carry-over	−12		5			6	6					
Distribution												
M^S	−84	−60	119	−59	−61	61	61	−122	119	−118	−120	−168
M kip-in	−39	30	−46	16	21	−21	320	−122	487	−118	−120	−168
Member	1″2″	2″1″	2″2′	2″3″	3″2″	3″3′	3′3″	3′2″	2′2″	2′3′	2′1′	1′2′
M kip-in	−129	−150	284	−134	−143	143	−198	−249				

Case (ii) consists of a frame that satisfies the principle of multiples and is subjected to lateral loading at the joints. This can be replaced by the two identical substitute frames shown at (v) and (vi). The moments in the interior columns of the actual frame are equal to twice the moments in the columns of the substitute frame, and the remaining moments in thes actual frame are equal and of the same sense as the corresponding moments in the substitute frames. The sway equations for each substitute frame and the initial fixed-end moments du to sway are:

top story:

$$2(M_{32} + M_{23}) = -2 \times 10 \times 12$$
$$M_{32}^F = M_{23}^F$$
$$= -60 \text{ kip-in}$$

bottom story:

$$2(M_{12} + M_{21}) = -2 \times 12 \times 12$$
$$M_{12}^F = M_{21}^F$$
$$= -72 \text{ kip-in}$$

The distribution procedure for the non-sway and sway cases is tabulated in Table 7.20, and the final moments are obtained as shown.

Example 7.19

Determine the sway displacement of the tops of the columns of the multi-bay frame shown in Figure 7.43.

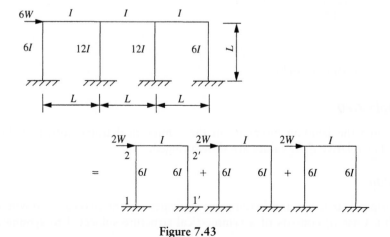

Figure 7.43

Solution

The frame satisfies the principle of multiples and can be replaced by the three identical substitute frames shown. The sway equation for each substitute frame and the initial fixed-end moments due to sway are:

$$2(M_{12} + M_{21}) = -2WL$$
$$M_{12}^F = M_{21}^F$$
$$= -WL/2$$

The distribution procedure is in Table 7.21.

Table 7.21 Distribution of moments in Example 7.1

Joint	1		2
Member	12	21	22'
Relative EI/L	6	6	1
Modified stiffness		6	6
Distribution factor	0	½	½
Carry-over factor	←	−1	
M^F, sway	−$WL/2$	−$WL/2$	
Distribution and carry-over	−$WL/4$	$WL/4$	$WL/4$
M	−3$WL/4$	−$WL/4$	$WL/4$

The sway displacement may be obtained from the final moments as shown at (i) and (ii) in Figure 7.44. Then:

$$M_{21} = 24EI\theta/L - 36EI\delta/L^2$$
$$= -WL/4$$

$$M_{12} = 12EI\theta/L - 36EI\delta/L^2$$
$$= -3WL/4$$

and

$$\delta = 5WL^2/144EI$$

Example 7.20

Determine the bending moments in the symmetrical frame shown in Figure 7.45. The relative EI/L values are shown ringed.

Solution

The original structure and loading are equivalent to the two cases shown at (i) and (ii). Case (i) consists of a symmetrical structure subjected to symmetrical

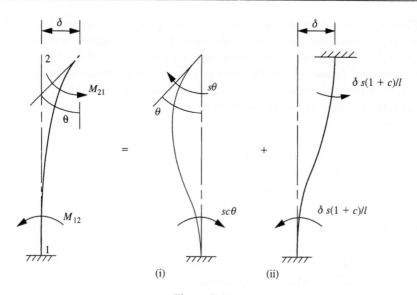

Figure 7.44

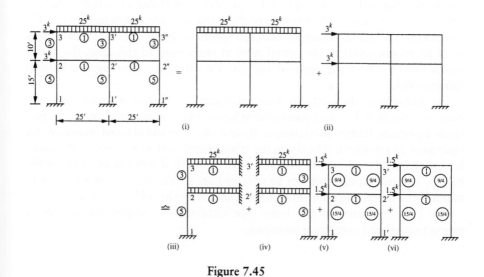

Figure 7.45

loads and may be replaced by the two identical substitute frames shown at (iii) and (iv). The fixed-end moments due to the applied loading are:

$$M^F_{3'3} = -M^F_{33'}$$
$$= -25 \times 25 \times 12/12$$
$$= -625 \text{ kip-in}$$

Case (ii) does not satisfy the principle of multiples, but an approximate solution may be obtained by means of the two identical substitute frames shown at (v) and (vi). The sum of the second moments of area for the columns of the two substitute frames equals the sum of the second moments of area for the columns of the real frame. The sway equations for each substitute frame and the initial fixed-end moments due to sway are:

top story:

$$2(M_{32} + M_{23}) = -3 \times 10 \times 12 / 2$$
$$M_{32}^{F} = M_{23}^{F}$$
$$= -45 \text{ kip-in}$$

bottom story:

$$2(M_{12} + M_{21}) = -3 \times 15 \times 12$$
$$M_{12}^{F} = M_{21}^{F} = -135 \text{ kip-in}$$

These sway moments are distributed as shown in Table 7.22. The column moments obtained are multiplied by the ratio (stiffness of actual column)/ (stiffness of substitute column) to give the actual column moments since, if the columns in the actual structure undergo identical deformations, the moments must be proportional to the actual column stiffnesses. In this case, the multiplying ratio for both the external and interior columns in the top and bottom stories is 4/3.

The moments at the joints are now unbalanced, as shown in Figure 7.46, and must be distributed. An approximate solution may be obtained from the previous substitute frames as shown at (i) and (ii). The unbalanced moments are equivalent to symmetrical loads on the symmetrical substitute frames, and distribution proceeds as shown in the table with a modified stiffness of $2EI/l$ for the beams. The correcting moments so obtained are small, and it is unnecessary to multiply the correcting moments for the columns by the stiffness ratios.

The final moments in the frame are obtained by adding the non-sway, proportioned, and correcting moments.

7.10   Symmetrical multi-story frames with inclined columns

The analysis of a single bay frame, with all beams parallel to the base and with the two columns in each story of equal stiffness and subjected to lateral loads at the joints, consists of a skew symmetrical distribution. The frame need not necessarily satisfy conditions of geometrical symmetry but must be symmetrical as regards stiffness of the columns. Only the left half of the frame

Joint	1	2			3			3'		2'			1'
Member	12	21	22'	23	32	33'	33	3'3"	3'2'	2'2"	2'1'	2'3'	1'2'
Relative E/l	5	5	1	3	3	1	1	1	3	1	5	3	5
Distribution factor	0 ←	5/9	1/9	1/3	3/4	1/4	0 →			0 →			
Carry-over factor			½	½ ←	½	½							
M^F, non-sway						−625	625 →						
Distribution and carry-over	−65	−130	−26	234	469	156	78						
Distribution and carry-over				−78	−39					−13			
Distribution and carry-over				15	29	10	5			−1			
Distribution and carry-over	−4	−8	−2	−5	−1	1							
M^W	−69	−138	−28	166	458	−458	708	0	−708	−14	0	0	14
E/l, substitute frame		15/4	1	9/4	9/4	1							
Modified stiffness (skew)		15	24	9	9	24							
Distribution factor	0 ←	5/16	1/2	3/16	3/11	8/11							
Carry-over factor	←	−1		−1 ↓		−1							
M^F, sway	−135	−135		−45	−45								
Distribution		56	90	34	12	33							
Carry-over	−56			−12	−34	25							
Distribution		4	6	2	9								
Carry-over	−4			−9	−2								
Distribution		3	4	2	1	1							
Carry-over and distribution	−3		1	−1	−2	2							

(Continued)

Table 7.22 (Continued)

Joint	1		2			3		3'		2'			1'	
M^S	-198	-72	101	-29	-61	61	61	-81	61	101	-39	-96	101	-264
Proportioned moments	-264	-96	101	-39	-81	61	61	-81	61	101	-39	-96	101	-264
Modified stiffness (symm.)		15	2	9	9	2								
Distribution factor	0	15/26	1/13	9/26	9/11	2/11								
Carry-over factor	←	-½		←	½	→	½							
Distribution	10	20	2	12	16	4								
Carry-over		-5	0	8	6	-1								
Distribution	-3	2		-3	-5	-1								
Carry-over	1			-3	-1	1								
Distribution and carry-over	1	2		1	1	1								
Correcting moments	8	17	2	15	17	3	-3	-35	-3	-2	-30	-33	-2	-16
M kip-in	-325	-217	75	142	394	-394	766	-116	-650	85	-69	-129	113	-280
Member	1"2"		2"1"	2"2'	2"3"	3"2"	3"3'							
M kip-in	-187		59	131	-190	-522	522							

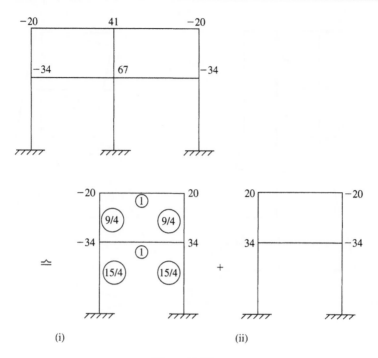

Figure 7.46

is considered, with a modified stiffness of $6EI/l$ applied to the beams and no carry-over between the two halves. As in the analysis of frames with vertical columns, fixed-end moments that satisfy the sway equations are imposed on the columns, and modified stiffness and carry-over factors are applied to the columns, which will ensure that the sway equations remain balanced throughout the distribution.

The sway mechanism for panel 1234 of the frame shown in Figure 7.47 is shown at (i). The displacement diagram, for a horizontal displacement of ϕh_1 units imposed at 2, is shown at (ii), and, since 2 must move perpendicularly to the original direction of 12, the point 2' is obtained. Similarly 4 moves perpendicularly to 24 and 34, and the point 4' is obtained. The member rotations are:

$$\phi_1 = 22'/l_{12}$$
$$= \phi$$
$$\phi_2 = 44'/l_{34}$$
$$= \phi$$
$$\phi_3 = -2'4'/l_{24}$$
$$= -\phi(rl_{24} - l_{24})/l_{24}$$
$$= -(r - 1)\phi$$

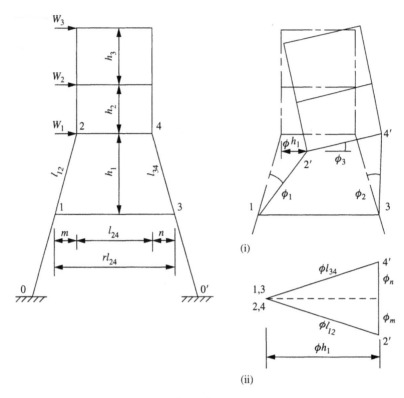

Figure 7.47

During the sway displacement, beams 13 and 24 are held infinitely rigid so that no deformations and no moments occur in them. Hence, as shown in Figure 7.48, the columns rotate through a sway angle ϕ at their bases and a sway angle $r\phi$ at their tops, and the frame above 24 rotates through an angle $-(r - 1)\phi$ and translates a distance ϕh_1 to the right. The sway equation for panel 1234 is:

$$-(M_{12} + M_{34})\phi - r(M_{21} + M_{43})\phi = (W_1 + W_2 + W_3)\phi h_1$$
$$- \{W_3(h_3 + h_2)(r - 1)\phi + W_2 h_1(r - 1)\phi\}$$

and

$$M_{12} + M_{34} + r(M_{21} + M_{43}) = M(r - 1) - Qh_1$$

where M is the clockwise moment of the external loads above the story about the top of the story and Q is the shear to the right due to all external forces above the bottom of the story. The columns are of equal stiffness, and the initial fixed-end moments required to satisfy the sway equation are:

$$2M_{12}^F + 2rM_{21}^F = M(r-1) - Qh_1$$

From Figure 7.48, the fixed-end moments due to an arbitrary rotation ϕ of the columns, while the beams are maintained infinitely rigid, are:

$$M_{12}^F/M_{21}^F = s\phi(1+rc)/s\phi(r+c)$$

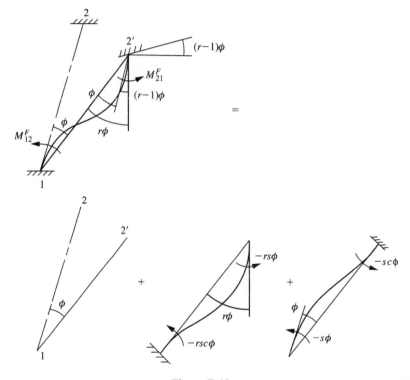

Figure 7.48

Hence:

$$2M_{21}^F(1+rc)/(r+c) + 2rM_{21}^F = M(r-1) - Qh_1$$

and:

$$M_{21}^F = \{M(r-1) - Qh_1\}(r+c)/2(1+2r+r^2)$$

$$M_{12}^F = \{M(r-1) - Qh_1\}(1+rc)/2(1+2rc+r^2)$$

For straight, prismatic columns these values reduce to:

$$M_{12}^F = \{M(r-1) - Qh_1\}(2+r)/4(1+r+r^2)$$
$$M_{21}^F = \{M(r-1) - Qh_1\}(1+2r)/4(1+r+r^2)$$

The out of balance moment at joint 1 is distributed while allowing joint 2 to translate laterally and maintaining the beams as infinitely rigid. This procedure is shown in Figure 7.49, where s'_{12} and c'_{12} refer to the modified stiffness of 12 and the modified carry-over factor from 1 to 2 required to leave the sway equation is left undisturbed. Then, substituting in the sway equation with the right-hand side set at zero:

$$s'_{12} + rs'_{12}\, c'_{12} = 0$$

and

$$c'_{12} = -1/r$$

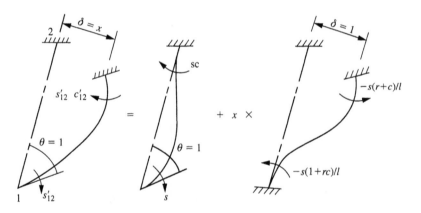

Figure 7.49

From Figure 7.49:

$$s'_{12} = s - xs(1+rc)/l$$
$$rs'_{12}c'_{12} = -s'_{12} = rsc - rxs(r+c)/l$$

Hence:

$$x = l(1+rc)/(1+2rc+r^2)$$

and

$$s'_{12} = sr^2(1-c^2)/(1+2rc+r^2)$$

For straight prismatic columns this reduces to:

$$s'_{12} = 3EIr^2/l(1 + r + r^2)$$

Expressions for s'_{21} and c'_{21} may be obtained by substituting $r = 1/r$ in the expressions for s'_{12} and c'_{12}.
Then:

$$s'_{21} = 3EI/l(1 + r + r^2)$$

and

$$c'_{21} = -r$$

In the bottom story of the frame, the stiffness and carry-over factor at the top of the column are as above, but there is no carry-over from the fixed base.

Should the column feet be hinged at the foundations, the initial fixed-end moment required at the top of the column to satisfy the sway equation is:

$$M^F_{10} = \{M(r - 1) - Qh_1\}/2r$$

and the modified stiffness of 10 is zero.

The Vierendeel girder shown in Figure 7.50, which has inclined top chords and the two chords in any panel of equal stiffness, may be analyzed in a similar manner. The end panel 012 has a value of infinity for r. Then:

$$\begin{aligned}
M^F_{10} &= M^F_{01} \\
&= c'_{10} \\
&= 0
\end{aligned}$$

and

$$s'_{10} = 3EI/l$$

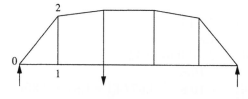

Figure 7.50

Example 7.21

Determine the bending moments in the frame shown in Figure 7.51. All members have the same second moment of area.

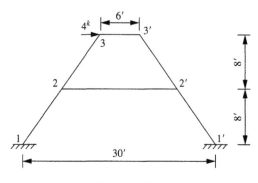

Figure 7.51

Solution

For the top story:

$$r = 18/6 = 3$$
$$M(r - 1) - Qh_1 = 0 - 32$$
$$= -32$$
$$M^F_{23} = -32(2 + 3)/4(1 + 3 + 9)$$
$$= -3.08$$
$$M^F_{32} = -32(1 + 6)/4(1 + 3 + 9)$$
$$= -4.31$$
$$c'_{23} = -1/3$$
$$= -0.333$$
$$c'_{32} = -3$$
$$s'_{23} = 27EI/10(1 + 3 + 9)$$
$$= 0.207EI$$
$$s'_{32} = 3EI/10(1 + 3 + 9)$$
$$= 0.023EI$$

For the bottom story:

$$r = 30/18$$
$$= 1.67$$
$$M(r - 1) - Qh_1 = 32 \times 0.67 - 32$$
$$= -10.67$$
$$M^F_{12} = -10.67 \times 3.67/4(1 + 1.67 + 2.80)$$
$$= -1.80$$

$$M_{21}^F = -10.67 \times 4.33/4(1 + 1.67 + 2.80)$$
$$= -2.12$$
$$c_{21}' = -1.67$$
$$s_{21}' = 3EI/10(1 + 1.67 + 2.80)$$
$$= 0.055EI$$

For the beams:

$$s_{33'}' = 6EI/6$$
$$= EI$$
$$s_{22'}' = 6EI/18$$
$$= 0.333EI$$

The distribution procedure and the final moments are shown in Table 7.23.

Table 7.23 Distribution of moments in Example 7.21

Joint	1		2			3
Member	12	21	22'	23	32	33'
Modified stiffness		0.055	0.333	0.207	0.023	1
Distribution factor	0	0.092	0.56	0.348	0.023	0.977
Carry-over factor	←	−1.670	←	−3.000		
				−0.333	→	
$M^F \times 100$ kip-ft	−180	−212		−308	−431	
Distribution		48	291	181	10	421
Carry-over	−80			−30	−60	
Distribution		3	17	10	1	59
Carry-over	−5			−3	−3	
Distribution		0	2	1	0	3
Carry-over and distribution	0			0	−1	1
Final moments × 100 kip-ft	−265	−161	310	−149	−484	484

7.11 Frames with non-prismatic members subjected to sway

The fixed-end moments produced in non-prismatic members by a unit displacement are determined by the column analogy method given in Section 6.5. In addition, the stiffness, carry-over factors, and fixed-end moments due to lateral loads on the members are required and are also determined by the column analogy method.

Example 7.22

Determine the bending moments in the frame shown in Figure 7.52. The second moments of area of the members are shown ringed.

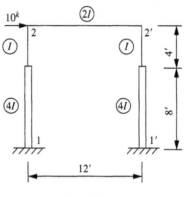

Figure 7.52

Solution

The stiffness, carry-over factors, and fixed-end moments due to an arbitrary displacement have been obtained for member 21 in Examples 6.6 and 6.7 and are:

$$s'_{21} = 5EI/12$$
$$c'_{21} = 4/5$$
$$M^F_{21} = -EI\delta/16$$
$$\quad = -170x$$
$$M^F_{12} = -EI\delta/8$$
$$\quad = -340x$$

The sway equation is:

$$-M_{12} - M_{21} = 1440/2$$
$$\quad = 720 \text{ kip-in}$$

Due to the skew symmetry, the modified stiffness of member 22' is:

$$s'_{22'} = 6E(2I)/12$$
$$\quad = EI$$

The distribution procedure is shown in Table 7.24, and substituting the final sway moments in the sway equation gives:

$$(300 + 120)x = 720$$
$$x = 1.71$$

The final moments are shown in Table 7.24.

Table 7.24 Distribution of moments in Example 7.22

Joint	1		2	
Member	12	21		22'
Modified stiffness		5/12		1
Distribution factor	0	5/17		12/17
Carry-over factor	$\leftarrow$	4/5		
M^F, sway	-340	-170		
Distribution and carry-over	40	50		120
M^S	-300	-120		120
$M = 1.71 \times M^S$	-513	-207		207

7.12 Frames with curved members

A single bay frame with the two columns connected by means of a straight beam has one degree of sway freedom. The single bay frame with a curved beam, shown in Figure 7.53, has the two degrees of sway freedom shown at (ii) and (iii), corresponding to the different translations that occur at the top of each column. The analysis proceeds in the three stages as shown at (i), (ii), and (iii), and the two sway equations required are obtained by considering a unit

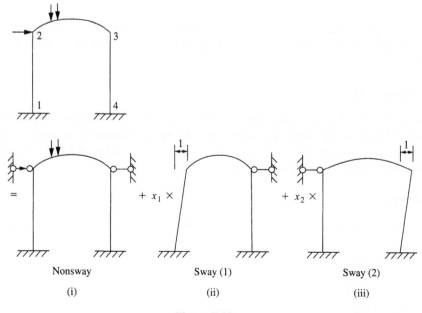

Figure 7.53

horizontal displacement at the top of each column. Considering sway (1) and (2) in turn, the sway equations are:

$$-(M_{12} + M_{21})/l_{12} + H_{23} = W$$
$$-(M_{34} + M_{43})/l_{34} - H_{32} = 0$$

where H_{23} and H_{32} are the outward thrusts of the arch at 2 and 3. The outward thrusts are determined by considering the arch to be initially fixed-end and subjected to the applied loads and translations and allowing for the change in thrust produced as the ends of the arch rotate to their positions of final equilibrium. Then:

$$H_{23} = H_{32}$$
$$= H^F + H^{\delta 2=1} - H^{\delta 3=1} + \psi_2 H^{\theta 2=1} - \psi_3 H^{\theta 3=1}$$

where H^F is the initial thrust due to the applied loads, $H^{\delta 2=1}$ and $H^{\delta 3=1}$ are the initial thrusts due to unit translations to the right at 2 and 3, $H^{\theta 2=1}$ and $H^{\theta 3=1}$ are the thrusts produced by unit rotations at 2 and 3, and ψ_2 and ψ_3 are the final clockwise rotations of 2 and 3. The final rotations are given by:

$$\psi_2 = \Sigma M_{23}^B / s_{23}$$

and

$$\psi_3 = \Sigma M_{32}^B / s_{32}$$

where ΣM_{23}^B is the sum of the balancing moments distributed to joint 23 from its initial to its final equilibrium position and s_{23} *is* the rotational stiffness of the arch.

Example 7.23

Determine the bending moments in the frame shown in Figure 7.54. The curved beam is parabolic in shape, and its second moment of area varies directly as the secant of the slope, with a value at the crown of I_0. The second moment of area of the columns is $2I_0$.

Solution

The characteristics of the arch, which were determined in Example 6.9, are:

$$s_{23} = s_{32}$$
$$= 9EI_0/l$$
$$= EI_0/5 \text{ kip-ft}$$
$$c_{23} = c_{32}$$
$$= -1/3$$

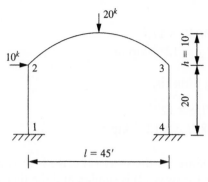

Figure 7.54

$$H^{\theta=1} = 15EI_0 / 2h^2l$$
$$= 0.0167EI_0 \text{ kip/rad}$$

$$H^{\delta=1} = 45EI_0 / 4h^2l$$
$$= 0.0025EI_0 \times 12^{-3} \text{ kip/in}$$

$$M^{\delta=1} = 15EI_0 / 2hl$$
$$= 0.0167EI_0 \times 12^{-2} \text{ kip-in/in}$$

$$M^F_{23} = -M^F_{32}$$
$$= Wl / 32$$
$$= 337 \text{ kip-in}$$

$$H^F = 15Wl / 64h$$
$$= 21.10 \text{ kip outward}$$

The stiffness and carry-over factors for the columns are:

$$s_{21} = s_{34}$$
$$= 4E(2I_0)/20$$
$$= 2EI_0 / 5 \text{ kip-ft}$$

$$c_{21} = c_{34}$$
$$= 1/2$$

The distribution procedure for the non-sway stage is shown in Table 7.25. The initial fixed-end moments due to sway (1) are:

$$M^F_{23} = -M^F_{32}$$
$$= 0.0167EI_0\delta \times 12^{-2}$$
$$= 0.0167EI_0\delta \times 12^{-2} \text{ kip-in}$$

$$M_{12}^F = M_{21}^F$$
$$= -6E(2I_o)\delta/(20 \times 12)^2$$
$$= 0.0300EI_o\delta \times 12^{-2} \text{ kip-in}$$

and the corresponding thrust is:

$$H^{\delta=\delta} = 0.0025EI_0\delta \times 12^{-3}$$
$$= 0.000208EI_0\delta \times 12^{-2} \text{ kip}$$

The distribution procedure for the sway (1) stage is shown in Table 7.25. The distribution procedure for sway (2) is similar, and the final moments are shown in the table.

Table 7.25 Distribution of moments in Example 7.23

Joint	1		2		3	4
Member	12	21	23	32	34	43
Relative stiffness	2	2	1	1	2	2
Distribution factor	0	2/3	1/3	1/3	2/3	0
Carry-over factor	$\leftarrow$	½	$\leftarrow$	1/3	½	$\rightarrow$
			$-1/3$	$\rightarrow$		
M^F, non-sway			337	-337		
Distribution		-225	-112	112	225	
Carry-over	-112		-37	37		112
Distribution		25	12	-12	-25	
Carry-over	12		4	-4		-12
Distribution and carry-over	-2	-3	-1	1	3	2
M^W	-102	-203	203	-203	203	102
ΣM^B			-101	101		
M^F, sway (1)	-300	-300	167	-167		
Distribution		89	44	56	111	
Carry-over	44		-19	-15		56
Distribution		13	6	5	10	
Carry-over	6		-2	-2		5
Distribution		1	1	1	1	
M^{S1}	-250	-197	197	-122	122	61
ΣM^B			51	62		
M^{S2}	61	122	-122	197	-197	-250
ΣM^B			62	51		
Final moments	-213	223	-223	905	-905	-1506

Substituting the final non-sway, sway (1), and sway (2) moments in sway equation (1) gives:

$$\begin{aligned}
C_1^W &= -(M_{12} + M_{21})/l_{12} + (H^F + \psi_2 H^{\theta 2=1} - \psi_3 H^{\theta 3=1}) \\
&= 305/240 + 21.10 - 2 \times 5 \times 0.0167 \times 101/12 \\
&= 20.97
\end{aligned}$$

$$\begin{aligned}
C_1^{S1} &= -(M_{12} + M_{21})/l_{12} + (H^{\delta 2=1} + \psi_2 H^{\theta 2=1} - \psi_3 H^{\theta 3=1}) \\
&= 447/240 + 2.08 + 5 \times 0.0167(51 - 62)/12 \\
&= 3.86
\end{aligned}$$

$$\begin{aligned}
C_1^{S2} &= -(M_{12} + M_{21})/l_{12} + (H^{\delta 3=1} + \psi_2 H^{\theta 2=1} - \psi_3 H^{\theta 3=1}) \\
&= -183/240 - 2.08 + 5 \times 0.0167(62 - 51)/12 \\
&= -2.76
\end{aligned}$$

Then:

$$20.97 + 3.86 x_1 - 2.76 x_2 = 10$$

Substituting the final non-sway, sway (1), and sway (2) moments in sway equation (2) gives:

$$-20.97 - 2.76 x_1 + 3.86 x_2 = 0$$

Solving these two equations simultaneously:

$$x_1 = 2.14$$
$$x_2 = 6.95$$

and the final moments are given by:

$$M^W + x_1 M^{S1} + x_2 M^{S2}$$

7.13 Rectangular grids

Rectangular grids, which have a large number of degrees of sway freedom, are most readily analyzed by a method of successive sway corrections[13].

The load W, applied to the grid shown in Figure 7.55, produces bending moments and torsion in all the members. The sign convention adopted is that, in a member parallel to the x-axis, bending moment is positive if clockwise

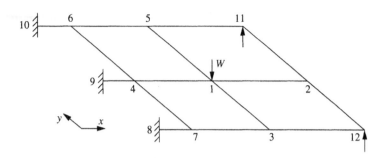

Figure 7.55

when viewed in the positive direction of the y-axis, and torsion is positive if clockwise when viewed in the positive direction of the x-axis. Each unsupported node of the grid undergoes a vertical deflection, and sway equations must be developed for each one. The sway equation for node 1 is obtained by considering unit vertical deflection of 1 and is:

$$(M_{12} + M_{21})/l_{12} + (M_{13} + M_{31})/l_{13} - (M_{14} + M_{41})/l_{14} \\ - (M_{15} + M_{51})/l_{15} - W = 0$$

Similar equations may be developed for nodes, 2, 3, 4, 5, 6, and 7.

An initial estimate is made of the vertical deflection at each node, and these deflections are imposed on the grid, with no joint rotations permitted. The initial fixed-end moments due to all the sway modes applied simultaneously are:

$$M_{12}^{F} = M_{21}^{F} \\ = 6EI(y_1 - y_2)/(l_{12})^2$$

$$M_{15}^{F} = M_{51}^{F} \\ = -6EI(y_1 - y_5)/(l_{15})^2, \text{ etc}$$

where $y_1, y_2, y_3,...$, are the estimated deflections at nodes 1, 2, 3,....

These initial fixed-end moments are distributed and the resulting moments substituted in the sway equations. Any residual that is produced means that the sway equation is not satisfied, as the initial estimate of the deflections was incorrect and additional sway moments must be applied and distributed. The procedure continues until all the sway equations are satisfied.

Distribution is required for all beams in the x and y directions, and the distribution factor at each node must allow for the torsional stiffness of the cross-members framing into the node from the y and x directions. The torsional stiffness of a member is GJ/l, where G is the modulus of torsional rigidity, J is the torsional inertia, and l is the length of the member. The carry-over factor for torsion is -1. Any out of balance moment at node 4 that is distributed to cross-member 46 during the distribution of moments in beam 92, is carried over to 64 with the sign changed. This produces an out of balance moment at node 6 during the distribution of moments in beam 1011, which must be distributed to the three members meeting there. Thus, the distribution of moments must proceed simultaneously in beams 1011, 92, 812, 67, 53, and 1112, and the balancing operation at any node affects the adjacent nodes on parallel beams.

When loading is applied between the nodes, the fixed-end moments due to this loading must be distributed at the same time as the initial fixed-end moments due to the sway deflections.

Example 7.24

Determine the moments in the grid shown in Figure 7.56. All members are of uniform section, and the flexural rigidity is twice the torsional rigidity.

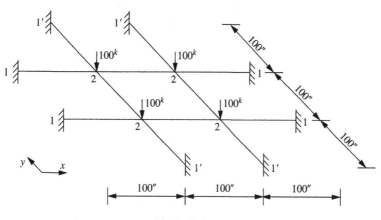

Figure 7.56

Solution

Due to the symmetry of the structure and loading, equal deflections occur at nodes 2 and there is no torsion in members 22. Distribution is required in only half of beam 11, and the flexural and torsional stiffness of each member is:

$$\text{flexural stiffness of } 12 = 4EI/l$$
$$= 8GJ/l$$
$$\text{flexural stiffness of } 22 = 2EI/l$$
$$= 4GJ/l$$
$$\text{torsional stiffness of } 21' = GJ/l$$

The sway equation is obtained by considering unit vertical deflection imposed simultaneously at all nodes 2 and is:

$$-2(M_{12} + M_{21})/100 - 100 = 0$$
$$-M_{12} - M_{21} - 5000 = 0$$

The initial sway moments are:

$$M_{12}^F = M_{21}^F = -6EIy_2/1000 = -4000, \text{ say}$$
$$M_{22}^F = 0$$

The distribution proceeds as shown in Table 7.26, and the residual is indicated at several stages. Equal correcting moments M_{21}^F and M_{12}^F are imposed at each stage, and the final moments are shown in the table.

Table 7.26 Distribution of moments in Example 7.24

Joint	1		2		Residual
Member	12	21	21'	22	
Relative stiffness		8	1	4	
Distribution factor		8/13	1/13	4/13	
Carry-over factor	←	½	−1		
M^F, sway	−4000	−4000			
Distribution		2460	310	1230	
Carry-over	1230				−690
1st correction	−600	−600			
Distribution		370	45	185	
Carry-over	185				− 45
2nd correction	−43	−43			
Distribution		27	3	13	
Carry-over	13				1
Final moments, kip-in	−3215	−1786	358	1428	

7.14 Direct distribution of moments and deformations

Direct moment distribution was introduced by Lin[14] as a means of eliminating the iteration required in the standard moment distribution procedure. The parameters required for the direct distribution of fixed-end moments require considerable preliminary effort, but, once these parameters have been obtained, alternative loading conditions can be quickly investigated. To assist in the calculation of these parameters, a number of graphs and charts[15,16,17] are available. Several alternative methods[18,19,20] have been developed for the direct distribution of moments but will not be considered here.

The direct distribution of deformation provides the most convenient way of determining influence lines for rigid frames[21]. The basic parameters required are obtained in a similar manner to the parameters required for the direct distribution of moments, and both techniques may be readily applied to a particular structure.

(a) Direct distribution of moment

The actual stiffness at the end 1 of any member 12 of a frame is the moment s_{12}^a required to produce unit rotation at 1, with all members meeting at 1 and 2 having their actual stiffnesses. As indicated in Figure 7.57, this may be obtained as the sum of the two operations shown at (i) and (ii). In operation (i), unit rotation is imposed at 1 with 2 clamped. In operation (ii), joint 2 is released and allowed to rotate to its position of final equilibrium while joint 1 is clamped. The balancing moment required at 2 is $-c_{12}s_{12}$, and this is distributed to each member in accordance with its distribution factor. The stiffness of 21 is the restrained stiffness s_{21} since joint 1 is clamped, while all the other n members meeting at 2 have their actual stiffnesses. The distribution factor for 21 is:

$$a_{21} = s_{21}/(s_{21} + \Sigma s_{2n}^a - s_{21}^a)$$

where s_{21}^a is the actual stiffness of member 21.

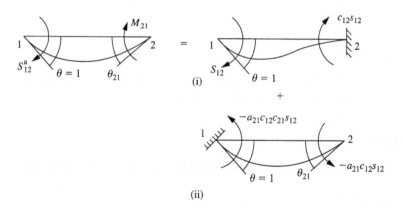

(i)

(ii)

Figure 7.57

The balancing moment distributed to 21 is $-a_{21}c_{12}s_{12}$, and the moment transmitted to 1 is:

$$-a_{21}c_{12}c_{21}s_{12} = b_{21}c_{12}s_{12}$$

Then the actual stiffness at 1 is:

$$s_{12}^a = s_{12}(1 - a_{21}c_{12}c_{21})$$

and the actual distribution factor for member 12 is:

$$d_{12}^a = s_{12}^a / \Sigma s_{1n}^a$$

The proportion of the out-of-balance moment at joint 2 which is transmitted to the fixed-end 1 when 2 is balanced, is:

$$b_{21} = -a_{21}c_{21}$$

For any member 12, the value of a_{12} is zero when joint 1 is fixed and unity when joint 1 is hinged.

The direct distribution procedure may be summarized as:

- determine the fixed-end moments produced by the applied loads
- starting with the extreme left-hand-side joint of the frame, transmit the out-of-balance moments at each joint to the next joint on the right, which is clamped
- repeat this operation starting at the extreme right-hand-side joint and transmit to the left with moments transmitted previously not included in the out-of-balance moments
- at each joint, balance the fixed-end moments and the transmitted moments using the actual distribution factors

Example 7.25

Determine the bending moments in the members of the frame shown in Figure 7.16. The relative EI values are shown ringed.

Solution

The parameters required are shown in Table 7.27 and are derived as follows:

$$a_{12} = 1$$

and:

$$s_{21}^a = 3(1 - 1/4)$$
$$= 2.25$$
$$s_{23}^a = 9$$

Table 7.27 Distribution of moments in Example 7.25

Joint	1		2			4		6
Member	12	21	23	24	42	45	46	64
Relative $4EI/l$	3	3	9	4	4	12	3	3
s^a		2.25	9	3.74	3.74	9	2.25	
a	1	0.19		0.26	0.26		0.19	1
b	-0.50	-0.10		-0.13	-0.13		-0.10	-0.50
d^a	1	0.15		0.25	0.25		0.15	1
M^F, kip-in	-100	100					-115	173
Transmission L to R		50			-20			13
Transmission R to L	-13			28			-87	
Distribution	113	-27		-45	56		33	-173
Final moments	0	123	-106	-17	36	133	-169	0

and

$$a_{24} = 4/(4 + 9 + 2.25)$$
$$= 0.26$$
$$s_{42}^a = 4(1 - 0.26/4)$$
$$= 3.74$$
$$s_{45}^a = 9$$

and

$$a_{46} = 3/(3 + 9 + 3.74)$$
$$= 0.19$$

The remaining parameters are readily obtained, and the transmission and distribution proceeds as shown in the table.

(b) Direct distribution of deformations

The rotation θ_{21} produced at the end 2 of member 12 shown in Figure 7.57 is given by:

$$\theta_{21} = -c_{12}s_{12}/(s_{21} + \Sigma s_{2n}^a - s_{21}^a)$$
$$= -a_{12}'c_{12}$$
$$= b_{12}'$$

Thus, the final rotation at joint 2 when a rotation θ_{12} is imposed at 1 is:

$$\theta_{21} = b'_{12}\theta_{12}$$

For symmetrical members:

$$s_{12} = s_{21}$$
$$a'_{12} = a_{21}$$
$$b'_{12} = b_{21}$$

Using the principles of Section 5.3, the direct distribution of deformation provides a rapid method of determining influence lines. The influence line for M_{23} for the frame shown in Figure 7.58 is obtained by imposing a unit clockwise rotation on member 23 with respect to members 20 and 21. The deflected form of structure is, from Müller–Breslau's principle, the influence line for M_{23}. The absolute rotation of member 23 is:

$$\theta_{23} = (s^a_{20} + s^a_{21})/\Sigma s^a_{2n}$$

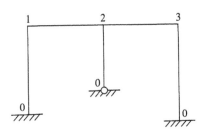

Figure 7.58

The rotation of member 21 is:

$$\theta_{21} = \theta_{23} - 1$$
$$= -s^a_{23}/\Sigma s^a_{2n}$$
$$= -d^a_{23}$$

These two imposed rotations may be readily transmitted through the structure. Thus, the final rotations of all the members in the structure are obtained, and the ordinates of the elastic curve are obtained from tabulated values of fixed-end moments.

A correction must be applied to these ordinates because of the sway that occurs in the frame. A unit sway displacement is imposed on the frame, as shown in Figure 7.59 (i), and the final moments are determined by the direct

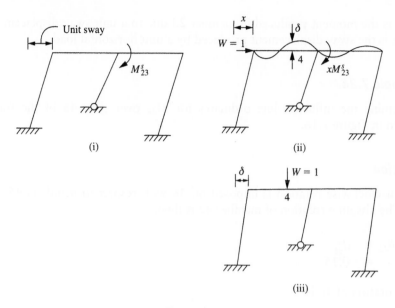

Figure 7.59

distribution procedure. The rotations θ^S at each joint are readily obtained by dividing the balancing moment distributed to any member at the joint by its actual stiffness, and the moment in member 23 is M_{23}^s.

A unit horizontal load applied at 1, as shown at (ii), produces a sway displacement x, a vertical deflection of δ at any point 4, and a moment in member 23 of xM_{23}^S. The value of x is determined by substituting the M^S values obtained in system (i) in the sway equation. Thus:

$$-x(M_{10}^S + M_{01}^S) - xM_{20}^S - x(M_{30}^S + M_{03}^S) = 1$$

A unit vertical load applied at 4 will, from Maxwell's reciprocal theorem, produce a sway displacement of δ, as shown at (iii). The moment in member 23, due only to this displacement δ, is δM_{23}^S. This value, then, is the sway correction that must be added to the non-sway moment at 23 due to a unit load at any point 4. Thus, the corrections are given by the ordinates of the elastic curve, with rotations at each joint of $\theta^S \times xM_{23}^S$.

The corrections are best applied to the non-sway rotations, and the true influence line ordinates for M_{23} are given by the deflected form of the structure with joint rotations of:

$$\theta^{NS} + \theta^S \times xM_{23}^S$$

where θ^{NS} is the rotation at a joint due to the imposed rotation with sway prevented, θ^S is the rotation at the same joint due to a unit sway displacement,

M_{23}^S is the moment produced in member 23 due to a unit sway displacement, and x is the sway displacement produced by a unit horizontal load.

Example 7.26

Determine the influence line ordinates for M_{46} over span 46 of the frame shown in Figure 7.16.

Solution

A unit clockwise rotation is imposed on 46 with respect to members 45 and 42. The absolute rotation of member 42 is then:

$$\theta_{42} = -d_{46}^a$$
$$= -0.15$$

The rotation of 46 is:

$$\theta_{46} = 1 + \theta_{42}$$
$$= 0.85$$

These imposed rotations are transmitted in Table 7.28 to obtain the final rotations shown.

Table 7.28 Distribution of deformations in Example 7.26

Joint	1	2		4		6
Member	12	21	24	42	46	64
b'		−0.50		−0.13	−0.50	
Imposed deformations				−0.15	0.85	
Transmitted deformations	−0.00975	0.0195				−0.425
Final rotations	−0.00975	0.0195		−0.15	0.85	−0.425

The influence line ordinates, at intervals of $0.2 \times$ the span of $10\,\text{ft}$, are obtainable in Table 7.29 from:

$$M_{46} = 0.85 \times M_{46}^F - 0.425 \times M_{64}^F$$

Table 7.29 Influence line ordinates for Example 7.26

Section	4	4.2	4.4	4.6	4.8	6
M_{46}^F ft units	0	−1.28	−1.44	−0.96	−0.32	0
M_{64}^F ft units	0	0.32	0.96	1.44	1.28	0
$M_{46}^F \times 0.85$	0	−1.09	−1.22	−0.82	−0.27	0
$M_{64}^F \times -0.425$	0	−0.14	−0.41	−0.61	−0.54	0
Ordinates, ft units	0	−1.23	−1.63	−1.43	−0.81	0

7.15 Elastically restrained members

Modified stiffness, carry-over factors, and fixed-end moments are required for members that are elastically restrained at their ends. These may be obtained by the methods given in Section 6.7.

Example 7.27

Determine the moments in the uniform beam shown in Figure 7.60. The beam is elastically restrained at end 1 with a rotation-moment ratio of $\eta_{12} = l/8EI$.

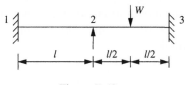

Figure 7.60

Solution

The modified stiffness and carry-over factor for member 21 are:

$$s_{21}' = 4EI(l + 3EI\eta_{12})/l(l + 4EI\eta_{12})$$
$$= 11EI/3l$$
$$c_{21}' = l/2(l + 3EI\eta_{12})$$
$$= 4/11$$

The initial fixed-end moments are:

$$M_{23}^F = -M_{32}^F$$
$$= -Wl/8$$
$$= -100, \text{ say}$$

The distribution proceeds as shown in Table 7.30.

Table 7.30 Distribution of moments in Example 7.27

Joint	1	2		3
Member	12	21	23	32
Modified stiffness		11/3	4	
Distribution factor		11/23	12/23	
Carry-over factor	←	4/11	1/2	→
M^F			−100	100
Distribution and carry-over	17	48	52	26
Final moments	17	48	−48	126

Supplementary problems

Use the moment distribution technique to solve the following problems.
S7.1 The continuous beam shown in Figure S7.1 has a second moment of area for member 34, which is 3 × that for members 12 and 23. Determine the reactions at supports 2, 3, and 4.

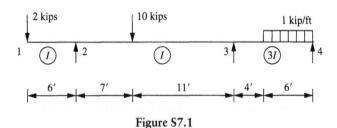

Figure S7.1

S7.2 Figure S7.2 shows a propped rigid frame with the relative EI/L values shown ringed alongside the members. Determine the reactions at supports 1 and 4.

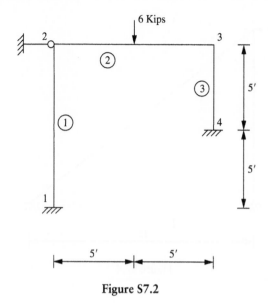

Figure S7.2

S7.3 The continuous beam shown in Figure S7.3 has a second moment of area of $376\,in^4$ and a modulus of elasticity of $29,000\,kips/in^2$. Determine the bending moments produced in the beam by a settlement of support 1 by 1 in, support 2 by 2 in, and support 3 by 1 in.

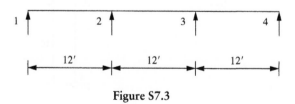

Figure S7.3

S7.4 Determine the moments produced in the members of the rigidly jointed frame shown in Figure S7.4 by the indicated load of 10 kips. The second moment of area of all members is $20\,in^4$. The cross-sectional area of all members is $2\,in^2$. The modulus of elasticity is constant for all members.

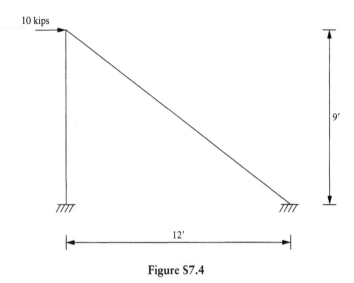

Figure S7.4

S7.5 Figure S7.5 indicates a sway frame, with pinned supports, with the relative *EI/L* values shown ringed alongside the members. Determine the bending moments at joints 2 and 3 caused by the lateral load.

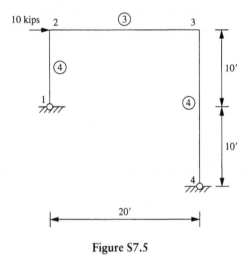

Figure S7.5

S7.6 Determine the forces produced at the joints of the rigidly jointed frame shown in Figure S7.6 by the indicated loads. The *EI/L* value is constant for all members.

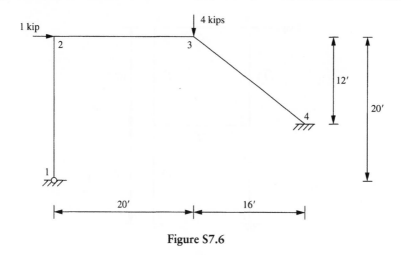

Figure S7.6

S7.7 Determine the bending moments produced at the joints of the rigidly jointed frame shown in Figure S7.7 by the indicated loads. The relative EI/L values are shown ringed alongside the members.

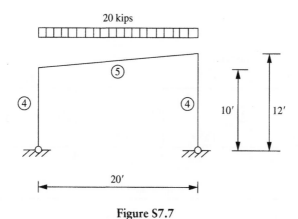

Figure S7.7

S7.8 The dimensions and loading on a symmetrical, single-bay, two-story frame are shown in Figure S7.8. The relative second moment of area values are shown ringed alongside the members. Determine the bending moments, shears, and axial forces in the members.

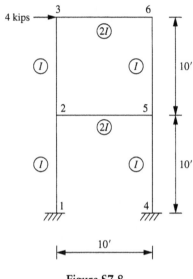

Figure S7.8

S7.9 Determine the bending moments produced at the joints of the Vierendeel girder shown in Figure S7.9 by the indicated loads. The relative EI/L values are shown ringed alongside the members.

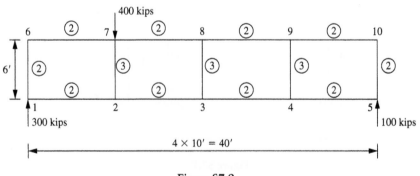

Figure S7.9

S7.10 Determine the bending moments produced at the joints of the rigidly jointed frame shown in Figure S7.10 by the indicated loads. The EI/L value is constant for all members.

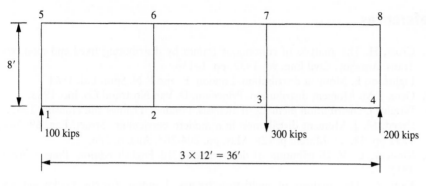

Figure S7.10

S7.11 Determine the bending moments produced in the left-hand columns of the rigidly jointed frame shown in Figure S7.11 by the indicated loads. The relative EI/L values are shown ringed alongside the members.

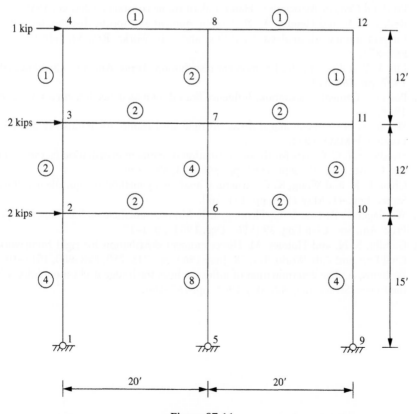

Figure S7.11

References

1. Cross, H. The analysis of continuous frames by distributing fixed end moments. Trans. Am. Soc. Civil Eng. 96. 1932. pp. 1–156.
2. Lightfoot, E. Moment distribution. London. E. and F. N. Spon Ltd. 1961.
3. Gere, J. M. Moment distribution. Princeton. D. Van Nostrand Co. Inc. 1963.
4. Wright, J. Distribution methods of analysis. London. Thames and Hudson. 1963.
5. Creed, M. J. Moment distribution in a modern curriculum. Struct. Eng. 69. Nov. 1991 pp. 48, 52. March. p. 128. May. pp. 204-205. Aug. p. 286.
6. Kloucek, C. V. Distribution of deformation. 3rd English edition. Prague. Artia. 1955.
7. Kani, G. The analysis of multistory frames. London. Crosby Lockwood and Son Ltd. 1957.
8. Glatz, R. General iteration method for translatory rigid frames. New York. Frederick Ungar. 1963.
9. Gray, C. S., Kent, L. E., Mitchell, W. A., Godfrey, G. B. Steel designers' manual. London. Crosby Lockwood and Son Ltd. 1960. pp. 29–35.
10. Northern Aluminum Company. Structural aluminum. London. 1959. pp. 137–139.
11. Portland Cement Association. Continuity in concrete building frames. Chicago, 1951. pp. 10–13.
12. Portland Cement Association. Handbook of frame constants. Chicago. 1948.
13. Reddy, D. V. and Hendry, A. W. A rapid moment and torque distribution method for grid framework analysis. Civil Eng. and Pub. Works. Rev. 54. July 1959. pp. 867–872.
14. Lin, T. Y. A direct method of moment distribution. Trans. Am. Soc. Civil Eng. 102. 1937. pp. 561–571.
15. Portland Cement Association. Influence lines drawn as deflection curves. Chicago. 1948.
16. Wood, R. H. An economical design of rigid steel frames for multistory buildings. London. HMSO. 1951.
17. Steedman, J. C. Charts for the determination of moment distribution factors. Conc. and Constr. Eng. 57. Sept. 1962. pp. 348–353, 395–396.
18. Chen, L. K. and Wong, K. C. Structural analysis by method of equivalent stiffness. Struct. Eng. 41. May 1963. pp. 151–160.
19. Wiesinger, F. P., Lee, S. L., Fleming, J. F. Moment coefficients for frame analysis. Proc. Am. Soc. Civil Eng. 89 (ST6). Dec. 1963. pp. 1–12.
20. Gandhi, S. N. and Holmes, M. Direct moment distribution for rigid frameworks. Civil Eng. and Pub. Works Rev. 58. June 1963. pp. 745–750, 898–900, 1013–1015.
21. Williams, A. The determination of influence lines for bridge decks monolithic with their piers. Struct. Eng. 42. May 1964. pp. 161–166.

8 Model analysis

Notation

a	area displaced by a deformed model
A	cross-sectional area
d	depth of member
E	Young's modulus
f	stress
g	gravitational acceleration
G	modulus of torsional rigidity
H	horizontal reaction
I	second moment of area
J	torsional inertia
k	scale factor
l	length of a member
L	dimension of length
m	model (used as suffix)
M	dimension of mass
p	prototype (used as suffix)
P	axial force
Q	shear force
r_p	radius of gyration of prototype member
t_m	thickness of model member
T	dimension of time
V	vertical reaction
w	distributed load
W	applied load
W'	knife-edge load
y	influence line ordinate
α	equals f_p/f_m
β	equals ϵ_p/ϵ_m
δ	deflection, displacement imposed on a model
ϵ	strain
θ	rotation imposed on a model
ρ	density
υ	Poisson's ratio

8.1   Introduction

A structural model is required to reproduce an actual structure (the prototype) to some convenient scale and in such a manner that it is possible to predict the behavior of the prototype subjected to its applied loads from the behavior of the model subjected to a system of proportional loads. Model analysis may be employed in the design of a complex structure for which a mathematical analysis is not possible or may be used to check the validity of a new theoretical method of analysis. In addition, simple model techniques may be utilized[1,2] to develop an appreciation of structural behavior in young engineers. No simplifying approximations are necessary for a model analysis, and all secondary effects are automatically included in the analysis.

Model analysis may be classified under the headings *direct methods* and *indirect methods*. The loading applied to an indirect model is unrelated to the applied loading on the prototype. An arbitrary deformation is applied to the model, and the deflected form obtained represents, to some scale, the influence line for the force corresponding to the applied deformation. The indirect method is applicable only to linear structures. The loading applied to a direct model is similar to the applied loads on the prototype. Strains and deflections of the model are recorded and stresses and deflections of the prototype deduced. The direct method is applicable to structures in both the elastic and inelastic states. For both direct and indirect methods, knowledge of the laws of similitude is necessary for the correct proportioning of the model and its loads and the correct interpretation of results.

Methods of constructing and testing models will not be considered here but may be referred to in several comprehensive textbooks[3,4,5,6].

8.2   Structural similitude

The required similitude between model and prototype quantities may be obtained by the method of dimensional analysis[7]. The relationship between the n fundamental variables affecting structural behavior can be expressed as a function of $(n - i)$ dimensionless products, where i is the number of dimensions involved. The dimensionless products are referred to as the *pi terms*.

The variables affecting structural behavior are given in Table 8.1 together with the exponents of their dimensions of mass, length, and time. Only the first eight of these variables are fundamental; the remainder have been included in order to derive their interdependence. The fifteen variables contain three dimensions, and thus the structural behavior can be expressed as the relationship between twelve pi terms. The pi terms are obtained by considering the

Table 8.1 Variables involved in structural similitude

Variable		ϵ	l	I	E	W	ρ	v	g	δ	f	w	W'	A	J	G
Dimension	M	0	0	0	1	1	1	0	0	0	1	1	1	0	0	1
	L	0	1	4	-1	1	-3	0	1	1	-1	-1	0	2	4	-1
	T	0	0	0	-2	-2	0	0	-2	0	-2	-2	-2	0	0	-2

exponents of the dimensions of the product formed by g, W, l, and one other variable at a time. Thus:

$$\pi_1 = \epsilon l^a W^b g^c$$
$$= \epsilon l^0 W^0 g^0$$
$$= \epsilon$$
$$\pi_2 = I l^a W^b g^c$$
$$= I l^{-4} W^0 g^0$$
$$= I/l^4$$
$$\pi_3 = l^2 E/W$$
$$\pi_4 = l^3 \rho g/W$$
$$\pi_5 = v$$
$$\pi_6 = \delta/l$$
$$= \delta EI/Wl^3$$
$$= \delta EA/Wl$$
$$= \delta GJ/Wl^3$$
$$= \delta GA/Wl$$
$$\pi_7 = fl^2/W$$
$$\pi_8 = wl^2/W$$
$$\pi_9 = W'l/W$$
$$\pi_{10} = A/l^2$$
$$\pi_{11} = J/l^4$$
$$\pi_{12} = Gl^2/W$$

Any number of the dimensionless products may be combined to give a different expression for the pi term, as shown by the five different expressions given for π_6.

For complete similitude, each variable affecting the behavior of the model and the prototype must be in a fixed ratio, which is dimensionless and is referred to as the *scale factor*. In addition, the pi terms of the model and the prototype must be in a fixed ratio and establish the relationships between the scale factors. Thus, for geometrical similarity, lengths on the model must be in a fixed ratio to the corresponding lengths on the prototype and to $l_p/l_m = k_l$, the linear scale factor. Considering each pi term in turn:

$$\epsilon_p/\epsilon_m = k_\epsilon$$
$$= 1$$
$$k_I = k_l^4$$
$$k_W = k_E k_l^2$$
$$k_\rho = k_W/k_g k_l^3$$
$$k_v = 1$$
$$k_\delta = k_l$$
$$= k_W k_l^3/k_E k_I$$
$$= k_W k_l/k_E k_A$$
$$= k_W k_l^3/k_G k_J$$
$$= k_W k_l/k_G k_A$$
$$k_f = k_w k_l^2$$
$$k_w = k_W/k_l^2$$
$$k_{W'} = k_W/k_l$$
$$k_A = k_l^2$$
$$k_J = k_l^4$$
$$k_G = k_W/k_l^2$$

When it is known that some variables have a negligible effect on the behavior of a particular structure, it is unnecessary to fulfill all the above conditions of similitude, and a distorted model may prove satisfactory and be easier to construct[8].

8.3   Indirect models

(a) Similitude requirements

Indirect model methods make use of the Müller-Breslau principle presented in Section 5.2. The influence line for any restraint in a structure is the deformed shape of the structure produced by a unit displacement replacing the restraint. The displacement corresponds to and is applied in the same direction as the restraint. The influence line ordinates obtained are positive when the deformation produced moves in opposition to the direction of an applied load at any point. It is impracticable to apply the displacements to the prototype, and a model is constructed to a suitable linear scale such that the deformations in the model and the prototype, due to a given displacement, are in a constant ratio at all corresponding points. Thus:

$$\delta_p/\delta_m = k_\delta$$
$$= k_W k_I^3/k_E k_I$$
$$= k_W k_I/k_E k_A$$
$$= k_W k_I^3/k_G k_J$$
$$= k_W k_I/k_G k_A$$

If only flexural deformations are significant in the prototype, the expression reduces to:

$$k_\delta = k_W k_I^3/k_E k_I$$

Since loads are not measured or reproduced, the only requirement to be observed is that the ratio $(EI)_p/(EI)_m$ is maintained constant for all members. It is not essential for $k_I = k_I^4$ or $k_v = 1$. Hence any suitable material may be used and any shape of cross-section chosen for ease of fabrication and prevention of buckling. Thus, the model members may be cut from a sheet of material of uniform thickness such that the ratio $(d_m)^3/(EI)_p$ is constant for all members.

If only axial deformations are significant in the prototype, the expression reduces to:

$$k_\delta = k_W k_I/k_E k_A$$

The only requirement to be observed is that the ratio $(EA)_p/(EA)_m$ is maintained constant for all members. Thus, the model members may be cut from a sheet of material of uniform thickness such that the ratio $d_m/(EA)_p$ is constant for all members.

If both flexural and axial deformations are significant in the prototype, the expression reduces to:

$$k_\delta = k_W k_I^3/k_E k_I$$
$$= k_W k_I/k_E k_A$$

and:

$$k_I^2 = k_I/k_A$$

Using rectangular model sections, this reduces to:

$$k_I^2 = A_p r_p^2/A_p \times 12 t_m d_m/t_m d_m^3$$
$$= 12 r_p^2/d_m^2$$

Thus, the model members may be of rectangular section with $d_m = r_p(12)^{0.5}/k_l$ and with the thickness adjusted such that the ratio $d_m t_m/(EA)_p$ is constant for all members.

If both flexural and shearing deformations are significant in the prototype, the expression reduces to:

$$k_\delta = k_W k_I^3/k_E k_I$$
$$= k_W k_I/k_G k_A$$

using a model material such that $k_v = 1$, $k_E = k_G$. Thus $k_I^2 = k_I/k_A$, as in the case of significant flexural and axial deformations.

If both flexural and torsional deformations are significant in the prototype, the expression reduces to:

$$k_\delta = k_W k_I^3/k_E k_I$$

$$= k_W k_I^3/k_G k_J$$

using a model material such that $k_v = 1$, $k_E = k_G$. Thus, $k_I = k_J$, and the model and prototype member cross-sections must be geometrically similar. Using a model material such that $k_v \neq 1$, the model members may be of any suitable section such that the ratio $k_E k_I/k_G k_J$ is constant for all members.

(b) Testing technique

To determine the reactions H_1, V_1, M_1 of the arch shown in Figure 8.1 (i), a model is constructed to a suitable linear scale factor, k_l, with the necessary similitude conditions satisfied. A convenient horizontal displacement, δ, is imposed on the model at 1, as shown at (ii), and the ordinate y and the area a measured. Had the displacement, δ, been applied to the prototype; the corresponding area displaced would have been ak_l. The horizontal reaction in the prototype is given by:

$$H_1 = Wy/\delta + wak_l/\delta$$

A convenient vertical displacement, δ, is imposed on the model at 1, as shown at (iii), and the ordinate y and the area a measured. The vertical reaction in the prototype is given by:

$$V_1 = Wy/\delta + wak_l/\delta$$

A convenient anticlockwise rotation, θ, is imposed on the model at 1, as shown at (iv), and the ordinate y and the area a measured. Had the rotation, θ, been applied to the prototype, the corresponding ordinate and displaced area

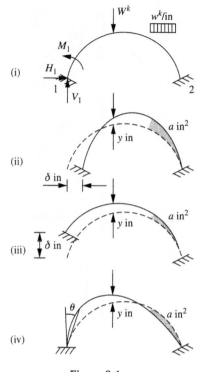

Figure 8.1

would have been yk_l and ak_l^2, respectively. The moment in the prototype is given by:

$$M_1 = Wyk_l/\theta - wak_l^2/\theta$$

To determine the internal forces P_c, Q_c, M_c at the crown of the arch shown in Figure 8.2 (i), the model must be cut at the crown. A convenient relative horizontal displacement, δ, is imposed on the cut ends, as shown at (ii), and the ordinate y and the area a measured. The axial thrust in the prototype at the crown is given by:

$$P_c = Wy/\delta + wak_l/\delta$$

A convenient relative vertical displacement, δ, is imposed on the cut ends, as shown at (iii), and the ordinate y and the area a measured. The shear force in the prototype at the crown is given by:

$$Q_c = -Wy/\delta + wak_l/\delta$$

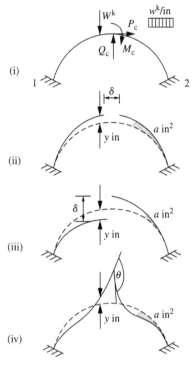

Figure 8.2

A convenient relative rotation, θ, is imposed on the cut ends, as shown at (iv), and the ordinate y and the area a measured. The internal moment in the prototype at the crown is given by:

$$M_c = Wyk_l/\theta - wak_l^2/\theta$$

The technique, as described above, may be readily applied to space frames[9].

An alternative technique for obtaining internal moments that avoids cutting members involves the use of a moment deformeter[3,4]. However, it may be applied only to members that are initially straight and prismatic. To determine the internal moment, M_1, in the frame shown in Figure 8.3 (i), a model is constructed to a suitable linear scale factor with the necessary similitude conditions satisfied. The moment deformeter applies the reaction system shown at (ii) to the model. This is equivalent to applying equal and opposite moments, of magnitude, $V \delta x$, to two sections in the model a short distance, δx, apart. Thus, as shown in Section 5.2(b), the ordinates of the elastic curve produced by the deformeter represent the value:

$$V \delta x \, \delta\theta = V(\delta x)^2/R$$
$$= VM_m(\delta x)^2/E_m I_m$$

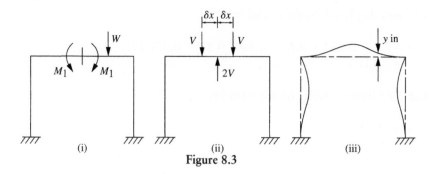

Figure 8.3

with the exception of the short length $2 \times \delta x$. The deformed shape of the model is shown at (iii) and the internal moment at 1 in the prototype is given by:

$$M_1 = W y k_l E_m I_m / V (\delta x)^2$$

This technique may be readily modified[10] and used to determine the influence surface for the bending moment at any point in a slab.

Example 8.1

The value of the horizontal thrust at the springings is required for the two-hinged arch shown in Figure 8.4 (i). A model is constructed to a suitable scale and a horizontal displacement of 0.375 in imposed on the end 1 produces the displacements shown at (ii).

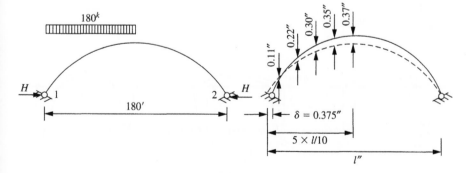

Figure 8.4

Solution

The linear scale factor is:

$$k_l = 2160/l$$

The area displaced on the model is:

$$a = l\{0 + 2(0.11 + 0.22 + 0.30 + 0.35) + 0.37\}/20$$
$$= 0.1165l$$

The distributed loading on the model is:

$$w = 2/12$$
$$= 1/6 \text{ kip/in}$$

The horizontal thrust in the prototype is given by:

$$H = wak_l/\delta$$
$$= 2160 \times 0.1165l/(6l \times 0.375)$$
$$= 112 \text{ kips}$$

Example 8.2

The values of the moments M_1 and M_2 are required for the symmetrical frame shown in Figure 8.5 (i). The column feet are initially hinged; an anti-clockwise rotation of 0.004 radian imposed at 1 produces the displacements shown at (ii).

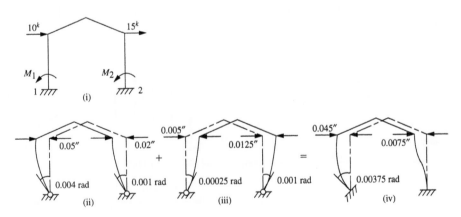

Figure 8.5

Solution

Because of the symmetry of the structure, a clockwise rotation of 0.001 radian imposed at 2 will produce the displacements shown at (iii). Thus, the displacements produced by a clockwise rotation of 0.00375 radian at 1 with end 2 clamped are shown at (iv). The values of the moments are given by:

$$M_1 = 10 \times 0.045/0.00375 + 15 \times 0.0075/0.00375$$
$$= 150 \text{ kip-in}$$
$$M_2 = 15 \times 0.045/0.00375 + 10 \times 0.0075/0.00375$$
$$= 200 \text{ kip-in}$$

8.4 Direct models

A direct model must be true to linear scale and, for complete similitude, must satisfy all the similitude conditions listed in Section 8.2. The linear scale factor should not be so large that the internal texture of the model material affects the results. Where a large linear scale factor is essential, the existence of scale effects can be distinguished by constructing several models to different linear scales. As with indirect models, complete geometrical similitude of cross-sections may often be disregarded, provided that the loading scale factor is modified. Thus, if only flexural effects are significant:

$$k_\delta = k_l$$
$$= k_W k_l^3 / k_E k_I$$

And:

$$k_I \neq k_l^4$$

Hence, the modified loading scale factor is:

$$k_W = k_E k_I / k_l^2$$

If only axial effects are significant:

$$k_\delta = k_l$$
$$= k_W k_l / k_E k_A$$

And:

$$k_A \neq k_l^2$$

Hence, the modified loading scale factor is:

$$k_W = k_E k_A$$

When the self weight of the structure appreciably affects the behavior of the prototype, an essential condition is:

$$k_\rho = k_W/k_g k_l^3$$
$$= k_E/k_g k_l$$

Thus, the linear scale factor is automatically fixed by the properties of the model material. When a linear scale factor larger than this is required, the density of the model material may be artificially increased by suspending weights from the model, or a centrifuge may be used to provide an increased gravitational field.

When torsional and shearing effects are significant, an essential condition is:

$$k_v = 1$$

To investigate the elastic behavior of flexible structures in which deflections are large and stress is not proportional to the applied loads, the two essential conditions are:

$$k_w = k_E k_l^2$$

and:

$$k_f = k_w/k_l^2$$
$$= k_E$$

Then:

$$k_\delta = k_l$$

And:

$$k_\epsilon = 1$$

The condition $k_W = k_E k_l^2$ fixes the relationship between applied loads and elastic critical loads. If this condition is ignored and an arbitrary value is assigned to k_W, it is possible to magnify model strains and deflections. However, changes in geometry of the model under load are no longer similar to those in the prototype, and errors in predicting elastic critical loads are incurred.

To investigate the elastic behavior of linear structures in which deflections are small, stress is proportional to the applied loads, and the principle of superposition is valid, arbitrary values may be assigned to k_W and k_E:

$$k_W \neq k_E k_l^2$$

Then:

$$k_\epsilon = k_f/k_E$$

And:

$$k_\delta = k_f k_l/k_E$$

since stress is proportional to strain in both model and prototype.

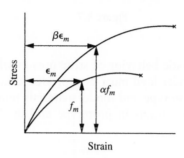

Figure 8.6

To investigate the plastic behavior of structures in which the deformations at failure have little influence on the structural behavior, the stress-strain curves for the prototype and model materials must be related as shown in Figure 8.6. Then:

$$k_f = \alpha$$

and:

$$k_\epsilon = \beta$$

and:

$$k_\delta = \beta k_l$$

and changes in geometry of the model under load are not similar to those in the prototype. In addition, the following scale factors must be observed:

$$
\begin{aligned}
k_W &= k_f k_l^2 \\
    &= \alpha k_l^2 \\
k_\rho &= k_W/k_g k_l^3 \\
       &= \alpha/k_g k_l \\
k_w &= k_W/k_l^2 \\
    &= \alpha
\end{aligned}
$$

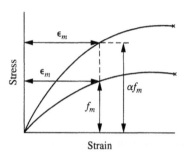

Figure 8.7

To investigate the plastic behavior of structures in which changes in geometry of the prototype under load must be reproduced in the model, the stress-strain curves for the prototype and model materials must be related as shown in Figure 8.7. Thus, the strains in prototype and model are equal at corresponding stresses and:

$$k_f = \alpha$$

and:

$$k_\epsilon = 1$$

and:

$$k_\delta = k_l$$

and:

$$k_W = \alpha k_l^2$$

and:

$$k_\rho = \alpha / k_g k_l$$

and:

$$k_w = \alpha$$

To investigate composite structures, the model materials used must all have the same scale factors α and β. Alternatively, for reinforced and prestressed concrete structures, it is possible to use steel reinforcement for the model and ensure that the scale of forces in the reinforcement is correct[11]. Thus, if k_S represents the scale factor for reinforcement areas in the prototype and model:

$$\alpha k_l^2 = k_S f_p / f_m$$
$$= k_S E_S \epsilon_m \beta / E_S \epsilon_m$$

And:

$$k_S = \alpha k_l^2 / \beta$$

In addition, it is necessary for the model reinforcement to yield at a strain of ϵ_S / β, where ϵ_S is the strain in the prototype reinforcement at yield.

Example 8.3

A model is constructed of a prototype in which only axial effects are significant. The model is constructed from the same material as the prototype, using the scale factors $k_l = 20$ and $k_A = 100$. Determine the required scale factor for loading if geometrical similitude is to be maintained during loading.

Solution

$$k_\delta = k_l$$
$$= k_W k_l / k_E k_A$$

Thus:

$$k_W = k_A$$
$$= 100$$

Example 8.4

A model is constructed of a prototype, in which only flexural effects are significant, using the scale factors $k_l = 20$ and $k_E = 100$. The model members are constructed to a distorted scale so that $k_l = 1000$ and $k_d = 10$. Determine the scale factors k_W, k_f, and k_ϵ if the required scale factor for deflection is $k_\delta = 10$.

Solution

$$k_\delta = k_W k_l^3 / k_E k_l$$

Thus:

$$k_W = 1000/8$$
$$= 125$$
$$k_f = k_W k_l k_d / k_I$$
$$= 25$$
$$k_\epsilon = k_f / k_E$$
$$= 0.25$$

Example 8.5

A model is constructed of a prototype, in which both axial and flexural effects are significant, using the scale factors $k_l = 20$ and $k_{EI} = 2 \times 10^6$. Determine the required scale factors k_W and k_{EA} if geometrical similitude is to be maintained during loading.

Solution

$$k_\delta = k_l$$
$$= k_W k_l^3 / k_{EI}$$
$$= k_W k_l / k_{EA}$$

Thus:

$$k_{EA} = k_W$$
$$= 5000$$

Example 8.6

A model is constructed of an arch dam using a linear scale factor $k_l = 40$. Determine the stress scale factor k_f if hydrostatic loads are produced in the model with mercury.

Solution

$$k_\rho = 1/13.6$$

and:

$$k_g = 1$$

Thus:

$$k_f = k_W / k_l^2$$
$$= k_\rho k_g k_l$$
$$= 40/13.6$$

Supplementary problems

S8.1 Figure S8.1 shows a symmetrical, rigid frame bridge with a 20 ft wide deck. The bridge is designed for loading, which consists of a uniformly distributed load of intensity 220 lb/ft² of any continuous length, plus a knife-edge load of 2700 lb/ft placed anywhere on the deck[12].

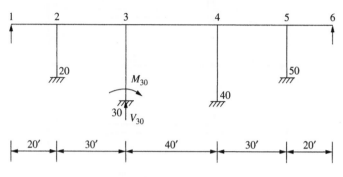

Figure S8.1

A model is constructed to a scale of 1 in = 2 ft, and the following displacements are imposed:

(i) a vertical displacement of 1 in on column base 40
(ii) a vertical displacement of +0.2 in and a clockwise rotation of 0.2 radians on column base 30

The measured vertical displacements over the central span at intervals of 0.2 × span are shown in the Table S8.1, with upward displacements positive.

Table S8.1 Displacements of the model

Section	3.0	3.2	3.4	3.6	3.8	4.0
Displacements (i)	0.00	0.10	0.35	0.75	0.95	1.00
Displacements (ii)	0.20	0.27	0.26	0.16	0.06	0.00

Determine the maximum value of M_{30} that can be produced in the prototype by the applied loading and the value of V_{30} that occurs simultaneously.

S8.2 A prototype shown in Figure S8.2 (i) consists of an unsymmetrical, fixed-ended, haunched beam, 12, with a concentrated load of 10 kips located at point 3 a distance of one-third its length from end 1. A model is constructed to a scale of 1 in = 1 ft, as shown in Figure S8.2(ii), with hinged ends. The following displacements are imposed on the model:

(i) a clockwise rotation of 0.1 radians at end 1′, which produces an anticlockwise rotation of 0.04 radians at end 2′ and a downward deflection of 0.15 in at point 3′

(ii) an anticlockwise rotation of 0.1 radians at end 2', which produces a clockwise rotation of 0.05 radians at end 2' and a downward deflection of 0.12 in at point 3'

Determine the bending moment produced at end 1 in the prototype by the 10 kips load.

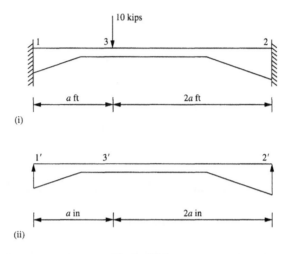

Figure S8.2

S8.3 An aluminum model with a modulus of elasticity of 10×10^6 lb/in^2 is constructed of a steel W section with a modulus of elasticity of 29×10^6 lb/in^2. The second moment of area of the W section is 5900 in^4 and of the model is 0.9 in^4. The linear scale factor adopted is $l_p/l_m = k_l = 10$. Determine the load scale factor, k_W, required if geometric similarity is to be maintained after loading the model (i.e., $\delta_p/\delta_m = k\delta = k_l$).

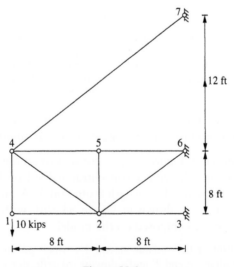

Figure S8.3

S8.4 Figure S8.3 shows a pin-jointed frame with a vertical load of 10 kips applied to node 1. A model is constructed to a linear scale factor of $l_p/l_m = k_l = 10$. In the model, member 47 is shortened by 0.5 in and the vertical displacement of node 1 is 0.43 in. Determine the force in the prototype in member 47 caused by the 10 kip load.

S8.5 The symmetrical, cable-stayed bridge shown in Figure S8.4 (i) consists of a continuous main girder, which is supported on rollers where it crosses the piers. The ties are anchored to the girder and the tops of the towers, and the dimensions of the prototype structure are indicated. The main girder has a second moment of area I_b and a modulus of elasticity E_b. Each tie has a cross-sectional area A_c and a modulus of elasticity E_c.

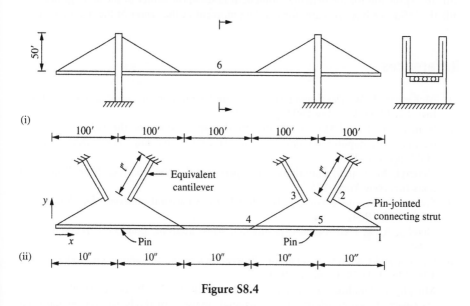

Figure S8.4

A model is constructed as shown at (ii). The stiffness of the ties is represented by equivalent cantilevers[13] of length l in, cut from the same sheet of Perspex and finished to the same section as the main beam. Axial effects in the beam and towers and flexural effects in the towers are neglected. The following displacements are imposed:

(i) strut 12 is shortened by 1 in, and the vertical displacements y_{12} of the main girder are recorded
(ii) strut 34 is shortened by 1 in, and the vertical displacements y_{34} of the main girder are recorded
(iii) a vertical displacement of +1 in is imposed at hinge 5 and the vertical displacements y_5 of the main girder are recorded.

The measured vertical displacements, over the left half of the main girder, at intervals of 5 in, are shown in the Table S8.2, with upward displacements positive.

Table S8.2 Displacements of the main girder

x, in	0	5	10	15	20	25
y_{12}, in	0.13	0.06	0	−0.06	−0.13	−0.21
y_{34}, in	−0.05	−0.03	0	0.03	0.04	0.01
y_5, in	−0.36	−0.18	0	0.18	0.37	0.59

Derive the following:

(i) the length of the equivalent cantilever, l, in terms of the properties of the main beam and ties
(ii) an expression for the bending moment at node 6, the center of the main girder
(iii) the influence line ordinates for bending moment at the center of the main girder

References

1. McKenzie, I. W. Developing structural understanding in young engineers. Proc. Inst. Civil Eng. 97. May 1993. pp. 94–99.
2. Romero, M. L. and Museros, P. Structural analysis education through model experiments and computer simulation. Proc. Am. Soc. Civil Eng. 128 (PIE4). October 2002. pp. 170–175.
3. Hetényi, M. I. (ed.) Handbook of experimental stress analysis. John Wiley and Sons Inc. New York. 1957.
4. Preece, B. W. and Davies, J. D. Models for structural concrete. London. C. R. Books Ltd. 1964.
5. Hendry, A. W. Elements of experimental stress analysis. London. Pergamon Press. 1964.
6. Cement and Concrete Association. Proceedings of the conference on model testing, 17th March 1964. London. 1964.
7. Murphy, G. Similitude in engineering. New York. Ronald Press Co. 1950.
8. Jasiewicz, J. Applications of dimensional analysis methods to civil engineering problems. Civil Eng. and Pub. Works. Rev. 58. Sept., Oct., Nov., Dec., 1963. pp. 1125–1128, 1279-1280, 1409-1410, 1547-1549.
9. Balint, P. S. The model analysis of rigid space frames. Civil Eng. and Pub. Works. Rev. 57. April 1962. pp. 481–483.
10. Kuske, A. Influence fields for transversely bent plates. Civil Eng. and Pub. Works. Rev. 58. May 1963. pp. 627–628.
11. Mattock, A. H. Structural model testing-- theory and applications. Skokie, IL. Portland Cement Association. 1962.
12. Williams, A. The determination of influence-lines for bridge decks monolithic with their piers. Struct. Eng. 42. May 1964. pp. 161–166.
13. Stephens, L. F. A simple model solution of the suspension bridge problem. Proc. Inst. Civil Eng. 34. July 1966. pp. 369–382, 36. April 1967. pp. 883–886.

9 Plastic analysis and design

Notation

A	cross-sectional area of a member
c'	carry-over factor for plastic moment distribution
D	Degree of indeterminacy of a structure
E	Young's modulus
f_y	yield stress
I	second moment of area of a member
m	bending moment in a member due to a virtual load
m_i	number of independent mechanisms
M	bending moment in a member due to the applied load
M_p	plastic moment of resistance of a member
M'_p	plastic moment of resistance modified for axial compression
M_y	moment at which the extreme fibers of a member yield in flexure
$M^{\max}$	maximum theoretical elastic moment
$M^{\min}$	minimum theoretical elastic moment
M^r	residual moment
N	load factor against collapse due to proportional loading $= W_u/W$
N_a	load factor against collapse due to alternating plasticity $= W_a/W$
N_s	load factor against incremental collapse equals $= W_s/W$
p	number of possible hinge positions
P	axial force in a member
S	elastic section modulus
V	vertical reaction
W	applied load
W_a	load producing collapse by alternating plasticity
W_s	load producing incremental collapse
W_u	ultimate load for proportional loading
Z	plastic section modulus $= M_p/f_y$
δ	deflection
θ	relative rotation at a plastic hinge during motion of collapse mechanism
λ	shape factor $= Z/S$
ϕ	total rotation at a plastic hinge during loading

9.1 Introduction

The plastic method of structural analysis is concerned with determining the maximum loads that a structure can sustain before collapse. The collapse load is known variously as the failure load, the ultimate load, and the limit load. A number of comprehensive textbooks dealing with plastic design are available[1,2,3,4,5].

The plastic method is applicable to structures constructed with an ideal elastic-plastic material that exhibits the stress-strain relationship shown in Figure 9.1 (i). The moment-curvature relationship for any section of the structure is assumed to have the ideal form shown in Figure 9.1 (ii). Thus, on applying a uniform sagging moment to a member the moment-curvature relationship is linear until the applied moment reaches the value of M_p the plastic moment of resistance of the section. At this stage, all material above the zero-strain axis of the section has yielded in compression and all material below has yielded in tension, and a plastic hinge has formed. Then the section can offer no additional resistance to deformation, and increase in curvature continues at a constant applied moment. In addition, in determining the collapse load of a structure, it is assumed that elastic deformations are negligible and do not affect the geometry of the structure. Thus, the structure behaves in a rigid-plastic manner with zero deformation until the formation of sufficient plastic hinges to produce a mechanism.

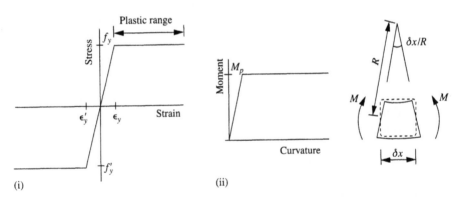

Figure 9.1

9.2 Formation of plastic hinges

A fixed-ended beam is subjected to a uniformly distributed working load of total magnitude W, as shown in Figure 9.2. The elastic distribution of bending moment in the member, drawn on the tension side of the member, is as shown at (i). The moment at the ends of the beam is twice the moment at the center, and, if the applied load is increased, at some stage the plastic moment of resistance will be reached simultaneously at both ends of the member; the distribution of bending moment is as shown at (ii). Further increase in the applied load causes the two plastic hinges to rotate, while the moment at the ends remains constant at the value M_p. Thus, the system is equivalent to a simply supported beam with an applied load and restraining end moments of value M_p,

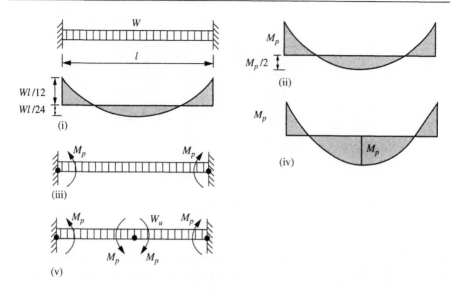

Figure 9.2

as shown at (iii). Finally, as the applied load is increased still further, a plastic hinge forms in the center of the beam, and the distribution of bending moment is as shown at (iv). The beam has now been converted to the unstable collapse mechanism shown at (v), and collapse is imminent under the ultimate load W_u The ratio of the collapse load to the working load is:

$$W_u/W = N$$

where N is the load factor. Since the structure is statically determinate at the point of collapse, the collapse load is readily determined as:

$$W_u = 16M_p/l$$

and this value is unaffected by settlement of the supports or elastically restrained end connections.

9.3 Plastic moment of resistance

After the formation of a plastic hinge in the section shown in Figure 9.3, the rectangular stress distribution shown at (i) is produced. Equating horizontal compressive and tensile forces:

$$P_c = P_t$$
$$f_y'A_c = f_y A_t$$

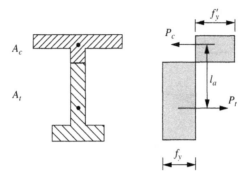

Figure 9.3

where f_y and f_y' are the yield stresses in tension and compression (which may be assumed to be equal) and A_t and A_c are the cross-sectional areas in tension and compression. Thus, the plastic moment of resistance is:

$$M_p = f_y A_t l_a$$

where l_a is the lever arm and equals the distance between the centroids of A_t and A_c. Also:

$$M_p = f_y Z$$

where Z is the plastic section modulus. The ratio of the plastic moment of resistance to the moment producing yield in the extreme fibers is:

$$M_p/M_y = f_y Z/f_y S = Z/S = \lambda$$

where λ is the shape factor.

An axial force applied to the section shown in Figure 9.4 reduces the value of the plastic moment of resistance that the section can develop. Assuming the axial force P is compressive, the rectangular stress distribution shown at (i) is produced after the formation of a plastic hinge with the cross-sectional area A_c' resisting the axial force P. Equating horizontal forces gives:

$$P = \Sigma C - \Sigma T$$
$$= f_y' A_c + f_y' A_c' - f_y A_t$$
$$= f_y' A_c'$$

and:

$$f_y' A_c = f_y A_t$$

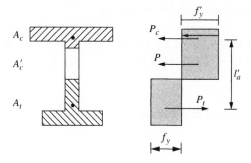

Figure 9.4

Thus, the modified plastic moment of resistance is:

$$M'_p = f_y A_t l'_a$$

where l'_a is the distance between the centroids of areas A_c and A_t.

Example 9.1

Determine the shape factor for the built-up steel beam shown in Figure 9.5.

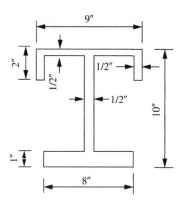

Figure 9.5

Solution

The area of the section is:

$$A = 10 \times 8 + 2 \times 1 - 8.5 \times 7.5$$
$$= 18.25 \, \text{in}^2$$

The height of the centroid is:

$$\bar{y} = (80 \times 5 + 2 \times 9 - 63.75 \times 5.25)/18.25$$
$$= 4.56 \, \text{in}$$

The second moment of area of the section is:

$$I = 80 \times 100/12 + 2 \times 4/12 - 63.75 \times 72/12 + 80 \times 0.46^2$$
$$+ 2 \times 4.46^2 - 63.75 \times 0.71^2$$
$$= 309 \text{ in}^4$$

The smaller elastic section modulus is:

$$S = 309/5.44$$
$$= 56.8 \text{ in}^3$$

The areas in tension and compression after the formation of a plastic hinge are:

$$A_t = A_c$$
$$= 9.125 \text{ in}^2$$

The zero strain axis is at a height above the base of:

$$1 + 2.25 = 3.25 \text{ in}$$

The centroid of A_t is at a height above the base of:

$$(8 \times 0.5 + 1.125 \times 2.125)/9.125 = 0.70 \text{ in}$$

The centroid of A_c is at a height above the base of:

$$(4 \times 9.75 + 2 \times 9 + 3.125 \times 6.375)/9.125 = 8.42 \text{ in}$$

The lever arm is:

$$l_a = 8.42 - 0.7$$
$$= 7.72 \text{ in}$$

The plastic section modulus is:

$$Z = 9.125 \times 7.72$$
$$= 70.4 \text{ in}^3$$

The shape factor is:

$$\lambda = 70.4/56.8$$
$$= 1.24$$

Example 9.2

The T-section shown in Figure 9.6 consists of a material that has a yield stress in compression 25% greater than the yield stress in tension. Determine the plastic moment of resistance when the section is subjected to (i) a bending moment producing tension in the bottom fiber and (ii) a bending moment producing tension in the top fiber.

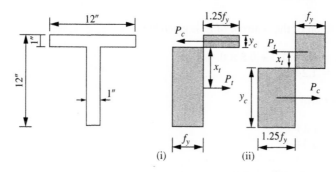

Figure 9.6

Solution

(i) The stress distribution, after the formation of a plastic hinge, is shown at (i). Equating horizontal forces:

$$1.25 \times 12y_c = 11 + 12(1 - y_c)$$

and:

$$y_c = 0.85 \text{ in}$$

The position of the centroid of A_t is given by:

$$x_t = (11 \times 5.65 + 1.8 \times 0.075)/12.8$$
$$= 4.87 \text{ in}$$

The lever arm is:

$$l_a = 4.87 + 0.425$$
$$= 5.295 \text{ in}$$

The plastic moment of resistance is:

$$M_p = 5.295 \times 12.8 f_y$$
$$= 67.6 f_y$$

(ii) The stress distribution after the formation of a plastic hinge is shown at (ii). The depth of the compressive stress block is:

$$1.25y_c = 12 + (11 - y_c)$$

and:

$$y_c = 10.2 \text{ in}$$

The position of the centroid of A_t is given by:

$$x_t = (12 \times 1.3 + 0.8 \times 0.4)/12.8$$
$$= 1.24 \text{ in}$$

The lever arm is:

$$l_a = 1.24 + 5.1$$
$$= 6.34 \text{ in}$$

The plastic moment of resistance is:

$$M_p = 6.34 \times 12.8f_y$$
$$= 80f_y$$

9.4 Statical method of design

The statical method may be used to determine the required plastic modulus for continuous beams.

The continuous beam shown in Figure 9.7 is first cut back to a statically determinate condition and the applied loads multiplied by the load factor as shown at (i). The statical bending moment diagram for this condition is shown at (ii). The fixing moment line due to the redundants is now superimposed on the static moment diagram, as shown at (iii), so that the collapse mechanism, shown at (iv), is formed. Collapse occurs simultaneously in the two end spans, and the required plastic modulus is $M_p = 0.686M$. An alternative fixing moment line is shown at (v) for a non-uniform beam section. Then $M_{p1} = 0.5M$ and $M_{p2} = 0.766M$, and collapse occurs simultaneously in all spans, as shown at (vi).

A similar procedure may be adopted for determining the collapse load of a given structure. A bending moment distribution is drawn in which the given value of M_p is not exceeded and the collapse load is computed. A possible moment diagram for the two-span beam of Figure 9.8 is shown at (i), and the computed collapse load is $W_u = 4M_p/l$. This is equivalent to the system shown at (ii) and is clearly not a collapse mechanism, which indicates that the assumed moment diagram is safe. An alternative moment diagram is shown at

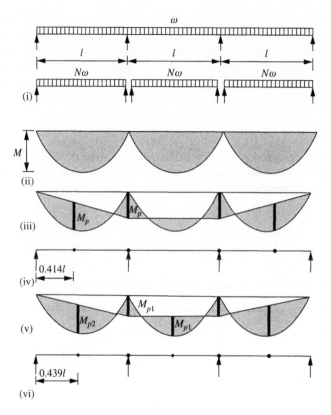

Figure 9.7

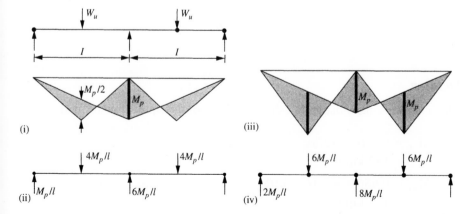

Figure 9.8

(iii) and is equivalent to the collapse mechanism shown at (iv). The computed collapse load is $W_u = 6M_p/l$ and is the maximum possible value. Thus, collapse loads computed by the statical method are either equal to or less than the correct values. That is, the statical method provides a lower bound on the collapse load, and the correct collapse load is that producing a collapse mechanism.

Example 9.3

Determine the required plastic modulus in each span of the continuous beam shown in Figure 9.9. The collapse loads are indicated, and collapse is to occur in all spans simultaneously.

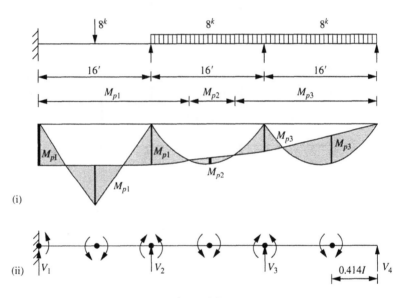

Figure 9.9

Solution

The collapse moment diagram is shown at (i), and the required plastic moduli are:

$$M_{p1} = 8 \times 16 \times 12/8$$
$$= 192 \text{ kip-in}$$
$$M_{p3} = 0.686 \times 8 \times 16 \times 12/8$$
$$= 132 \text{ kip-in}$$
$$M_{p2} = 8 \times 16 \times 12/8 - (192 + 132)/2$$
$$= 30 \text{ kip-in}$$

The reactions at collapse may be obtained from the collapse mechanism shown at (ii) and are:

$$V_1 = 8/2$$
$$= 4.0 \text{ kip}$$
$$V_2 = 8 + (192 - 132)/(16 \times 12)$$
$$= 8.31 \text{ kip}$$
$$V_3 = 8 - 0.31 + 132/(16 \times 12)$$
$$= 8.38 \text{ kip}$$
$$V_4 = 4 - 0.69$$
$$= 3.31 \text{ kip}$$

9.5 Mechanism method of design

The mechanism, or kinematic, method may be used to determine the plastic modulus required for the members of rigid frames[6,7] and grids[8].

The fixed-ended beam shown in Figure 9.10 is assumed to collapse when the mechanism shown at (i) has formed. A virtual displacement δ is imposed on the mechanism, as shown at (ii), and the internal work equated to the external work. Thus:

$$4M_p\theta = W_u\delta$$

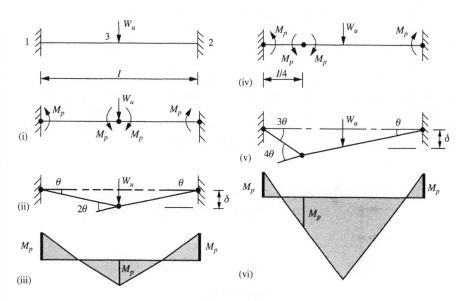

Figure 9.10

and:

$$M_p = W_u l/8$$

The moment diagram for this mechanism is shown at (iii) and is satisfactory, since M_p is nowhere exceeded.

A similar procedure may be adopted for determining the collapse load of a given structure. A collapse mechanism is assumed, as shown at (iv), and the collapse load determined in terms of the plastic moment of resistance of the members. A virtual displacement is imposed on the mechanism, as shown at (v), and equating internal and external work gives:

$$8M_p\theta = W_u\delta$$

and:

$$W_u = 16M_p/l$$

The moment diagram for this assumed mechanism is shown at (vi) and is unsafe, since M_p is exceeded over the central portion of the beam. The maximum possible collapse load corresponds to the correct mechanism shown at (ii) and is:

$$W_u = 8M_p/l$$

Thus, collapse loads computed by the mechanism method are either equal to or greater than the correct values. That is, the virtual work method provides an upper bound on the collapse load, and the correct collapse load is that producing a moment diagram in which M_p is nowhere exceeded.

The correct location of plastic hinges is a prior requirement of the virtual work method of analysis. Hinges are usually formed at the positions of maximum moment, which occur at the ends of a member, under a concentrated load, and at the position of zero shear in a prismatic member subjected to a distributed load. In the case of two members meeting at a joint, a plastic hinge forms in the weaker member. In the case of three or more members meeting at a joint, plastic hinges may form at the ends of each of the members. Possible locations of plastic hinges are shown in Figure 9.11.

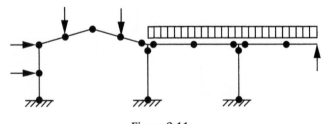

Figure 9.11

The position of the plastic hinge in a member subjected to a distributed load depends on the values of the fixing moments M_{p1} and M_{p2} at the ends of the member, as shown in Figure 9.12. Equating internal and external work for the virtual displacement shown at (ii):

$$M_{p1}\theta + M_{p2}\theta x/(l - x) + M_{p3}\theta l/(l - x) = W_u\delta/2$$

and:

$$W_u = 2(M_{p1}l - M_{p1}x + M_{p2}x + M_{p3}l)/x(l - x)$$

The value of x is such as to make W_u a minimum, or $\partial W_u/\partial x = 0$. Thus:

$$l^2(M_{p1} + M_{p3}) - 2lx(M_{p1} + M_{p3}) + x^2(M_{p1} - M_{p2}) = 0$$

and the value of x may be determined in any particular instance. This value may also be readily determined by means of charts[4]. As a first approximation, the hinge may be assumed in the center of the member and subsequently adjusted if necessary when the collapse mechanism is known.

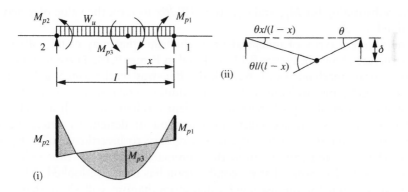

Figure 9.12

The position of the plastic hinge in a non-prismatic member subjected to a distributed load depends on the variation of the plastic moment M_x along the member in addition to the fixing moments. A fixed-ended beam, in which the plastic moment of resistance varies linearly with the distance from the end, is shown in Figure 9.13. The plastic hinge occurs at the point where the static

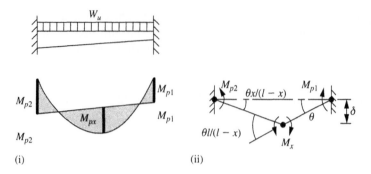

Figure 9.13

moment diagram touches the M_x diagram, as shown at (i). Equating internal and external work for the virtual displacement shown at (ii):

$$W_u = 2(M_{p1}l - M_{p1}x + M_{p2}x + M_{px}l)/x(l - x)$$

and

$$M_{px} = M_{p1} + (M_{p2} - M_{p1})x/l$$

Thus, substituting for M_{px} and equating $\partial W_u/\partial x$ to zero, the value of x may be obtained.

The different types of independent mechanisms that may cause collapse of the structure shown in Figure 9.11 are listed in Figure 9.14. These may be classified as beam mechanisms $B1$, $B2$, $B3$, $B4$, $B5$; sway mechanism S; gable mechanism G; and joint mechanisms $J1$, $J2$. In addition, collapse may occur through the combination of any of these independent mechanisms, such as $(B4 + J2)$ and $(B4 + J2 + S)$. In the mechanism method of design, to determine the maximum required value for M_p, it is necessary to investigate all independent mechanisms and combinations that eliminate a hinge and thus reduce the internal work. To ensure that no combination has been overlooked, a bending moment diagram for the assumed collapse mechanism will show that M_p is nowhere exceeded when the assumed mechanism is the critical one.

An independent mechanism corresponds to a condition of unstable equilibrium in the structure. A structure that is indeterminate to the degree D becomes stable and determinate when D plastic hinges have formed and the formation of one more hinge will produce a collapse mechanism. Thus, in a structure in which there are p possible hinge positions, the number of independent mechanisms is given by:

$$m_i = p - D$$

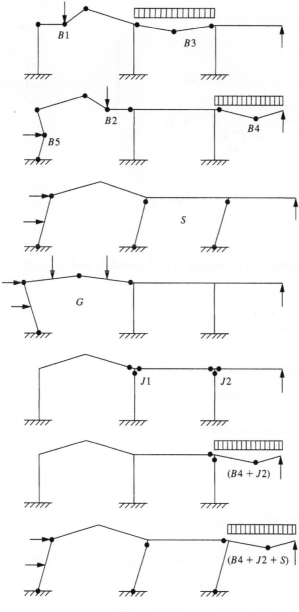

Figure 9.14

Example 9.4

The propped cantilever shown in Figure 9.15 is fabricated from two 12 in × 1 in flange plates and a thin web plate, which may be neglected in determining the plastic moment of resistance. The yield stress of the flange plates is 36 kips/in². Determine the collapse load.

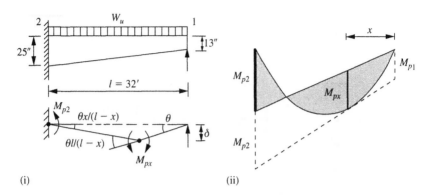

Figure 9.15

Solution

The plastic moments of resistance at end 1, end 2, and at a section a distance x ft from 1 are:

$$M_{p1} = 12 \times 12 f_y$$
$$= 144 f_y$$
$$M_{p2} = 24 \times 12 f_y$$
$$= 288 f_y$$
$$M_{px} = 12 f_y (12 + 3x/8)$$

Equating internal and external work for the mechanism shown at (i):

$$M_{p2} \theta x/(l - x) + M_{px} \theta l/(l - x) = W_u \delta/2 = W_u x \theta/2$$

and:

$$W_u = 288 f_y (32 + 3x)/x(32 - x)$$

The minimum value of W_u is obtained by equating $\partial W_u/\partial x$ to zero. Thus:

$$3x(32 - x) = (32 - 2x)(32 + 3x)$$
$$3x^2 + 64x - 1024 = 0$$
$$x = 10.67 \text{ ft}$$

Then:

$$W_u = 288 \times 36 \times 64/(10.67 \times 12 \times 21.33)$$
$$= 243 \text{ kips}$$

Example 9.5

The rigid frame shown in Figure 9.16 has columns with a plastic section modulus of 15.8 in³ and a beam with a plastic section modulus of 30.8 in³. The shape factor of both sections is 1.125, and the yield stress of the steel is 50 kips/in². Neglecting the effects of axial loads and instability, determine the ratio of W to H for all possible modes of plastic collapse of the frame. Determine the values of W and H when $W = 1.8\,H$.

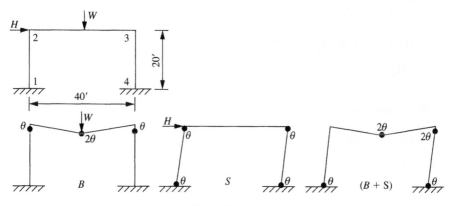

Figure 9.16

Solution

The plastic moment of resistance of the columns is:

$$M_p = 15.8 \times 1.125 \times 50/12$$
$$= 74.1 \text{ kip-ft}$$

The plastic moment of resistance of the beam is:

$$M_{pb} = 30.8M_p/15.8$$
$$= 1.95M_p$$

The number of independent collapse mechanisms is:

$$m_i = p - D$$
$$= 5 - 3$$
$$= 2$$

and these are shown in the figure together with the combined mechanism $(B + S)$. The hinge rotations and displacements of mechanism $(B + S)$ are obtained as the sum of the rotations and displacements of mechanism B and mechanism S. Thus, the hinge at 2 is eliminated since the rotation there is θ closing in the case of mechanism B and θ opening in the case of mechanism S Similarly, the hinge at 3 rotates 2θ closing, since the rotation there is θ closing in the case of mechanism B and θ closing in the case of mechanism S. In addition, the external work done is the sum of the external work done in the mechanism B and mechanism S cases since the displacements of the applied loads are the sum of the corresponding displacements in the mechanism B and mechanism S cases.

For mechanism B:

$$2M_p + 2 \times 1.95 M_p = 20W$$

$$W/M_p = 0.295$$

For mechanism S:

$$4M_p = 20H$$

$$H/M_p = 0.20$$

For mechanism $(B + S)$:

$$4M_p + 2 \times 1.95 M_p = 20W + 20H$$

$$W/M_p + H/M_p = 0.396$$

These three expressions may be plotted on the interaction diagram shown in Figure 9.17. The mode of collapse for a particular ratio of W to H is determined by plotting a line with a slope equal to this ratio. The mechanism expression that this line first intersects gives the collapse mode.

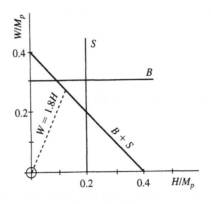

Figure 9.17

When $W = 1.8 H$, mechanism $(B + S)$ controls and:

$$W + H = 0.396 \times 74.1$$
$$= 2.8H$$

Thus:

$$H = 10.48 \text{ kips}$$

and:

$$W = 18.85 \text{ kips}$$

Example 9.6

The two-bay rigid frame shown in Figure 9.18 is fabricated from members of a uniform section having a shape factor of 1.15 and a yield stress of $50\,\text{kips/in}^2$. Neglecting the effects of axial loads and instability, determine the required plastic section modulus to provide a load factor against collapse of $N = 1.75$.

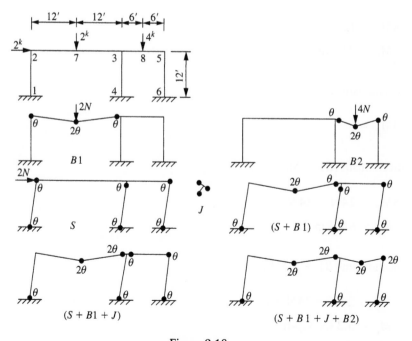

Figure 9.18

Solution

The number of independent collapse mechanisms is:

$$m_i = p - D$$
$$= 10 - 6$$
$$= 4$$

and these are shown in the figure together with a number of combined mechanisms.

For mechanism $B1$:

$$4M_p = 12 \times 2N$$
$$M_p = 6N \text{ kip-ft}$$

For mechanism $B2$:

$$4M_p = 6 \times 4N$$
$$M_p = 6N \text{ kip-ft}$$

For mechanism S:

$$6M_p = 12 \times 2N$$
$$M_p = 4N \text{ kip-ft}$$

For mechanism J:

$$3M_p = 0$$

For mechanism $(B1 + S)$:

$$8M_p = 24N + 24N$$
$$M_p = 6N \text{ kip-ft}$$

For mechanism $(B1 + S + J)$:

$$9M_p = 24N + 24N + 0$$
$$M_p = 5.33N \text{ kip-ft}$$

For mechanism $(B1 + B2 + S + J)$:

$$11M_p = 24N + 24N + 24N + 0$$
$$M_p = 6.55N \text{ kip-ft}$$

and this mechanism controls, since M_p is nowhere exceeded in the structure, as is shown by Example 9.11. The required plastic section modulus is:

$$S = 6.55 \times 1.75 \times 12/(50 \times 1.15)$$
$$= 2.4 \text{ in}^3$$

Example 9.7

The members of the rigid frame shown in Figure 9.19 have the relative plastic moments of resistance shown ringed, and the frame is to collapse under the loading shown. Assuming as a first approximation that plastic hinges occur either at the joints or at the mid-span of the members and neglecting the effects of axial loads and instability, determine the required plastic moments of resistance.

Solution

The number of independent collapse mechanisms is:

$$m_i = p - D$$
$$= 16 - 9$$
$$= 7$$

and these are shown in the figure together with a number of combined mechanisms.

For mechanism $B1$:

$$4M_p = 10 \times 10/2$$
$$M_p = 12.5 \text{ kip-ft}$$

For mechanism $B2$ and $B3$:

$$8M_p = 20 \times 10/2$$
$$M_p = 12.5 \text{ kip-ft}$$

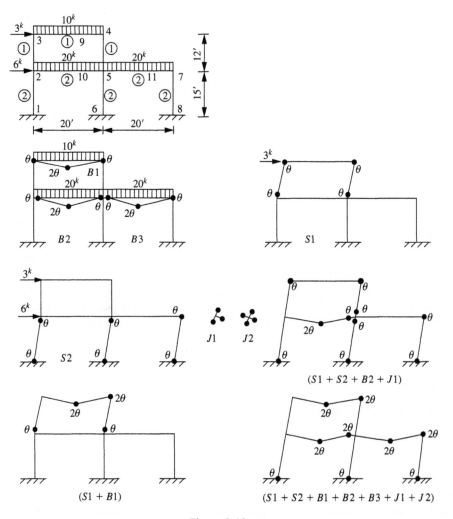

Figure 9.19

For mechanism $S1$:

$$4M_p = 3 \times 12$$
$$M_p = 9 \text{ kip-ft}$$

For mechanism $S2$:

$$12M_p = 9 \times 15$$
$$M_p = 11.25 \text{ kip-ft}$$

For mechanism $J1$:

$$5M_p = 0$$

For mechanism $J2$:

$$7M_p = 0$$

For mechanism $(S1 + S2 + B2 + J1)$:

$$19M_p = 36 + 135 + 100 + 0$$
$$M_p = 14.25 \text{ kip-ft}$$

For mechanism $(S1 + B1)$:

$$6M_p = 36 + 50$$
$$M_p = 14.33 \text{ kip-ft}$$

For mechanism $(S1 + S2 + B1 + B2 + B3 + J1 + J2)$:

$$26M_p = 36 + 135 + 50 + 100 + 100$$
$$M_p = 16.2 \text{ kip-ft}$$

and this mechanism controls as the plastic moments of resistance are nowhere exceeded, as shown by Example 9.12.

Example 9.8

The members of the Vierendeel girder shown in Figure 9.20 have their relative plastic moments of resistance shown ringed. The shape factor of the section is 1.15, the yield stress of the steel is $50\,\text{kips/in}^2$, and the required load factor against collapse is $N = 1.75$. Neglecting the effects of axial loads and instability, determine the required elastic section moduli.

Solution

The number of independent collapse mechanisms is:

$$m_i = p - D$$
$$= 16 - 9$$
$$= 7$$

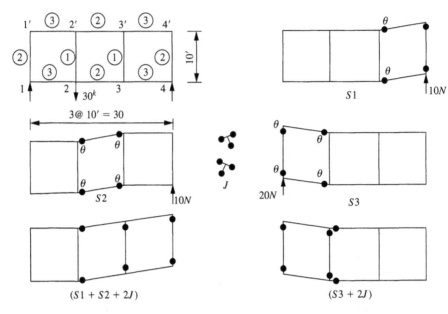

Figure 9.20

and these are shown in the figure together with a number of combined mechanisms.

For mechanism $S1$:

$$10M_p = 10 \times 10N$$
$$M_p = 10N \text{ kip-ft}$$

For mechanism $S2$:

$$8M_p = 10 \times 10N$$
$$M_p = 12.5N \text{ kip-ft}$$

For mechanism $S3$:

$$10M_p = 20 \times 10N$$
$$M_p = 20N \text{ kip-ft}$$

For mechanism $(S1 + S2 + 2J)$:

$$10M_p = 100N + 100N + 0$$
$$M_p = 20N \text{ kip-ft}$$

For mechanism $(S3 + 2J)$:

$$10M_p = 200N + 0$$
$$M_p = 20N \text{ kip-ft}$$

and this mechanism controls since the plastic moments of resistance are nowhere exceeded, as shown by Example 9.13.

The required elastic section modulus for the central post members is:

$$S = 20 \times 1.75 \times 12/(50 \times 1.15)$$
$$= 7.30 \text{ in}^3$$

Example 9.9

The ridged portal frame shown in Figure 9.21 is fabricated from members of a uniform section and is to collapse under the loading shown. Neglecting the effects of axial loads and instability, determine the plastic moment of resistance.

Solution

The independent and combined mechanisms are shown in the figure. In the case of the gable mechanism, it is necessary to construct the displacement diagram shown to determine the relative rotations of the members. A rotation θ is imposed on member 23, and 4 moves a horizontal distance 44'. The point 3 must move perpendicularly to the original directions of 23 and 34, and the point 3' is obtained. The rotation of member 45 is:

$$44'/l_{45} = 20\theta/20$$
$$= \theta$$

The rotation of member 34 is:

$$3'4'/l_{34} = \theta$$

Thus, the rotation of the hinge at 4 is 2θ and of the hinge at 3 is 2θ.

For mechanism B:

$$4M_p = 10 \times 2W$$
$$M_p = 5W$$

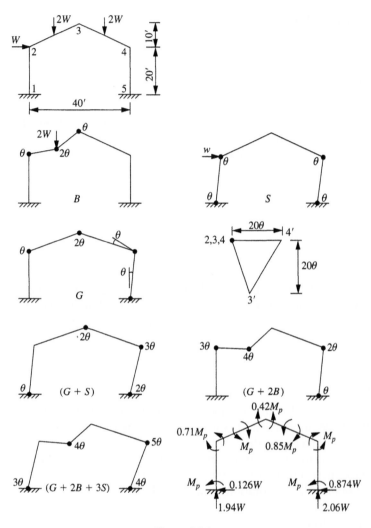

Figure 9.21

For mechanism S:

$$4M_p = 20 \times W$$
$$M_p = 5W$$

For mechanism G:

$$6M_p = 10 \times 2W + 10 \times 2W$$
$$M_p = 6.67W$$

For mechanism $(G + S)$:

$$8M_p = 40W + 20W$$
$$M_p = 7.5W$$

For mechanism $(G + 2B)$:

$$10M_p = 40W + 40W$$
$$M_p = 8W$$

For mechanism $(G + 2B + 3S)$:

$$16M_p = 40W + 40W + 60W$$
$$M_p = 8.75W$$

and this mechanism controls since this value of M_p is nowhere exceeded.

Example 9.10

The single-bay multi-story frame shown in Figure 9.22 consists of n storys each of height l. The plastic moment of resistance of each beam is M_p, and the columns are infinitely rigid. A load factor of $N_1 = 1.75$ is required against collapse due to vertical loads only, and a load factor of $N_2 = 1.4$ is required against collapse involving wind loads. Determine the value of n, in terms of W/H, for all the possible modes of collapse of the frame.

Solution

The beam, sway, and combined mechanisms are shown in the figure.
For mechanism B:

$$4M_p = N_1 Wl$$
$$M_p = N_1 Wl/4$$
$$= 7Wl/16$$

For mechanism S:

$$2nM_p = N_2 lH\{1 + 2 + \cdots + (n - 1) + n/2\}$$
$$= N_2 lH[(n - 1)\{1 + (n - 1)\}/2 + n/2]$$
$$= N_2 lHn^2/2$$
$$M_p = N_2 lHn/4$$
$$= 5.6lHn/16$$

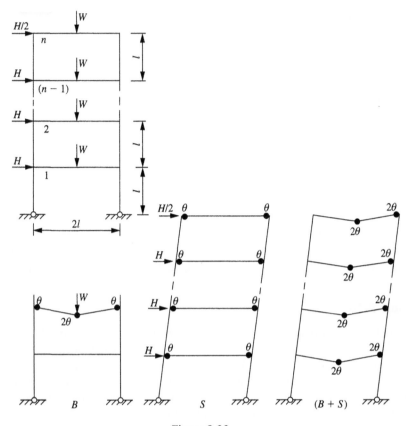

Figure 9.22

For mechanism $(B + S)$:

$$4nM_p = N_2Wln + N_2lHn^2/2$$
$$M_p = N_2l(2W + Hn)/8$$
$$= 2.8l(2W + Hn)/16$$

Thus, mechanism B controls when $n < W/2H$; mechanism S controls when $n > 2W/H$; and mechanism $(B + S)$ controls when $W/2H < n < 2W/H$.

9.6   Plastic moment distribution

The plastic moment distribution, or moment balancing, method[9,10] may be used to determine the bending moment diagram for an assumed collapse mechanism.

The assumed mechanism is the critical mechanism when the plastic moment of resistance is nowhere exceeded in the structure. The method may also be used as a direct method of designing a structure subjected to several loading combinations[9]. As the distribution proceeds simultaneously for all loading combinations, suitable plastic moments of resistance may be assigned to each member in the structure. In addition, the method is particularly suitable for the design of grids, in both the no-torsion[11] and torsion[12,13] cases. An alternative approach to the design of grids is the yield line analysis technique[14].

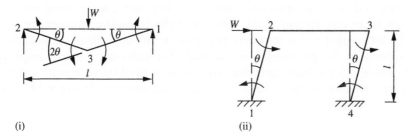

(i)   (ii)

Figure 9.23

The sign convention adopted for the bending moments in plane frames is that clockwise end moments acting from the support on the member are positive and moments within a member that produce tension on the bottom fibers of the member are positive. Knowledge is required of the equations of equilibrium for a structure, and these may be obtained from Figure 9.23. The equilibrium equation for the beam shown at (i) may be obtained by considering it to deform as the beam mechanism indicated. Equating internal and external work, the equilibrium equation is given by:

$$M_{12} + 2M_3 - M_{21} = Wl/2$$

The equilibrium equation for the frame shown at (ii) is obtained by equating internal and external work for the sway mechanism indicated and:

$$-M_{12} - M_{21} - M_{34} - M_{43} = Wl$$

The equilibrium equation for any joint in a structure is that the moments in the members meeting at the joint must sum to zero.

To determine the bending moment diagram for a particular collapse mechanism, the known plastic moments of resistance are first inserted at the hinge positions. The remaining moments are selected arbitrarily to satisfy

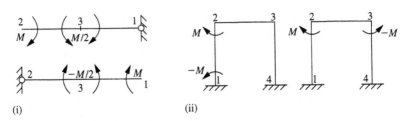

Figure 9.24

the beam and sway equations of equilibrium. In general, these initial moments will not be in equilibrium at the joints, and the balancing moments required may be distributed to the members at a joint in any convenient proportions— the stiffness of the members is immaterial. The carry-over factors employed must ensure that the beam and sway equations of equilibrium remain satisfied. The carry-over factors for a beam are derived as shown in Figure 9.24 (i) and are:

$$c'_{23} = \tfrac{1}{2}$$
$$c'_{21} = 0$$
$$c'_{13} = -\tfrac{1}{2}$$
$$c'_{12} = 0$$

The carry-over factor for a column is derived as shown at (ii) and is:

$$c'_{21} = -1$$

Alternatively, the carry-over may be made to adjacent columns in any convenient proportions, and the sum of the carry-over moments must equal the distribution moment with change of sign.

The sign convention adopted for the bending moments and torsions in grids is given in Section 7.13. A sway equation may be developed for each unsupported node, and the initial moments are selected to satisfy these equations and the hinge positions of the assumed collapse mechanism. The balancing moment required at any node may be distributed to the members at the node in any convenient way, and any moment or torque distributed to one end of a member is carried over to the other end with sign reversed.

Example 9.11

Determine the bending moments at collapse in the rigid frame shown in Figure 9.18.

Solution

In Example 9.6 it is shown that mechanism $(B + B2 + S + J)$ controls, and the required plastic moment of resistance is:

$$M_p = 6.55 \times 1.75 \times 12$$
$$= 138 \text{ kip-in}$$

The equilibrium equations are obtained by considering the independent mechanisms.

For mechanism $B1$:

$$-M_{23} + 2M_7 + M_{32} = 12 \times 2 \times 1.75 \times 12 = 504$$

For mechanism $B2$:

$$-M_{36} + 2M_8 + M_{53} = 6 \times 4 \times 1.75 \times 12 = 504$$

For mechanism S:

$$-M_{12} - M_{21} - M_{34} - M_{43} - M_{56} - M_{65} = 12 \times 2 \times 1.75 \times 12 = 504$$

The initial moments are:

$$
\begin{aligned}
M_7 &= M_{32} \\
&= M_8 \\
&= M_{53} \\
&= -M_{12} \\
&= -M_{43} \\
&= -M_{65} \\
&= 138 \\
-M_{23} &= 504 - 3 \times 138 \\
&= 90 \\
-M_{35} &= 504 - 3 \times 138 \\
&= 90 \\
-M_{21} &= -M_{34} \\
&= -M_{56} \\
&= (504 - 3 \times 138)/3 \\
&= 30
\end{aligned}
$$

These are inserted in Table 9.1, and the distribution proceeds and the final moments are obtained as shown.

Table 9.1 Plastic moment distribution in Example 9.11

Joint	1	4	6	2				3			5	
Moment	12	43	65	21	23	7	32	34	35	8	53	56
Initial moments	−138	−138	−138	−30	−90	138	138	−30	−90	138	138	−30
Distribution				120								
Sway equilibrium								−12				−108
Final moments	−138	−138	−138	90	−90	138	138	−42	−90	138	138	−138

Example 9.12

Determine the bending moments at collapse in the rigid frame shown in Figure 9.19.

Solution

In Example 9.7 it is shown that mechanism $(S1 + S2 + B1 + B2 + B3 + J1 + J2)$ controls, and the required plastic moments of resistance are 194 kip-in and 389 kip-in.

For mechanism $B1$:

$$-M_{34} + 2M_9 + M_{43} = 600$$

For mechanism $B2$:

$$-M_{25} + 2M_{10} + M_{52} = 1200$$

For mechanism $B3$:

$$-M_{57} + 2M_{11} + M_{75} = 1200$$

For mechanism $S1$:

$$-M_{23} - M_{32} - M_{45} - M_{54} = 432$$

For mechanism $S2$:

$$-M_{12} - M_{21} - M_{56} - M_{65} - M_{78} - M_{87} = 1620$$

The initial moments are:

$$M_9 = M_{43}$$
$$= -M_{45}$$
$$= 194$$
$$M_{10} = M_{52}$$
$$= M_{11}$$
$$= M_{75}$$
$$= -M_{78}$$
$$= -M_{12}$$
$$= -M_{65}$$
$$= -M_{87}$$
$$= 389$$
$$-M_{23} = -M_{54}$$
$$= -M_{32}$$
$$= (432 - 194)/3$$
$$= 79.33$$
$$-M_{21} = -M_{56}$$
$$= (1620 - 4 \times 389)/2$$
$$= 32$$
$$-M_{25} = -M_{57}$$
$$= 1200 - 3 \times 389$$
$$= 33$$
$$-M_{34} = 600 - 3 \times 194$$
$$= 18$$

These are inserted in Table 9.2, and the distribution proceeds and the final moments are obtained as shown.

Table 9.2 Plastic moment distribution in Example 9.12

Joint	1	6	8	3			4	
Moment	12	65	87	32	34	9	43	45
Initial moments	−389	−389	−389	−79	−18	194	194	−194
Distribution				97				
Sway equilibrium								
Final moments	−389	−389	−389	18	−18	194	194	−194

Joint		2				5			7		
Moment	21	23	25	10	52	56	54	57	11	75	78
Initial moments	−32	−79	−33	389	389	−32	−80	−33	389	389	−389
Distribution	144										
Sway equilibrium						−144	−97				
Final moments	112	−72	−33	389	389	−176	−177	−33	389	389	−389

Example 9.13

Determine the bending moments at collapse in the Vierendeel girder shown in Figure 9.20.

Solution

In Example 9.8 it is shown that mechanism $(S3 + 2J)$ controls, and the required plastic moments of resistance are:

For mechanism $S1$:

$$M_{34} + M_{43} = 1050$$

For mechanism $S2$:

$$M_{32} + M_{23} = 1050$$

For mechanism $S3$:

$$M_{21} + M_{12} = 2100$$

The initial moments are:

$$M_{12} = -M_{11'}$$
$$= -M_{23}$$
$$= -840$$
$$-M_{22'} = -M_{33'}$$
$$= -420$$
$$M_{21} = 840 - 2100$$
$$= -1260$$
$$M_{32} = 1050 - 840$$
$$= 210$$
$$M_{34} = M_{43}$$
$$= 1050/2$$
$$= 525$$

These are inserted in Table 9.3, and the distribution proceeds and the final moments are obtained as shown.

Table 9.3 Plastic moment distribution in Example 9.13

Joint	1		2				3		4	
Moment	11'	12	21	22'	23	32	33'	34	43	44'
Initial moments	840	−840	−1260	420	840	210	420	525	525	−525
Distribution							−840	−315		
Sway equilibrium							315			−315
Final moments	840	−840	−1260	420	840	210	−420	210	840	−840

Example 9.14

Determine the bending moments at collapse in the no-torsion grid shown in Figure 9.25. All the members of the grid are of uniform section, and collapse is to occur under the loading shown.

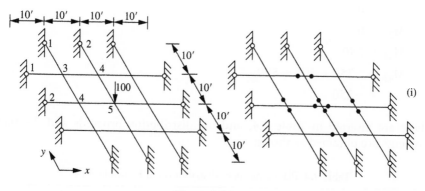

Figure 9.25

Solution

The assumed collapse mechanism is shown at (i), and, equating internal and external work:

$$4 \times 2M_p + 2 \times 4M_p = 100 \times 20 \times 2$$
$$M_p = 250 \text{ kip-ft}$$

The sway equations are obtained by considering a unit vertical deflection at nodes 3, 4 and 5.

For mechanism $S1$:

$$M_{34} + M_{43} - M_{31} = 0$$

For mechanism $S2$:

$$M_{45} + M_{54} - M_{42} - 2M_{34} - 2M_{43} = 0$$

For mechanism $S3$:

$$-M_{45} - M_{54} = 250$$

Due to the symmetry of the grid, distribution is required only in beams 14 and 25, and the initial moments are:

$$M_{43} = M_{54}$$
$$\quad = -250$$
$$M_{45} = 250 - 250$$
$$\quad = 0$$
$$M_{31} = 0$$
$$M_{34} = 250$$
$$M_{42} = -250 - 500 + 500$$
$$\quad = -250$$

These are inserted in Table 9.4, and the distribution proceeds and the final moments are obtained as shown.

Table 9.4 Plastic moment distribution in Example 9.14

Beam		14			25	
Joint		3	4	4		5
Moment	31	34	43	42	45	54
Initial moments	0	250	−250	−250	0	−250
Distribution	−125	−125				
Sway equilibrium				250		
Final moments, kip-ft	−125	125	−250	0	0	−250

9.7 Variable repeated loads

The collapse analysis of the preceding sections has been based on the assumption of proportional applied loads. That is, all the applied loads act simultaneously and are continuously increased in a constant ratio until collapse occurs. In practice, horizontal applied loads due to wind and seismic effects may reverse in direction and may be applied in a random manner.

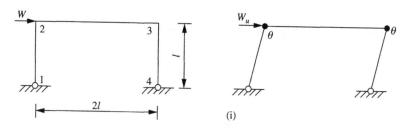

Figure 9.26

The uniform two-hinged portal frame shown in Figure 9.26 is subjected to a working load W applied horizontally at the top of the columns. If W is increased continuously to the value W_u the frame collapses as in the mechanism shown at (i). The required plastic moment of resistance is:

$$M_p = W_u l/2$$

and the load factor against collapse is $N = W_u/W$.

When the applied load may act in any direction, as shown in Figure 9.27, and the frame is subjected to the loading cycle shown at (i), (ii), (iii), and (iv), collapse will occur under a load W_a. At this value of the load, yielding has just been produced in the outer fibers of the frame at sections 2 and 3 at stage (ii) of the cycle. When the load is reversed, as at stage (iv), yielding is again just produced at sections 2 and 3 but in the opposite sense. Thus, sections 2 and 3 are subjected to alternating plasticity, and failure will occur, due to brittle fracture at these sections, after a sufficient number of loading cycles. The required yield moment is:

$$M_y = W_a l/2$$
$$= M_p/\lambda$$

and the load factor against failure due to alternating plasticity is:

$$N_a = W_a/W$$

For the frame to just collapse under an alternating load $W_a = W_u$ the required plastic moment of resistance is:

$$M_p = W_u l\lambda/2$$

which is greater than that required for proportional loading.

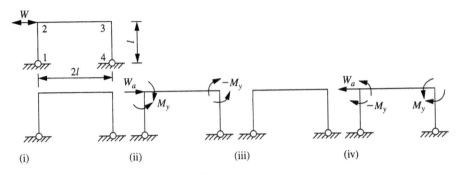

Figure 9.27

Under a general system of random loads, a plastic hinge may be produced at a section in the structure at one stage of the loading cycle, and yield in the opposite sense may be produced at the same section at another stage. The moment-curvature relationship for these two stages at this section is as shown by the solid lines in Figure 9.28.

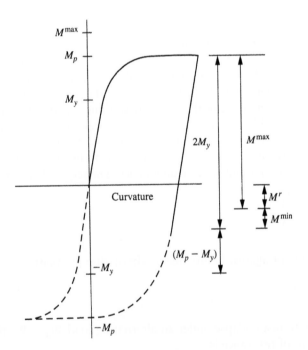

Figure 9.28

Thus, in general, the total range of bending moment that a section can sustain is $2M_y$. If the maximum theoretical elastic moment applied to the section exceeds M_p then, on unloading, an elastic change in bending moment will occur and a residual moment will remain at the section of:

$$M^r = M^{max} - M_p$$

where M^r is the residual moment and M^{max} is the maximum theoretical elastic moment for the worst loading combination. Hence:

$$M^{max} - M^{min} = 2M_y$$
$$= 2M_p/\lambda$$

where M^{min} is the minimum elastic moment for the worst loading combination. For an ideal elastic-plastic material and members with unit shape factor:

$$M^{max} - M^{min} = 2M_p$$

The uniform frame subjected to the proportional loading shown in Figure 9.29 will collapse as the mechanism shown at (i), and the required plastic moment of resistance is:

$$M_p = 0.75W_u l$$

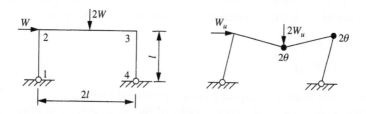

Figure 9.29

When the loads shown in Figure 9.30 are applied in a random manner and reach full magnitude independently, failure will occur under the loads W_a due to alternating plasticity. The magnitudes of the elastic bending moments due to each load applied independently are shown in Table 9.5, and the maximum range in moments at sections 2, 3, and 5 are obtained. The sign convention adopted is that moments producing tension on the inside of the frame are positive.

Thus, at sections 2 and 3:

$$11W_a l/8 = 2M_y$$
$$= 2M_p/\lambda$$
$$M_p = 11W_a l\lambda/16$$

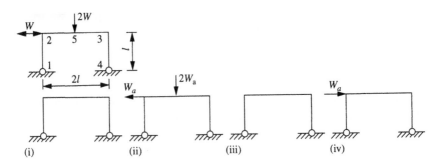

Figure 9.30

Table 9.5 Elastic moments for the frame shown in Figure 9.30

Loading	M_2	M_5	M_3
$2W_a$ vertical	$-3W_al/8$	$5W_al/8$	$-3W_al/8$
W_a horizontal	$W_al/2$	0	$-W_al/2$
$-W_a$ horizontal	$-W_al/2$	0	$W_al/2$
Maximum moment	$W_al/2$	$5W_al/8$	$W_al/2$
Minimum moment	$-7W_al/8$	0	$-7W_al/8$
Moment range	$11W_al/8$	$5W_al/8$	$11W_al/8$

The loading cycle causing failure at section 2 is shown at (i), (ii), (iii), and (iv). For the frame to just fail under the random loads W_u and $2W_u$ and with a shape factor of 1.15, the required plastic moment of resistance is:

$$M_p = 12.65W_u/16$$
$$= 0.79W_ul$$

which is greater than that required for proportional loading.

In practice, because of the dead weight of the structure, some proportion of the vertical load must be continuously applied, and this significantly increases the load factor against failure due to alternating plasticity[15,16,17,18]. Table 9.6 gives the magnitudes of the elastic bending moments when half of the vertical load remains permanently in position.

Thus, at sections 2 and 3:

$$19W_al/16 = 2M_y$$
$$= 2M_p/\lambda$$

The loading cycle causing failure at section 2 is shown in Figure 9.31. For the frame to just fail under loads of W_u and $2W_u$, and with a shape factor of 1.15, the required plastic moment of resistance is:

$$M_p = 0.68W_ul$$

Table 9.6 Elastic moments for the frame shown in Figure 9.30 with half of the vertical load permanently in position

Loading	M_2	M_5	M_3
W_a vertical	$-3W_al/16$	$5W_al/16$	$-3W_al/16$
W_a vertical $+$ W_a horizontal	$5W_al/16$	$5W_al/16$	$-11W_al/16$
W_a vertical $-$ W_a horizontal	$-11W_al/16$	$5W_al/16$	$5W_al/16$
Maximum moment	$5W_al/16$	$5W_al/8$	$5W_al/16$
Minimum moment	$-7W_al/8$	$5W_al/16$	$-7W_al/16$
Moment range	$19W_al/16$	$15W_al/16$	$19W_al/16$

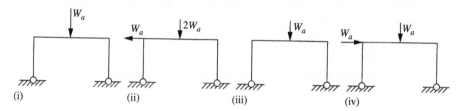

Figure 9.31

Thus, proportional loading is the more critical condition.

When the loads applied to a structure are random but nonreversible, collapse will occur under a load W_s. The application of some combination of the applied load to a magnitude W_s results in the formation of plastic hinges at some sections. However, insufficient hinges are formed to produce a collapse mechanism. The removal of this loading combination produces elastic changes in the bending moments and a residual bending moment pattern in the structure. The application of another loading combination to a magnitude W_s results in a moment pattern that, combined with the residual moment pattern, causes plastic hinges at some sections. Again, insufficient hinges are formed to produce a collapse mechanism. However, if the hinges produced by the two loading combinations are such that a collapse mechanism would be produced if they all occurred simultaneously, repetitions of this loading cycle will produce eventual collapse as regular increments of plastic yield occur during each cycle. Thus, the resulting deformed shape is the same as a collapse mechanism, and the correct mode of incremental collapse may be obtained by investigating each possible independent and combined mechanism. The application of the loading combinations to a magnitude less than W_s results after a few cycles in a residual stress pattern such that the applied loading causes purely elastic changes in the moments and the structure is said to have shaken down. The load factor against failure due to incremental collapse is:

$$N_s = W_s/W$$

and the criteria for shake-down to just occur is:

$$M^r + M^{max} = M_p$$

$$M^r + M^{min} = -M_p$$

The residual bending moment pattern for any assumed mode of collapse must satisfy the virtual work equilibrium equations with the external work set equal to zero, as no applied loads are acting. Thus, for the span 12 of the continuous beam shown in Figure 9.32 (i):

$$-M_1^r + 2M_3^r - M_2^r = 0$$

where moments producing tension in the bottom fibers are regarded as positive. For the frame shown in Figure 9.32 (ii):

$$-M_1^r + M_2^r - M_3^r + M_4^r = 0$$

where moments producing tension on the inside of the frame are regarded as positive.

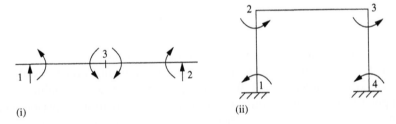

(i)   (ii)

Figure 9.32

Example 9.15

Determine the plastic moment of resistance required for the uniform frame shown in Figure 9.33 if incremental collapse is just to occur under the random loading shown.

Solution

The magnitudes of the elastic bending moments due to each load applied independently are shown in Table 9.7, and the maximum range in moments at sections 2, 3, and 5 are obtained.

For incremental failure in the beam mode shown at (i):

$$M_2^r = -M_p + 90$$

$$M_3^r = -M_p + 210$$

$$M_5^r = M_p - 150$$

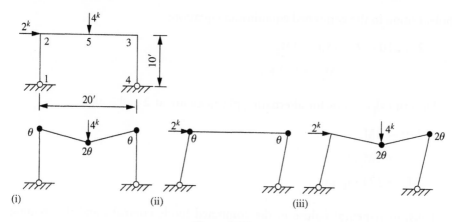

Figure 9.33

Table 9.7 Elastic moments in Example 9.15

Loading	M_2	M_5	M_3
4 kip vertical	−90	150	−90
2 kip horizontal	120	0	−120
Maximum moment	120	150	0
Minimum moment	−90	0	−210

Substituting in the beam equilibrium equation:

$$90 + 2 \times 150 + 210 = 4M_p$$
$$M_p = 150 \text{ kip-in}$$

For incremental failure in the sway mode shown at (ii):

$$M_2^r = M_p - 120$$
$$M_3^r = -M_p + 210$$

Substituting in the sway equilibrium equation:

$$120 + 210 = 2M_p$$
$$M_p = 165 \text{ kip-in}$$

For incremental failure in the combined mode shown at (iii):

$$M_3^r = -M_p + 210$$
$$M_5^r = M_p - 150$$

Substituting in the combined equilibrium equation:

$$2 \times 210 + 2 \times 150 = 4M_p$$
$$M_p = 180 \text{ kip-in}$$

The critical sections for alternating plasticity are at 2 and 3 and:

$$210 = 2M_y$$
$$= 2M_p/\lambda$$
$$M_p = 121 \text{ kip-in}$$

Thus, incremental failure in the combined mode controls, and the loading cycle causing failure is shown in Figure 9.34.

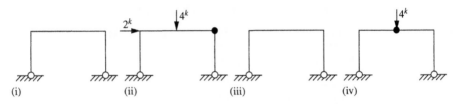

Figure 9.34

Example 9.16

Determine the plastic moment of resistance required for the uniform frame shown in Figure 9.35 if collapse is just to occur under the random loading shown. The members of the frame have a shape factor of unity.

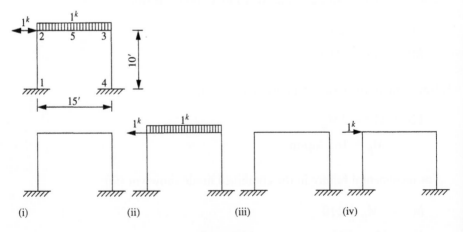

Figure 9.35

The magnitude of the elastic bending moments due to each load applied independently are shown in Table 9.8, and the maximum range in moments at sections 1, 2, 3, 4, and 5 are obtained.

Table 9.8 Elastic moments in Example 9.16

Loading	M_1	M_2	M_5	M_3	M_4
1 kip distributed	5.63	−11.25	11.25	−11.25	5.63
1 kip to right	−36.00	24.00	0.00	−24.00	36.00
1 kip to left	36.00	−24.00	0.00	24.00	−36.00
Maximum moment	41.63	24.00	11.25	24.00	41.63
Minimum moment	−36.00	−35.25	0.00	−35.25	−36.00
Moment range, kip-in	77.63	59.25	11.25	59.25	77.63

Collapse occurs under proportional loading in the sway mode, and the required plastic moment of resistance is:

$$M_p = 120/4$$
$$= 30 \text{ kip-in}$$

For incremental failure in the beam mode:

$$35.25 + 2 \times 11.25 + 35.25 = 4M_p$$
$$M_p = 23.25 \text{ kip-in}$$

For incremental failure in the sway mode:

$$36 + 24 + 35.25 + 41.63 = 4M_p$$
$$M_p = 34.22 \text{ kip-in}$$

For incremental failure in the combined mode:

$$36 + 2 \times 11.25 + 2 \times 35.25 + 41.63 = 6M_p$$
$$M_p = 28.46 \text{ kip-in}$$

The critical sections for alternating plasticity are at 2 and 3 and:

$$77.63 = 2M_y = 2M_p$$
$$M_p = 38.82$$

Thus, failure due to alternating plasticity controls, and the loading cycle causing failure is shown in Figure 9.35 at (i), (ii), (iii), and (iv).

9.8   Deflections at ultimate load

The deflections in a structure at incipient collapse may be readily determined. It is assumed that all sections of the structure have the ideal elastic-plastic moment-curvature relationship shown in Figure 9.1 (ii), that plastic yield is concentrated at the plastic hinge positions, and that the remaining portion of each member retains its original flexural rigidity. In addition, it is assumed that continuous rotation occurs at each hinge at a constant value of the plastic moment of resistance and that the loading is proportional. A conservative estimate of the deflections at working load may be obtained by dividing the deflection at incipient collapse by the load factor.

As the applied loading on a structure is progressively increased, plastic hinges are formed and discontinuities are produced in the structure due to the hinge rotations, ϕ. Eventually the last hinge that is required to produce a mechanism is formed. Immediately before the mechanism motion begins, the moment at the last hinge equals the plastic moment of resistance, and there is no discontinuity there as the hinge rotation equals zero. The required deflection must be calculated on the assumption that one particular hinge is the last to form. The calculation is then repeated in turn, assuming that the other hinges are the last to form and the largest value obtained is the correct one. The choice of an incorrect hinge as the last to form is equivalent to determining the deflection after a reversed mechanism motion such that the rotation that has occurred at this hinge before collapse is just eliminated. Thus, the deflection obtained using an incorrect assumption is necessarily smaller than the correct value. The deflection may be computed using slope-deflection[19,20],conjugate beam[21], virtual work[19,22], or moment distribution[23] methods. The virtual work method will be used here, as it leads to the quickest solution.

The collapse mechanism for the structure is first obtained, and the moments M in the structure at collapse are determined. Then the deflection at a particular point is given by:

$$\delta = \Sigma \int Mm \ dx/EI + \Sigma m\phi$$

where ϕ is the total rotation at a hinge during the application of the loads, m is the bending moment at any section due to a unit virtual load applied to the structure at the point in the direction of the required displacement, and the summations extend over all the hinges and all the members in the structure. It was shown in Section 3.2 that the unit virtual load may be applied to any cut-back structure that can support it. Thus, if a cut-back structure that gives a zero value for m at all plastic hinge positions except the last to form is selected, the term $\Sigma m\phi$ is zero and it is unnecessary to calculate the hinge rotations. A suitable form of the cut-back structure may be obtained by inserting frictionless hinges in the actual structure at all plastic hinge positions except the last to form. The term $\int Mm \ dx/EI$ may be determined by the method of volume integration, given in Section 2.5 if desired.

When failure occurs by partial collapse of the structure, plastic hinges additional to those predicted by the rigid-plastic collapse mechanism may be required for the elastic-plastic solution. The redundant moments in the frame at collapse are obtained from the equilibrium equations, and at those sections at which the plastic moment of resistance is exceeded, an additional plastic hinge is inserted. The remaining redundant moments are obtained by reapplying the equilibrium equations. The introduction of the additional plastic hinges does not affect collapse of the structure, as a mechanism is not produced until the last hinge predicted by the rigid-plastic solution forms. Virtual work equilibrium equations additional to those derived in Section 9.6 may be required. These are obtained by applying virtual internal forces, which are in equilibrium with zero external load, to a cut-back structure with frictionless hinges inserted at the plastic hinge positions. Thus, the external work and the term $\Sigma m\phi$ are zero and:

$$0 = \Sigma \int Mm\,dx/EI$$

where m is the bending moment at any section due to the applied internal forces.

Example 9.17

Determine the deflection at incipient collapse at the position of the final hinge in the propped cantilever shown in Figure 9.36.

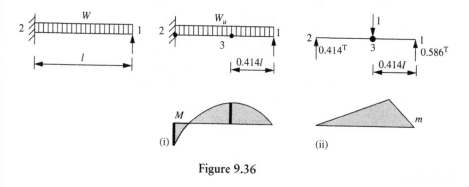

Figure 9.36

Solution

The collapse mechanism is shown at (i), and the ultimate load and reactions are:

$$W_u = 11.66M_p/l$$

$$V_1 = 4.83M_p/l$$

$$V_2 = 6.83M_p/l$$

The last hinge to form is a distance $0.414l$ from the prop, and the unit virtual load applied to the cut-back structure is shown at (ii). The deflection of 3 at incipient collapse is:

$$\delta = \int_1^3 Mm \ dx/EI + \int_2^3 Mm \ dx/EI$$

$$= \int_0^{0.414l} (4.83M_p x/l - 5.83M_p x^2/l^2)0.586x \ dx/EI$$

$$+ \int_0^{0.586l} (6.83M_p x/l - 5.83M_p x^2/l^2 - M_p)0.414x \ dx/EI$$

$$= 0.0412M_p l^2/EI + 0.0472M_p l^2/EI$$

$$= 0.0884M_p l^2/EI$$

Example 9.18

Determine the horizontal deflection at incipient collapse at joint 2 of the uniform frame shown in Figure 9.37.

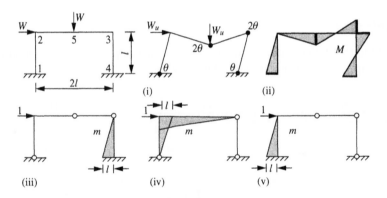

Figure 9.37

Solution

The collapse mechanism and final bending moment diagram are shown at (i) and (ii). Assuming the hinge at 4 forms last, the unit virtual load is applied to the cut-back structure shown at (iii). The deflection of 2 at incipient collapse is then:

$$\delta = -M_p \int_0^l (1 - 2x/l)x \ dx/EI$$

$$= M_p l^2/6EI$$

Assuming the hinge at 5 forms last, the unit virtual load is applied to the cut-back structure shown at (iv). The deflection of 2 at incipient collapse is then:

$$\delta = -M_p \int_0^l x(l-x)dx/lEI + M_p \int_0^l x(l-x/2)dx/lEI$$
$$\quad - M_p \int_0^l (1-2x/l)x \, dx/2EI$$
$$\quad = M_p l^2/4EI$$

Assuming the hinge at 1 forms last, the unit virtual load is applied to the cut-back structure shown at (v). The deflection of 2 at incipient collapse is then:

$$\delta = M_p \int_0^l x^2 \, dx/lEI$$
$$\quad = M_p l^2/3EI$$

and this value controls.

Example 9.19

Determine the horizontal deflection at incipient collapse at joint 2 of the uniform frame shown in Figure 9.38.

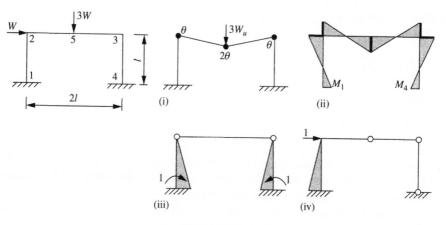

Figure 9.38

Solution

The collapse mechanism and final bending moment diagram are shown at (i) and (ii) and:

$$W_u l = 4M_p/3$$

From the sway equilibrium equation, using the convention that positive moments produce tension on the inside of the frame:

$$-M_1 - M_p + M_p + M_4 = W_u l = 4M_p/3$$
$$-M_1 + M_4 = 4M_p/3$$

An additional equilibrium equation is obtained by applying unit virtual moments to the column feet of the cut-back structure shown at (iii). Then:

$$0 = \int_0^l \{-M_p + (M_p + M_1)x/l\}x \, dx + \int_0^l \{-M_p + (M_p + M_4)x/l\}x \, dx$$
$$M_1 + M_4 = M_p$$

Solving these two equations for M_1 and M_4 gives:

$$M_4 = 7M_p/6$$

and:

$$M_1 = -M_p/6$$

The plastic moment of resistance is exceeded at 4, and hence a plastic hinge must be formed there in addition to the hinges predicted by the rigid-plastic collapse mechanism.
Then:

$$M_4 = M_p$$

and:

$$M_1 = -M_p/3$$

Thus, for the elastic-plastic solution, plastic hinges are required at 2, 3, 4, and 5, and the deflection at collapse is the maximum value obtained by assuming the rotations at 2, 3, and 5 are zero in turn.

Assuming the hinge at 2 forms last, the unit virtual load is applied to the cut-back structure shown at (iv). The deflection of 2 at incipient collapse is:

$$\delta = M_p \int_0^l (1 - 2x/3l)x \, dx/EI$$
$$= 5M_p l^2/18EI$$

and this value controls, as identical values are obtained by assuming the hinges at 3 and 5 form last.

Supplementary problems

S9.1 Figure S9.1 shows a continuous beam of total length l with a uniform plastic moment of resistance M_p supporting a distributed load w, including its own weight. Using a load factor of N, determine the ratio of l/a for plastic hinges to occur simultaneously in each span and at the supports. For this condition, calculate the maximum value of w that may be supported.

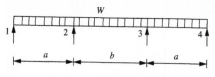

Figure S9.1

S9.2 A rectangular portal frame of uniform section is hinged at the base and subjected to the loads shown in Figure S9.2. The shape factor of the section is 1.15, the yield stress of the steel is $16\,\text{kips/in}^2$, and the required load factor is 1.75. Neglecting the effects of axial loads and instability, determine the required elastic section modulus of the members.

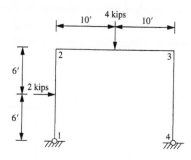

Figure S9.2

S9.3 The rigid frame shown in Figure S9.3 has a uniform plastic moment of resistance of M_p. Determine the ratio of W to H for the three possible modes of collapse of the frame and plot the relevant interaction diagram.

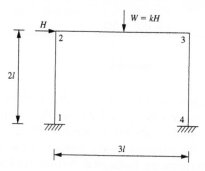

Figure S9.3

S9.4 The members of the rigid frame shown in Figure S9.4 have the relative plastic moments of resistance shown ringed, and the frame is to collapse under the loading shown. Neglecting the effects of axial loads and instability, determine the required plastic moments of resistance.

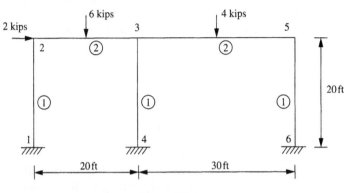

Figure S9.4

S9.5 The members of the Vierendeel girder shown in Figure S9.5 have the relative plastic moments of resistance shown ringed, and the frame is to collapse under the loading shown. Determine the required plastic moments of resistance.

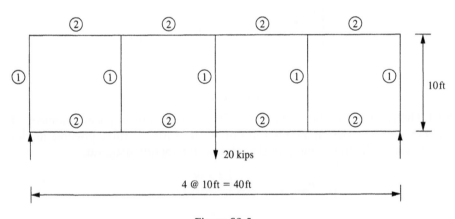

Figure S9.5

S9.6 The ridged portal frame shown in Figure S9.6 is fabricated from members of a uniform section and is to collapse in the combined mechanism (gable + sway) under the loading shown. Neglecting the effects of axial loads and instability, determine the required plastic moment of resistance. If the last hinge forms at joint 3, determine the horizontal deflection of joint 2 at collapse.

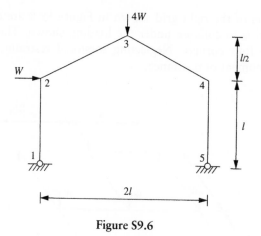

Figure S9.6

S9.7 The right bridge grid shown in Figure S9.7 consists of three simply supported main beams and five diaphragms. Determine the magnitude of the collapse load W for each of the following three situations.

(i) Main beams: plastic moment of resistance in flexure = M_p,
plastic moment of resistance in torsion = 0
Diaphragms: plastic moment of resistance in flexure = $0.25M_p$,
plastic moment of resistance in torsion = 0

(ii) Main beams: plastic moment of resistance in flexure = M_p,
plastic moment of resistance in torsion = $0.3M_p$
Diaphragms: plastic moment of resistance in flexure = $0.25M_p$,
plastic moment of resistance in torsion = 0

(iii) Main beams: plastic moment of resistance in flexure = M_p,
plastic moment of resistance in torsion = $0.15M_p$
Diaphragms: plastic moment of resistance in flexure = $0.25M_p$,
plastic moment of resistance in torsion = 0

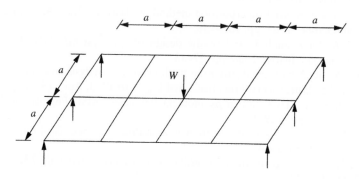

Figure S9.7

S9.8 The members of the right grid shown in Figure S9.8 are of uniform section, and the grid is to collapse under the loading shown. The grid is simply supported at the four corners. Neglecting torsional restraint, determine the required plastic moment of resistance.

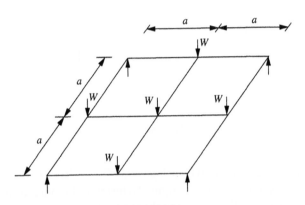

Figure S9.8

References

1. Beedle, L. S. Plastic design of steel frames. New York. John Wiley & Sons, Inc. 1958.
2. Hodge, P. G. Plastic analysis of structures. New York. McGrawHill Book Co. 1959.
3. Neal, B. G. The plastic methods of structural analysis. London. Chapman and Hall Ltd. 1959.
4. Baker, J. F., Horne, M. R., Heyman, J. The steel skeleton. Volume 2. Cambridge University Press. 1956.
5. Horne, M. R. and Morris, L. J. Plastic design of low rise frames. London. Granada Publishing. 1981.
6. Pincus, G. Method of combining mechanisms in plastic analysis. AISC Engineering Journal. July 1965. pp. 85–87.
7. Pincus, G. Generalized superposition method in plastic analysis. AISC Engineering Journal. October 1967. pp. 136–138.
8. Vukov, A. Limit analysis and plastic design of grid systems. AISC Engineering Journal. Second quarter 1986. pp. 77–83.
9. Horne, M. R. A moment distribution method for the analysis and design of structures by the plastic theory. Proc. Inst. Civil Eng. Part III. 3. April 1954. pp. 51–98.
10. Gaylord, E. H. Plastic design by moment balancing. AISC Engineering Journal. October 1967. pp. 129–135.
11. Shaw, F. S. Limit analysis of grid frameworks. Proc. Am. Soc. Civil Eng. 89 (ST5). October 1963. pp. 1–14.
12. Percy, J. H. A design method for grillages. Proc. Inst. Civil Eng. 23. November 1962. pp. 409–422.
13. Martin, J. B. Rigid plastic analysis of general structures. Proc. Am. Soc. Civil Eng. 89 (EM2). April 1963. pp. 107–132.

14. Cannon, J. P. Yield-line analysis and design of grids. AISC Engineering Journal. October 1969. pp. 124–129.
15. Popov, E. P. and McCarthy, R. E. Deflection stability of frames under repeated loads. Proc. Am. Soc. Civil Eng. 86 (EM1). January 1960. pp. 61–78.
16. Popov, E. P. and Becker, E. Shakedown analyses of some typical one storey bents. Proc. Am. Soc. Civil Eng. 90 (ST5). October 1964. pp. 1–9.
17. WRC ASCE Committee. Commentary on plastic design in steel: progress report no. 4. Proc. Am. Soc. Civil Eng. 85 (EM4). October 1959. pp. 157–185.
18. Xiaofeng, Y. Shakedown analysis in plastic design of steel structures. Proc. Am. Soc. Civil Eng. 115 (EM2). February 1989. pp. 304–314.
19. WRC ASCE Committee. Commentary on plastic design in steel: progress report no. 7. Proc. Am. Soc. Civil Eng. 86 (EM2). April 1960. pp. 141–160.
20. Perrone, N. and Soteriades, M. C. Calculating deflections of a partially collapsed frame. Proc. Am. Soc. Civil Eng. 91 (EM1). February 1965 pp. 1–5 and 91 (EM5). October 1965. pp. 218–223.
21. Lee, S. L. The conjugate frame method and its application in the elastic and plastic theories of structures. Journal of the Franklin Inst. 266. September 1958. pp. 207–222.
22. Heyman, J. On the estimation of deflections in elastic plastic framed structures. Proc. Inst. Civil Eng. 19. May 1961 pp. 39–60 and 23. October 1962. pp. 298–315.
23. Lavingia, K. P. Deformation analysis of structures near collapse load. AISC Engineering Journal. July 1968. pp. 128–134.

10 Matrix and computer methods

Notation

A	cross-sectional area of a member
c_{12}	carry-over factor for a member 12 from the end 1 to the end 2
D	degree of indeterminacy
E	Young's modulus
f_{ij}	element of $[F]$ = displacement produced at point i by a unit force replacing the redundant force at j
$[F]$	flexibility matrix of the cut-back structure
$[F_i]$	element of $[\bar{F}]$ = flexibility sub-matrix for member i
$[\bar{F}]$	diagonal matrix formed from the sub-matrices $[F_i]$
G	modulus of torsional rigidity
H	horizontal reaction
$\{H\}$	lack of fit vector
I	second moment of area of a member
J	torsional inertia
l	length of a member
M_{12}	moment produced at the end 1 of a member 12 by the joint displacements
M_{12}^F	moment produced at the end 1 of a member 12 by the external loads, all joints in the structure being clamped
P_i	element of $\{P\}$ = equals the total internal force at joint i
P_{12}	axial force produced at the end 1 of a member 12 by the joint displacements
P_{xi}	element of $\{P\}$ = total internal force, acting in the x-direction, produced at joint i by the joint displacements
P_{yi}	element of $\{P\}$ = total internal force, acting in the y-direction, produced at joint i by the joint displacements
$P_{\theta i}$	element of $\{P\}$ = total internal moment produced at joint i by the joint displacements
P_{12}^F	axial force produced at the end 1 of a member 12 by the external loads, all joints in the structure being clamped
P_{xi}^F	element of $\{P^F\}$ = equals the total internal force, acting in the x-direction, produced at joint i by the external loads, all joints in the structure being clamped
P_{yi}^F	element of $\{P^F\}$ = total internal force, acting in the y-direction, produced at joint i by the external loads, all joints in the structure being clamped
$P_{\theta i}^F$	element of $\{P^F\}$ = total internal moment produced at joint i by the external loads, all joints in the structure being clamped
P_i^R	element of $\{P^F\}$ = force produced in member i by the redundants applied to the cut-back structure

P_i^W element of $\{P^W\}$ equals the force produced in member i by the external loads acting on the cut-back structure

$\{P\}$ vector of total internal forces at the joints produced by the joint displacements

$\{P'\}$ vector of member forces, referred to the member axes, produced by the joint displacements

$\{P_{12}\}$ vector of member forces at the end 1 of member 12, referred to the x- and y-axes, produced by the joint displacements

$\{P^F\}$ vector of total fixed-end forces at the joints produced by the external loads

$\{P^{F'}\}$ vector of member forces, referred to the member axes, produced by the external loads, all joints in the structure being clamped

$\{P^R\}$ vector of support reactions, vector of forces produced in the cut-back structure by the redundants

$\{P^W\}$ vector of forces produced in the cut-back structure by the external loads

Q_{12} shear force produced at the end 1 of a member 12 by the joint displacements

Q_{12}^F shear force produced at the end 1 of a member 12 by the external loads, all joints in the structure being clamped

R_j redundant force acting at joint j

$\{R\}$ vector of redundant forces

s_{12} restrained stiffness at the end 1 of a member 12

S_{ij} element of $[S]$ = equals the force produced at point i by a unit displacement at point j, all other joints being clamped

$[S]$ stiffness matrix for the whole structure

$[S_{ij}']$ element of $[\bar{S}]$ = stiffness sub-matrix of member ij, referred to its own axis

$[\bar{S}]$ diagonal matrix formed from the sub-matrices $[S_{ij}']$

$[T]$ orthogonal transformation matrix

u_{ij} element of $[U]$ = force produced in member i by unit value of the redundant R_j acting on the cut-back structure

$[U]$ force matrix for unit value of the redundants

v_{ik} element of $[V]$ = force produced in member i by unit value of the external load W_k acting on the cut-back structure

V vertical reaction

$[V]$ force matrix for unit value of the external loads

W_k element of $\{W\}$ = external load applied at joint k

W_{xi} element of $\{W\}$ = external load applied at joint i in the x-direction

W_{yi} element of $\{W\}$ = external load applied at joint i in the y-direction

$W_{\theta i}$ element of $\{W\}$ = external moment applied at joint i

$\{W\}$ applied load vector

x horizontal displacement

y vertical displacement

α angle of inclination of member

δ_i element of $\{\Delta\}$, displacement at joint i

$\{\Delta\}$ displacement vector

$\{\Delta'\}$ displacement vector referred to member axes

$\{\Delta^I\}$ initial displacements at the releases due to lack of fit in the members

$\{\Delta^R\}$ vector of support displacements, vector of displacements at the releases produced by the redundants

$\{\Delta^W\}$ vector of displacements at the releases produced by the external loads

θ rotation

λ equals $\cos \alpha$

μ equals $\sin \alpha$

10.1   Introduction

Matrix algebra is a mathematical notation that simplifies the presentation and solution of simultaneous equations. It may be used to obtain a concise statement of a structural problem and to create a mathematical model of the structure. The solution of the problem by matrix structural analysis techniques[1,2,3,4] then proceeds in an entirely systematic manner. All types of structures, whether statically determinate or indeterminate, may be analyzed by matrix methods. In addition, matrix concepts and techniques, because of their systematic character, form the basis of the computer analysis and design of structures[5,6,7,8]. Highly indeterminate structures may be easily handled in this way and alternative loading conditions readily investigated. There are two general approaches to the matrix analysis of structures: the stiffness matrix method and the flexibility matrix method.

The stiffness method is also known as the displacement or equilibrium method. It obtains the solution of a structure by determining the displacements at its joints. The number of displacements involved equals the number of degrees of freedom of the structure. Thus, for a pin-jointed frame with j joints the solution of $2j$ equations is required, and for a rigid frame, allowing for axial effects, the solution of $3j$ equations is required. If axial effects in rigid frames are ignored, the number of equations involved reduces to j plus the number of degrees of sway freedom. Irrespective of the number involved, these equations may be formulated and solved automatically by computer. The stiffness matrix method is the customary method utilized in computer programs for the solution of building structures.

The flexibility method is also known as the force or compatibility method. It obtains the solution of a structure by determining the redundant forces. Thus, the number of equations involved is equal to the degree of indeterminacy of the structure. The redundants may be selected in an arbitrary manner, and their choice is not an automatic procedure. The primary consideration in the selection of the redundants is that the resulting equations are well conditioned.

10.2   Stiffness matrix method

(a) Introduction

The structure subjected to the applied loads shown in Figure 10.1 may be considered as the sum of system (i) and system (ii). In system (i) the external loads

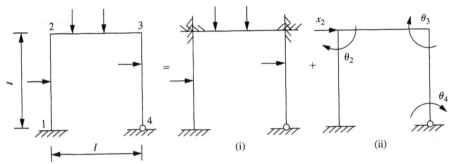

Figure 10.1

imposed between the joints are applied to the structure with all joints clamped. The total fixed-end forces $\{P^F\}$ at the joints are readily determined. In system (ii) the actual joint displacements $\{\Delta\}$ are imposed on the structure. These produce the total internal forces $\{P\}$ at the joints. The expressions force and displacement are used in their general sense and imply moment, shear, thrust, rotation, deflection, and axial deformation. Since the forces at any joint are in equilibrium, the applied loads at a joint are given by:

$$\{W\} = \{P\} + \{P^F\}$$

The total internal forces at a joint are given by the principle of superposition as:

$$P_1 = S_{11}\delta_1 + S_{12}\delta_2 + \cdots + S_{1n}\delta_n$$
$$P_2 = S_{21}\delta_1 + S_{22}\delta_2 + \cdots + S_{2n}\delta_n$$
$$\vdots \qquad \vdots \qquad \vdots \qquad \qquad \vdots$$
$$P_n = S_{n1}\delta_1 + S_{n2}\delta_2 + \cdots + S_{nn}\delta_n$$

where the stiffness coefficient S_{ij} is the force produced at point i by a unit displacement at point j, all other joints being clamped, and δ_j is the displacement at j in the actual structure. Thus:

$$\begin{bmatrix} P_1 \\ P_2 \\ \vdots \\ P_n \end{bmatrix} = \begin{bmatrix} S_{11} & S_{12} & \cdots & S_{1n} \\ S_{21} & S_{22} & \cdots & S_{2n} \\ \vdots & \vdots & \vdots & \vdots \\ S_{n1} & S_{n2} & \cdots & S_{nn} \end{bmatrix} \begin{bmatrix} \delta_1 \\ \delta_2 \\ \vdots \\ \delta_n \end{bmatrix}$$

or,

$$\{P\} = [S]\{\Delta\}$$

where $[S]$ is the complete stiffness matrix for the whole structure. This may be obtained from the stiffness sub-matrices of the individual members and is a symmetric matrix since $s_{ij} = s_{ji}$ by Maxwell's reciprocal theorem. Thus:

$$\{W\} = [S]\{\Delta\} + \{P^F\}$$

and:

$$\{\Delta\} = [S]^{-1}\{\{W\} - \{P^F\}\}$$

The internal forces in any member may now be obtained by back substitution in the stiffness sub-matrix of the member. The stiffness matrix depends solely on the geometrical properties of the members of the structure. Thus, once the stiffness matrix has been inverted, the displacements and internal forces due to alternative loading conditions may be quickly investigated.

The sign convention and notation used for rectangular frames is shown in Figure 10.2. The positive sense of the applied loads, joint displacements, fixed-end

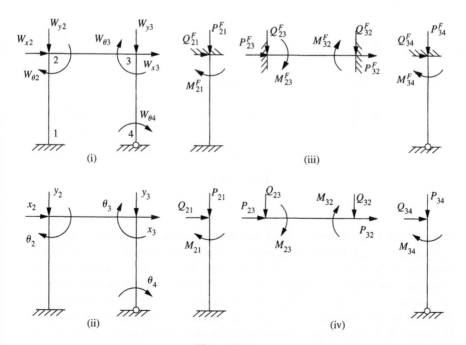

Figure 10.2

forces, and internal forces due to the joint displacements are shown at (i), (ii), (iii), and (iv). Thus, the total fixed-end forces at joint 2 are:

$$P_{x2}^F = Q_{21}^F + P_{23}^F$$
$$P_{y2}^F = P_{21}^F + Q_{23}^F$$
$$P_{\theta 2}^F = M_{21}^F + M_{23}^F$$

The total internal forces at joint 2 are:

$$P_{x2} = Q_{21} + P_{23}$$
$$P_{y2} = P_{21} + Q_{23}$$
$$P_{\theta 2} = M_{21} + M_{23}$$

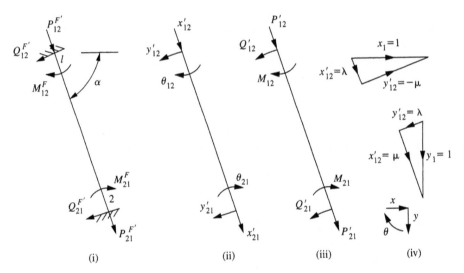

Figure 10.3

For structures containing inclined members, additional notation is required for fixed-end forces, joint displacements, and internal forces due to joint displacements referred to the member axes. In Figure 10.3 (i), the positive sense of the fixed-end forces $\{P^F\}$ acting on an inclined member 12 is shown, referred to the longitudinal axis of the member. In Figure 10.3 (ii), the positive sense of the joint displacements, $\{\Delta'\}$ is shown, referred to the inclined member axis. In Figure 10.3 (iii), the positive sense of the internal forces $\{P'\}$ due to the joint displacements is shown, referred to the inclined member axis.

(b) Rigid frames and grids

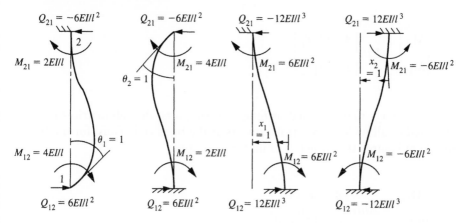

Figure 10.4

The stiffness matrix for rectangular frames in which axial effects are neglected may be readily assembled manually[9]. The elements of the stiffness sub-matrix for a vertical prismatic member may be obtained by applying unit displacements in turn to the ends of the member, as shown in Figure 10.4. Thus:

$$
\begin{bmatrix} M_{12} \\ M_{21} \\ lQ_{12} \\ lQ_{21} \end{bmatrix} = EI/l \begin{bmatrix} 4 & 2 & 6 & -6 \\ 2 & 4 & 6 & -6 \\ 6 & 6 & 12 & -12 \\ -6 & -6 & -12 & 12 \end{bmatrix} \begin{bmatrix} \theta_1 \\ \theta_2 \\ x_1/l \\ x_2/l \end{bmatrix}
$$

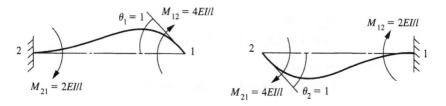

Figure 10.5

The elements of the stiffness sub-matrix for a horizontal prismatic member may be obtained by applying unit rotations to the ends, as shown in Figure 10.5.

Thus:

$$\begin{bmatrix} M_{12} \\ M_{21} \end{bmatrix} = EI/l \begin{bmatrix} 4 & 2 \\ 2 & 4 \end{bmatrix} \begin{bmatrix} \theta_1 \\ \theta_2 \end{bmatrix}$$

For members that are fixed or hinged to supports, the sub-matrices may be modified to allow for zero displacements at the fixed end. Thus, for the frame shown in Figure 10.1, which is of uniform section:

$$\begin{bmatrix} M_{21} \\ lQ_{21} \end{bmatrix} = EI/l \begin{bmatrix} 4 & -6 \\ -6 & 12 \end{bmatrix} \begin{bmatrix} \theta_2 \\ x_2/l \end{bmatrix}$$

$$\begin{bmatrix} M_{23} \\ M_{32} \end{bmatrix} = EI/l \begin{bmatrix} 4 & 2 \\ 2 & 4 \end{bmatrix} \begin{bmatrix} \theta_2 \\ \theta_3 \end{bmatrix}$$

$$\begin{bmatrix} M_{34} \\ M_{43} \\ lQ_{34} \end{bmatrix} = EI/l \begin{bmatrix} 4 & 2 & -6 \\ 2 & 4 & -6 \\ -6 & -6 & 12 \end{bmatrix} \begin{bmatrix} \theta_3 \\ \theta_4 \\ x_2/l \end{bmatrix}$$

The total internal forces at the joints are obtained by selecting the relevant elements from the sub-matrices.

Thus:

$$P_{\theta2} = M_{21} + M_{23}$$
$$= EI/l\{(4 + 4)\theta_2 + 2\theta_3 + 0 - 6x_2/l\}$$

$$P_{\theta3} = M_{32} + M_{34}$$
$$= EI/l\{(2\theta_2 + (4 + 4)\theta_3 + 2\theta_4 - 6x_2/l\}$$

$$P_{\theta4} = M_{43}$$
$$= EI/l\{0 + 2\theta_3 + 4\theta_4 - 6x_2/l\}$$

$$lP_{x2} + lP_{x3} = lQ_{21} + lQ_{34}$$
$$= EI/l\{-6\theta_2 - 6\theta_3 - 6\theta_4 + (12 + 12)x_2/l\}$$

or:

$$
\begin{bmatrix}
P_{\theta 2} \\
P_{\theta 3} \\
P_{\theta 4} \\
lP_{x2} + lP_{x3}
\end{bmatrix}
=
\begin{bmatrix}
W_{\theta 2} - P_{\theta 2}^F \\
W_{\theta 3} - P_{\theta 3}^F \\
W_{\theta 4} - P_{\theta 4}^F \\
l(W_{x2} + W_{x3} - P_{x2}^F - P_{x3}^F)
\end{bmatrix}
$$

$$
= EI/l
\begin{bmatrix}
8 & 2 & 0 & -6 \\
2 & 8 & 2 & -6 \\
0 & 2 & 4 & -6 \\
-6 & -6 & -6 & 24
\end{bmatrix}
\begin{bmatrix}
\theta_2 \\
\theta_3 \\
\theta_4 \\
x_2/l
\end{bmatrix}
$$

that is:

$$
\{P\} = \{W\} - \{P^F\} = [S]\{\Delta\}
$$

The force vector $\{\{W\} - \{P^F\}\}$ is readily obtained, and inversion of the stiffness matrix gives the values of θ_2, θ_3, θ_4, x_2. The member forces may then be obtained from the sub-matrices. Thus, the actual moment at end 2 of member 21 is:

$$
M_{21}^F + M_{21} = M_{21}^F + 4EI\theta_2/l - 6EIx_2/l^2
$$

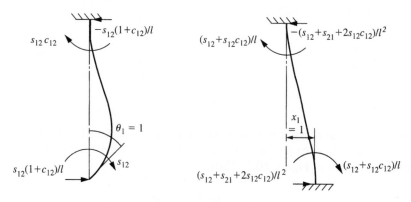

Figure 10.6

Similarly, frames with non-prismatic members and curved members may be analyzed if the stiffness, carry-over factors, and fixed-end forces are known for each member. The elements of the stiffness sub-matrix for a vertical non-prismatic member may be obtained by applying unit displacements in turn to the ends of the member, as shown in Figure 10.6. Since $s_{21}c_{21} = s_{12}c_{12}$, the stiffness sub-matrix is determined as:

$$
\begin{bmatrix} M_{12} \\ M_{21} \\ lQ_{12} \\ lQ_{21} \end{bmatrix}
$$

$$
= EI/l \begin{bmatrix} s_{12} & s_{12}c_{12} & s_{12} + s_{12}c_{12} & -(s_{12} + s_{12}c_{12}) \\ s_{12}c_{12} & s_{21} & s_{21} + s_{12}c_{12} & -(s_{21} + s_{12}c_{12}) \\ s_{12} + s_{12}c_{12} & s_{21} + s_{12}c_{12} & s_{12} + s_{21} + 2s_{12}c_{12} & -(s_{12} + s_{21} + 2s_{12}c_{12}) \\ -(s_{12} + s_{12}c_{12}) & -(s_{21} + s_{12}c_{12}) & -(s_{12} + s_{21} + 2s_{12}c_{12}) & s_{12} + s_{21} + 2s_{12}c_{12} \end{bmatrix}
$$

$$
\times \begin{bmatrix} \theta_1 \\ \theta_2 \\ x_1/l \\ x_2/l \end{bmatrix}
$$

The values of s_{12}, s_{21} and c_{12} may be obtained by the methods of Section 6.7(b).

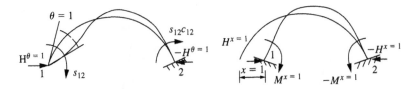

Figure 10.7

The elements of the stiffness sub-matrix for a symmetrical curved member may be obtained by applying unit displacements in turn to the ends of the member, as shown in Figure 10.7. Thus:

$$
\begin{bmatrix} M_{12} \\ M_{21} \\ P_{x1} \\ P_{x2} \end{bmatrix} = EI/l \begin{bmatrix} s_{12} & s_{12}c_{12} & M^{x=1} & -M^{x=1} \\ s_{12}c_{12} & s_{12} & -M^{x=1} & M^{x=1} \\ H^{\theta=1} & -H^{\theta=1} & H^{x=1} & -H^{x=1} \\ -H^{\theta=1} & H^{\theta=1} & -H^{x=1} & H^{x=1} \end{bmatrix} \begin{bmatrix} \theta_1 \\ \theta_2 \\ x_1 \\ x_2 \end{bmatrix}
$$

The values of s_{12}, c_{12}, $H^{\theta=1}$, and $H^{x=1}$ may be determined by the methods of Section 6.7(c). For a curved member, the carry-over factor is negative, and thus the term $s_{12}c_{12}$ is negative.

For hand computational purposes, it is desirable to reduce the order of the matrix that requires inversion. When the force vector $\{W - P^F\}$ contains zero elements, this may be achieved by partitioning the matrices in the form:

$$\left[\begin{array}{c} \{W\} - \{P^F\} \\ \hline 0 \end{array}\right] = \left[\begin{array}{c|c} [S_1] & [S_2] \\ \hline [S_3] & [S_4] \end{array}\right]\left[\begin{array}{c} \{\Delta_1\} \\ \hline \{\Delta_2\} \end{array}\right]$$

where $\{\Delta_1\}$ are the displacements at the joints subjected to an applied force and $\{\Delta_2\}$ are the displacements at the remaining joints. Thus:

$$\{\Delta_2\} = [S_4]^{-1}[S_3]\{\Delta_1\}$$

and:

$$\{\{W\} - \{P^F\}\} = [[S_1] - [S_2][S_4]^{-1}[S_3]]\{\Delta_1\}$$

Hence the displacements $\{\Delta_1\}$ and $\{\Delta_2\}$ may be obtained.

Support reactions may be determined by considering the equilibrium of the members at the supports after the internal forces are obtained. Alternatively, the internal forces at the supports may be included in the force matrix and the matrices partitioned in the form:

$$\left[\begin{array}{c} \{P\} \\ \hline \{P^R\} \end{array}\right] = \left[\begin{array}{c|c} [S_1] & [S_2] \\ \hline [S_3] & [S_4] \end{array}\right]\left[\begin{array}{c} \{\Delta\} \\ \hline 0 \end{array}\right]$$

where $\{\Delta\}$ and $\{P\}$ are the displacements and total internal forces at the joints and $\{P^R\}$ are the internal forces at the supports with zero corresponding displacements.

Thus:

$$\{\Delta\} = [S_1]^{-1}\{P\}$$
$$= [S]^{-1}\{\{W\} - \{P^F\}\}$$
$$\{P^R\} = [S_3][S_1]^{-1}\{P\}$$

and the reactions are given by:

$$\{P^R\} + \{P^F\}$$

For a structure that undergoes settlement at the supports, the internal forces at the supports that yield must be included in the force matrix and the matrices partitioned in the form:

$$\left[\begin{array}{c} \{P\} \\ \hline \{P^R\} \end{array}\right] = \left[\begin{array}{c|c} [S_1] & [S_2] \\ \hline [S_3] & [S_4] \end{array}\right]\left[\begin{array}{c} \{\Delta\} \\ \hline \{\Delta^R\} \end{array}\right]$$

where $\{P^R\}$ are the internal forces at the supports that yield and $\{\Delta^R\}$ are the corresponding known displacements.

Thus:

$$\{\Delta\} = [S_1]^{-1}\{\{P\} - [S_2]\{\Delta^R\}\}$$

and:

$$\{P^R\} = [S_3]\{\Delta\} + [S_4]\{\Delta^R\}$$

For a structure resting on flexible supports, the stiffness of the supports may be incorporated in the complete stiffness matrix as illustrated in Example 10.4.

In the case of symmetrical and skew symmetrical conditions, only half the structure may be considered and modified stiffness factors adopted for members that cross the axis of symmetry. Applied loads and the cross-sectional properties of members that lie along the axis of symmetry must be halved in value.

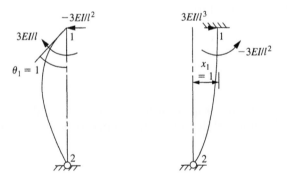

Figure 10.8

The stiffness sub-matrix for a straight prismatic member that is hinged at one end may be modified to allow for the hinge. The modified matrix is obtained from Figure 10.8 as:

$$\left[\begin{array}{c} M_{12} \\ lQ_{12} \end{array}\right] = EI/l\left[\begin{array}{cc} 3 & -3 \\ -3 & 3 \end{array}\right]\left[\begin{array}{c} \theta_1 \\ x_1/l \end{array}\right]$$

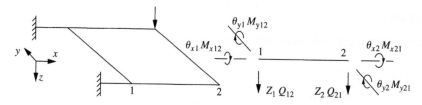

Figure 10.9

The stiffness matrix for rectangular grids may be assembled manually from the sub-matrices for each member[10,11]. The sub-matrix for a member parallel to the x-axis of the grid, shown in Figure 10.9, is obtained by applying unit displacements, in the positive directions shown, in turn to the ends of the member. The torsional stiffness of the member about the x-axis is GJ/l where G is the modulus of torsional rigidity, J is the torsional inertia, and l is the length of the member. The sub-matrix is given by:

$$
\begin{bmatrix} M_{x12} \\ M_{x21} \\ M_{y12} \\ M_{y21} \\ lQ_{12} \\ lQ_{21} \end{bmatrix} = 1/l \begin{bmatrix} GJ & -GJ & 0 & 0 & 0 & 0 \\ -GJ & GJ & 0 & 0 & 0 & 0 \\ 0 & 0 & 4EI & 2EI & 6EI & -6EI \\ 0 & 0 & 2EI & 4EI & 6EI & -6EI \\ 0 & 0 & 6EI & 6EI & 12EI & -12EI \\ 0 & 0 & -6EI & -6EI & -12EI & 12EI \end{bmatrix} \begin{bmatrix} \theta_{x1} \\ \theta_{x2} \\ \theta_{y1} \\ \theta_{y2} \\ z_1/l \\ z_2/l \end{bmatrix}
$$

Example 10.1

Determine the bending moments at joint 5 in the frame shown in Figure 10.10 for values of $W = 8$ kips and $l = 10$ ft. The relevant second moment of area values are shown ringed.

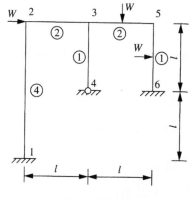

Figure 10.10

Solution

The member fixed-end forces are:

$$
\begin{aligned}
M_{35}^F &= M_{53}^F \\
&= -10 \text{ kip-ft} \\
M_{56}^F &= 10 \text{ kip-ft} \\
Q_{56}^F &= -4 \text{ kip}
\end{aligned}
$$

The total fixed-end forces at the joints are:

$$
\begin{aligned}
P_{\theta 2}^F &= P_{\theta 4}^F \\
&= 0 \\
P_{\theta 3}^F &= -10 \text{ kip-ft} \\
P_{\theta 5}^F &= 10 + 10 \\
&= 20 \text{ kip-ft} \\
P_{x5}^F &= -4 \text{ kips} \\
P_{x3}^F &= P_{x2}^F \\
&= 0
\end{aligned}
$$

The applied loads at the joints are:

$$
\begin{aligned}
W_{\theta 2} &= W_{\theta 3} = W_{\theta 4} = W_{\theta 5} \\
&= 0 \\
W_{x2} &= 8 \text{ kips} \\
W_{x3} &= W_{x4} \\
&= 0
\end{aligned}
$$

The sub-matrices for members 12, 23, 34, 35, and 56 are:

$$
\begin{bmatrix} M_{21} \\ lQ_{21} \end{bmatrix} = EI/l \begin{bmatrix} 4 \times 4/2 & -6 \times 4/4 \\ -6 \times 4/4 & 12 \times 4/8 \end{bmatrix} \begin{bmatrix} \theta_2 \\ x_2/l \end{bmatrix}
$$

$$
\begin{bmatrix} M_{23} \\ M_{32} \end{bmatrix} = EI/l \begin{bmatrix} 4 \times 2 & 2 \times 2 \\ 2 \times 2 & 4 \times 2 \end{bmatrix} \begin{bmatrix} \theta_2 \\ \theta_3 \end{bmatrix}
$$

$$
\begin{bmatrix} M_{34} \\ M_{43} \\ lQ_{34} \end{bmatrix} = EI/l \begin{bmatrix} 4 & 2 & -6 \\ 2 & 4 & -6 \\ -6 & -6 & 12 \end{bmatrix} \begin{bmatrix} \theta_3 \\ \theta_4 \\ x_2/l \end{bmatrix}
$$

$$
\begin{bmatrix} M_{35} \\ M_{53} \end{bmatrix} = EI/l \begin{bmatrix} 4 \times 2 & 2 \times 2 \\ 2 \times 2 & 4 \times 2 \end{bmatrix} \begin{bmatrix} \theta_3 \\ \theta_5 \end{bmatrix}
$$

$$
\begin{bmatrix} M_{56} \\ lQ_{56} \end{bmatrix} = EI/l \begin{bmatrix} 4 & -6 \\ -6 & 12 \end{bmatrix} \begin{bmatrix} \theta_5 \\ x_2/l \end{bmatrix}
$$

Collecting the relevant elements from the stiffness sub-matrices gives:

$$
\begin{bmatrix} M_{21} + M_{23} \\ M_{32} + M_{34} + M_{35} \\ M_{43} \\ M_{53} + M_{56} \\ l(Q_{21} + Q_{34} + Q_{56}) \end{bmatrix}
$$

$$
= \begin{bmatrix} P_{\theta 2} \\ P_{\theta 3} \\ P_{\theta 4} \\ P_{\theta 5} \\ l(P_{x2} + P_{x3} + P_{x5}) \end{bmatrix} = \begin{bmatrix} W_{\theta 2} - P_{\theta 2}^F \\ W_{\theta 3} - P_{\theta 3}^F \\ W_{\theta 4} - P_{\theta 4}^F \\ W_{\theta 5} - P_{\theta 5}^F \\ l(W_{x2} + W_{x3} + W_{x5} - P_{x2}^F - P_{x3}^F - P_{x5}^F) \end{bmatrix}
$$

$$
= \begin{bmatrix} 0 \\ 10 \\ 0 \\ -20 \\ 10(8+4) \end{bmatrix} = EI/l \begin{bmatrix} 16 & 4 & 0 & 0 & -6 \\ 4 & 20 & 2 & 4 & -6 \\ 0 & 2 & 4 & 0 & -6 \\ 0 & 4 & 0 & 12 & -6 \\ -6 & -6 & -6 & -6 & 30 \end{bmatrix} \begin{bmatrix} \theta_2 \\ \theta_3 \\ \theta_4 \\ \theta_5 \\ x_2/l \end{bmatrix}
$$

Inverting the stiffness matrix gives:

$$
\begin{bmatrix} \theta_2 \\ \theta_3 \\ \theta_4 \\ \theta_5 \\ x_2/l \end{bmatrix} = l/10EI \begin{bmatrix} 0.755 & & & & \\ -0.157 & 0.601 & & \text{symmetric} & \\ 0.433 & -0.329 & 4.109 & & \\ 0.174 & -0.210 & 0.591 & 1.065 & \\ 0.243 & -0.019 & 0.963 & 0.324 & 0.636 \end{bmatrix} \begin{bmatrix} 0 \\ 10 \\ 0 \\ -20 \\ 120 \end{bmatrix}
$$

The rotations at joints 3 and 5 are:

$$
\theta_3 = l(0 + 0.601 + 0 + 0.420 - 0.228)/EI
$$
$$
= 0.793l/EI
$$
$$
\theta_5 = l/(0 - 0.201 + 0 - 2.130 - 3.888)/EI
$$
$$
= 1.548l/EI
$$

The bending moment at joint 5 is:

$$
M^F_{53} + M_{53} = 10 + EI(4\theta_3 + 8\theta_5)/l
$$
$$
= 25.556 \text{ kip-ft}
$$

Example 10.2

Determine the bending moments in the frame shown in Figure 7.52. The second moments of area of the members are shown ringed.

Solution

The stiffness and carry-over factors for the non-prismatic columns were given in Example 7.22 as:

$$
s_{12} = 14EI/l
$$
$$
s_{21} = 5EI/l
$$
$$
c_{21} = 4/5
$$

Thus,

$$
(s_{21} + s_{12}c_{12})/l = 9EI/l^2
$$
$$
(s_{12} + s_{21} + 2s_{12}c_{12})/l = 27EI/l^3
$$

Due to the skew symmetry, the modified stiffness of member 22′ is:

$$s_{22'} = 6E(2I)/l$$
$$= 12EI/l$$

and:

$$W_{x2} = 10/2$$
$$= 5 \text{ kips}$$

The sub-matrices for members 12 and 22′ are:

$$\begin{bmatrix} M_{21} \\ lQ_{21} \end{bmatrix} = EI/l \begin{bmatrix} 5 & -9 \\ -9 & 27 \end{bmatrix} \begin{bmatrix} \theta_2 \\ x_2/l \end{bmatrix}$$

$$[M_{22'}] = EI/l[12][\theta_2]$$

Collecting the relevant elements from the stiffness sub-matrices gives:

$$\begin{bmatrix} P_{\theta 2} \\ lP_{x2} \end{bmatrix} = \begin{bmatrix} 0 \\ 60 \end{bmatrix} = EI/l \begin{bmatrix} 17 & -9 \\ -9 & 27 \end{bmatrix} \begin{bmatrix} \theta_2 \\ x_2/l \end{bmatrix}$$

Inverting the stiffness matrix gives:

$$\begin{bmatrix} \theta_2 \\ x_2/l \end{bmatrix} = l/EI \begin{bmatrix} 0.0714 & 0.0238 \\ 0.0238 & 0.0450 \end{bmatrix} \begin{bmatrix} 0 \\ 60 \end{bmatrix} = l/EI \begin{bmatrix} 1.428 \\ 2.700 \end{bmatrix}$$

The bending moment at joint 2 is:

$$0 + M_{22'} = 12EI\theta_2/l$$
$$= 12 \times 1.428$$
$$= 17.14 \text{ kip-ft}$$

The bending moment at support 1 is:

$$-60 + 17.14 = -42.86 \text{ kip-ft}$$

Example 10.3

Determine the bending moments in the arched frame shown in Figure 7.54.

Solution

The stiffness, carry-over factors, and fixed-end reactions were given in Example 7.23 as:

$$s_{21} = 8EI_o/20$$
$$c_{21} = 1/2$$
$$s_{23} = 9EI_o/l$$
$$c_{23} = -1/3$$
$$H^{x=1} = 0.1125EI_o/l$$
$$H^{\theta=1} = M^{x=1}$$
$$= 0.75EI_o/l$$
$$M_{23}^F = -M_{32}^F$$
$$= 28.125 \text{ kip-ft}$$
$$H_{23}^F = -H_{32}^F$$
$$= 21.09375 \text{ kip}$$

where:

$$l = 45 \text{ ft}$$

The sub-matrices for members 23, 21, and 34 are:

$$
\begin{bmatrix} M_{23} \\ M_{32} \\ P_{23} \\ P_{32} \end{bmatrix} = EI_o/l
\begin{bmatrix}
9 & -3 & 0.75 & -0.75 \\
-3 & 9 & -0.75 & 0.75 \\
0.75 & -0.75 & 0.1125 & -0.1125 \\
-0.75 & 0.75 & -0.1125 & 0.1125
\end{bmatrix}
\begin{bmatrix} \theta_2 \\ \theta_3 \\ x_2 \\ x_3 \end{bmatrix}
$$

$$
\begin{bmatrix} M_{21} \\ Q_{21} \end{bmatrix} = EI_o
\begin{bmatrix} 8/20 & -12/400 \\ -12/400 & 24/800 \end{bmatrix}
\begin{bmatrix} \theta_2 \\ x_2 \end{bmatrix}
$$

$$
= EI_o/l
\begin{bmatrix} 18 & -1.35 \\ -1.35 & 0.135 \end{bmatrix}
\begin{bmatrix} \theta_2 \\ x_2 \end{bmatrix}
$$

$$
\begin{bmatrix} M_{34} \\ Q_{34} \end{bmatrix} = EI_o/l
\begin{bmatrix} 18 & -1.35 \\ -1.35 & 0.135 \end{bmatrix}
\begin{bmatrix} \theta_3 \\ x_3 \end{bmatrix}
$$

Collecting the relevant elements from the sub-matrices gives:

$$
\begin{bmatrix} P_{\theta2} \\ P_{\theta3} \\ P_{x2} \\ P_{x3} \end{bmatrix} = \begin{bmatrix} -28.25 \\ 28.125 \\ -11.09375 \\ 21.09375 \end{bmatrix}
$$

$$
= EI_o/l \begin{bmatrix} 27 & -3 & -0.60 & -0.75 \\ -3 & 27 & -0.75 & -0.60 \\ -0.60 & -0.75 & 0.2475 & -0.1125 \\ -0.75 & -0.60 & -0.1125 & 0.2475 \end{bmatrix} \begin{bmatrix} \theta_2 \\ \theta_3 \\ x_2 \\ x_3 \end{bmatrix}
$$

Inverting the stiffness matrix gives:

$$
\begin{bmatrix} \theta_2 \\ \theta_3 \\ x_2 \\ x_3 \end{bmatrix} = l/EI_o \begin{bmatrix} 0.064 & & & \\ 0.031 & 0.064 & \text{symmetric} & \\ 0.469 & 0.483 & 9.857 & \\ 0.483 & 0.469 & 7.074 & 9.857 \end{bmatrix} \begin{bmatrix} -28.125 \\ 28.125 \\ -11.09375 \\ 21.09375 \end{bmatrix}
$$

$$
= l/EI_o \begin{bmatrix} 4.046 \\ 5.477 \\ 40.249 \\ 129.063 \end{bmatrix}
$$

The bending moments at joints 2 and 3 are:

$$
\begin{aligned}
M_{21} &= 18EI_o\theta_2/l - 1.35EI_ox_2/l \\
&= 18.6 \text{ kip-ft} \\
M_{34} &= 18EI_o\theta_3/l - 1.35EI_ox_3/l \\
&= -75.5 \text{ kip-ft}
\end{aligned}
$$

Example 10.4

Determine the support reactions of the longitudinal beam $11'$, which is simply supported on the five transverse beams shown in Figure 7.13. All the beams are of the same second moment of area.

Solution

The longitudinal beam may be considered as resting on flexible supports that have a stiffness of:

$$s = 3EI/4l^3$$

where:

$$l = 5 \text{ ft.}$$

Due to the symmetry, only half the longitudinal beam need be considered. There is no rotation at 3, and the applied load and the stiffness of the spring at 3 must be halved.

The stiffness sub-matrices for members 12 and 23 and supports 10, 20, and 30 are:

$$
\begin{bmatrix} M_{12} \\ M_{21} \\ lQ_{12} \\ lQ_{21} \end{bmatrix} = EI/l
\begin{bmatrix} 4 & 2 & 6 & -6 \\ 2 & 4 & 6 & -6 \\ 6 & 6 & 12 & -12 \\ -6 & -6 & -12 & 12 \end{bmatrix}
\begin{bmatrix} \theta_1 \\ \theta_2 \\ y_1/l \\ y_2/l \end{bmatrix}
$$

$$
\begin{bmatrix} M_{23} \\ lQ_{23} \\ lQ_{32} \end{bmatrix} = EI/l
\begin{bmatrix} 4 & 6 & -6 \\ 6 & 12 & -12 \\ -6 & -12 & 12 \end{bmatrix}
\begin{bmatrix} \theta_2 \\ y_2/l \\ y_3/l \end{bmatrix}
$$

$$
\begin{bmatrix} M_{10} \\ lP_{20} \end{bmatrix} = EI/l
\begin{bmatrix} 0 & 0 \\ 0 & 0.75 \end{bmatrix}
\begin{bmatrix} \theta_1 \\ y_1/l \end{bmatrix}
$$

$$
\begin{bmatrix} M_{20} \\ lP_{20} \end{bmatrix} = EI/l
\begin{bmatrix} 0 & 0 \\ 0 & 0.75 \end{bmatrix}
\begin{bmatrix} \theta_2 \\ y_2/l \end{bmatrix}
$$

$$
\begin{bmatrix} M_{30} \\ lP_{30} \end{bmatrix} = EI/l
\begin{bmatrix} 0 & 0 \\ 0 & 0.375 \end{bmatrix}
\begin{bmatrix} \theta_3 \\ y_3/l \end{bmatrix}
$$

Collecting the relevant elements from the sub-matrices gives:

$$
\begin{bmatrix} P_{\theta 1} \\ P_{\theta 2} \\ lP_{y1} \\ lP_{y2} \\ lP_{y3} \end{bmatrix} = \begin{bmatrix} 0 \\ 0 \\ 0 \\ 0 \\ Wl/2 \end{bmatrix}
$$

$$
= EI/l \begin{bmatrix} 4 & 2 & 6 & -6 & 0 \\ 2 & 8 & 6 & 0 & -6 \\ 6 & 6 & 12.75 & -12 & 0 \\ -6 & 0 & -12 & 24.75 & -12 \\ 0 & -6 & 0 & -12 & 12.375 \end{bmatrix} \begin{bmatrix} \theta_1 \\ \theta_2 \\ y_1/l \\ y_2/l \\ y_3/l \end{bmatrix}
$$

Inverting the stiffness matrix gives:

$$
\begin{bmatrix} \theta_1 \\ \theta_2 \\ y_1/l \\ y_2/l \\ y_3/l \end{bmatrix} = l/10EI \begin{bmatrix} 12.79 & & & & \\ 4.866 & 6.294 & \text{symmetric} & & \\ -5.543 & -3.806 & 8.704 & & \\ 2.938 & 1.536 & 3.740 & 6.210 & \\ 5.209 & 4.541 & 1.781 & 6.767 & 9.572 \end{bmatrix} \begin{bmatrix} 0 \\ 0 \\ 0 \\ 0 \\ Wl/2 \end{bmatrix}
$$

$$
= Wl^2/10EI \begin{bmatrix} 2.605 \\ 2.271 \\ 0.891 \\ 3.384 \\ 4.786 \end{bmatrix}
$$

The support reactions at 1, 2, and 3 are:

$$V_1 = 0.75EIy_1/l^3$$
$$= 0.0665W$$
$$V_2 = 0.75EIy_2/l^3$$
$$= 0.254W$$
$$V_3 = 2 \times 0.375EIy_3/l^3$$
$$= 0.359W$$

Example 10.5

Determine the bending moments and support reactions in the continuous beam shown in Figure 7.17 due to the loading shown and settlements of ½ in at support 2 and 1 in at support 3. The second moment of area of the beam is $120\,\text{in}^4$, and the modulus of elasticity is 29,000 kips/in².

Solution

The fixed-end reactions due to the applied loads were given in Example 7.8 as:

$$M_{21}^F = 138 \text{ kip-in}$$
$$M_{12}^F = -207 \text{ kip-in}$$
$$M_{32}^F = -M_{23}^F$$
$$= 100 \text{ kip-in}$$

$$Q_{21}^F = -4.225 \text{ kips}$$
$$Q_{12}^F = -7.775 \text{ kips}$$
$$Q_{23}^F = Q_{32}^F$$
$$= -5 \text{ kips}$$

Collecting relevant elements gives:

$$P_{\theta 2} = W_{\theta 2} - P_{\theta 2}^F$$
$$= 0 - (M_{21}^F + M_{23}^F)$$
$$= -(138 - 100)$$
$$= -38 \text{ kip-in}$$

The sub-matrices for members 12 and 23 are:

$$
\begin{bmatrix} M_{12} \\ M_{21} \\ lQ_{12} \\ lQ_{21} \end{bmatrix} = EI/l \begin{bmatrix} 4 & 2 & 6 & -6 \\ 2 & 4 & 6 & -6 \\ 6 & 6 & 12 & -12 \\ -6 & -6 & -12 & 12 \end{bmatrix} \begin{bmatrix} \theta_1 = 0 \\ \theta_2 \\ y_1/l = 0 \\ y_2/l = 0.5/l \end{bmatrix}
$$

$$
\begin{bmatrix} M_{23} \\ M_{32} \\ lQ_{23} \\ lQ_{32} \end{bmatrix} = EI/l \begin{bmatrix} 4 & 2 & 6 & -6 \\ 2 & 4 & 6 & -6 \\ 6 & 6 & 12 & -12 \\ -6 & -6 & -12 & 12 \end{bmatrix} \begin{bmatrix} \theta_2 \\ 0 \\ 0.5/l \\ 1.0/l \end{bmatrix}
$$

Collecting the relevant elements and partitioning the matrices give:

$$
\begin{bmatrix} P_{\theta 2} = -38 \\ \hline lP_{y2}^R \\ lP_{y3}^R \\ \hline P_{\theta 1}^R \\ P_{\theta 3}^R \\ lP_{y1}^R \end{bmatrix} = EI/l \left[\begin{array}{ccc|ccc} 8 & 0 & -6 & 2 & 2 & 6 \\ \hline 0 & 24 & -12 & -6 & 6 & -12 \\ -6 & -12 & 12 & 0 & -6 & 0 \\ \hline 2 & -6 & 0 & 4 & 0 & 6 \\ 2 & 6 & -6 & 0 & 4 & 0 \\ 6 & -12 & 0 & 6 & 0 & 12 \end{array} \right] \begin{bmatrix} \theta_2 \\ \hline 0.5/l \\ 1.0/l \\ \hline 0 \\ 0 \\ 0 \end{bmatrix}
$$

From Section 10.2(b), for the partitioned matrix:

$$\{\Delta\} = [S_1]^{-1}\{\{P\} - [S_2]\{\Delta^R\}\}$$

where $\{P\}$ and $\{\Delta\}$ are the internal forces and displacements at the supports that yield, and $\{\Delta^R\}$ are the corresponding known displacements. Thus:

$$
\begin{aligned}
\theta_2 &= (1/8)(-38l/EI + 6/l) \\
&= (1/8)[-38 \times 120/(29,000 \times 120) + 6/120] \\
&= 0.006086 \, \text{rad}
\end{aligned}
$$

From the sub-matrix for member 12, the bending moment in the beam at 2 is:

$$
\begin{aligned}
M_{21} &= M_{21}^F + (4\theta_2 - 3/l)EI/l \\
&= 119 \, \text{kip-in}
\end{aligned}
$$

From the partitioned matrix

$$\begin{bmatrix} lP^R_{y2} \\ lP^R_{y3} \end{bmatrix} = EI\theta_2/l \begin{bmatrix} 0 \\ -6 \end{bmatrix} + EI/l \begin{bmatrix} 24 & -12 \\ -12 & 12 \end{bmatrix} \begin{bmatrix} 0.5/l \\ 1.0/l \end{bmatrix} = \begin{bmatrix} 0 \\ 391 \end{bmatrix}$$

The vertical reactions at 2 and 3 are:

$$V_2 = 0 - 4.225 - 5.0$$
$$= -9.225 \text{ kips}$$
$$V_3 = 391/l - 5$$
$$= -1.741 \text{ kips}$$

From the partitioned matrix:

$$\begin{bmatrix} P^R_{\theta 1} \\ P^R_{\theta 3} \\ lP^R_{y1} \end{bmatrix} = EI\theta_2/l \begin{bmatrix} 2 \\ 2 \\ 6 \end{bmatrix} + EI/l \begin{bmatrix} -6 & 0 \\ 6 & -6 \\ -12 & 0 \end{bmatrix} \begin{bmatrix} 0.5/l \\ 1.0/l \end{bmatrix} = \begin{bmatrix} -372 \\ -372 \\ -391 \end{bmatrix}$$

The bending moments at 1 and 3 and the vertical reaction at 1 are:

$$M_{12} = M^F_{12} + P^R_{\theta 1}$$
$$= -207 - 372$$
$$= -579 \text{ kip-in}$$

$$M_{31} = M^F_{32} + P^R_{\theta 3}$$
$$= 100 - 372$$
$$= -272 \text{ kip-in}$$

$$V_1 = Q^F_{12} + P^R_{y1}$$
$$= -391/120 - 7.775$$
$$= -11.033 \text{ kips}$$

Example 10.6

Determine the moments in the grid shown in Figure 10.11: (i) due to a single load at 5, (ii) due to equal loads at 2, 3, 4, and 5. All members are of uniform section, and the flexural rigidity is twice the torsional rigidity.

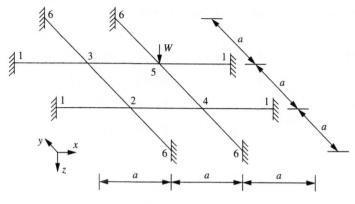

Figure 10.11

Solution

(i) The sub-matrices for members 24, 21, and 41 are:

$$
\begin{bmatrix}
M_{x24} \\
M_{y24} \\
aQ_{24} \\
M_{x42} \\
M_{y42} \\
aQ_{42}
\end{bmatrix}
= EI/2a
\begin{bmatrix}
1 & 0 & 0 & -1 & 0 & 0 \\
0 & 8 & 12 & 0 & 4 & -12 \\
0 & 12 & 24 & 0 & 12 & -24 \\
-1 & 0 & 0 & 1 & 0 & 0 \\
0 & 4 & 12 & 0 & 8 & -12 \\
0 & -12 & -24 & 0 & -12 & 24
\end{bmatrix}
\begin{bmatrix}
\theta_{x2} \\
\theta_{y2} \\
z_2/a \\
\theta_{x4} \\
\theta_{y4} \\
z_4/a
\end{bmatrix}
$$

$$
\begin{bmatrix}
M_{x21} \\
M_{y21} \\
aQ_{21}
\end{bmatrix}
= EI/2a
\begin{bmatrix}
1 & 0 & 0 \\
0 & 8 & -12 \\
0 & -12 & 24
\end{bmatrix}
\begin{bmatrix}
\theta_{x2} \\
\theta_{y2} \\
z_2/a
\end{bmatrix}
$$

$$
\begin{bmatrix}
M_{x41} \\
M_{y41} \\
aQ_{41}
\end{bmatrix}
= EI/2a
\begin{bmatrix}
1 & 0 & 0 \\
0 & 8 & 12 \\
0 & 12 & 24
\end{bmatrix}
\begin{bmatrix}
\theta_{x4} \\
\theta_{y4} \\
z_4/a
\end{bmatrix}
$$

and the sub-matrices for members 35, 13, and 51 are similar.

The sub-matrices for members, 23, 26, and 36 are:

$$
\begin{bmatrix} M_{x23} \\ M_{y23} \\ aQ_{23} \\ M_{x32} \\ M_{y32} \\ aQ_{32} \end{bmatrix} = EI/2a
\begin{bmatrix}
8 & 0 & -12 & 4 & 0 & 12 \\
0 & 1 & 0 & 0 & -1 & 0 \\
-12 & 0 & 24 & -12 & 0 & -24 \\
4 & 0 & -12 & 8 & 0 & 12 \\
0 & -1 & 0 & 0 & 1 & 0 \\
12 & 0 & -24 & 12 & 0 & 24
\end{bmatrix}
\begin{bmatrix} \theta_{x2} \\ \theta_{y2} \\ z_2/a \\ \theta_{x3} \\ \theta_{y3} \\ z_3/a \end{bmatrix}
$$

$$
\begin{bmatrix} M_{x26} \\ M_{y26} \\ aQ_{26} \end{bmatrix} = EI/2a
\begin{bmatrix}
8 & 0 & 12 \\
0 & 1 & 0 \\
12 & 0 & 24
\end{bmatrix}
\begin{bmatrix} \theta_{x2} \\ \theta_{y2} \\ z_2/a \end{bmatrix}
$$

$$
\begin{bmatrix} M_{x36} \\ M_{y36} \\ aQ_{36} \end{bmatrix} = EI/2a
\begin{bmatrix}
8 & 0 & -12 \\
0 & 1 & 0 \\
-12 & 0 & 24
\end{bmatrix}
\begin{bmatrix} \theta_{x3} \\ \theta_{y3} \\ z_3/a \end{bmatrix}
$$

and the sub-matrices for members 45, 64, and 56 are similar.

Collecting the relevant elements from the sub-matrices gives:

$$
\begin{bmatrix} P_{\theta x2} \\ P_{\theta y2} \\ aP_{z2} \\ P_{\theta x3} \\ P_{\theta y3} \\ aP_{z3} \\ P_{\theta x4} \\ P_{\theta y4} \\ aP_{z4} \\ P_{\theta x5} \\ P_{\theta y5} \\ aP_{z5} \end{bmatrix} = EI/2a
\begin{bmatrix}
18 & 0 & 0 & 4 & 0 & 12 & -1 & 0 & 0 & 0 & 0 & 0 \\
0 & 18 & 0 & 0 & -1 & 0 & 0 & 4 & -12 & 0 & 0 & 0 \\
0 & 0 & 96 & -12 & 0 & -24 & 0 & 12 & -24 & 0 & 0 & 0 \\
4 & 0 & -12 & 18 & 0 & 0 & 0 & 0 & 0 & -1 & 0 & 0 \\
0 & -1 & 0 & 0 & 18 & 0 & 0 & 0 & 0 & 0 & 4 & -12 \\
12 & 0 & -24 & 0 & 0 & 96 & 0 & 0 & 0 & 0 & 12 & -24 \\
-1 & 0 & 0 & 0 & 0 & 0 & 18 & 0 & 0 & 4 & 0 & 12 \\
0 & 4 & 12 & 0 & 0 & 0 & 0 & 18 & 0 & 0 & -1 & 0 \\
0 & -12 & -24 & 0 & 0 & 0 & 0 & 0 & 96 & -12 & 0 & -24 \\
0 & 0 & 0 & -1 & 0 & 0 & 4 & 0 & -12 & 18 & 0 & 0 \\
0 & 0 & 0 & 0 & 4 & 12 & 0 & -1 & 0 & 0 & 18 & 0 \\
0 & 0 & 0 & 0 & -12 & -24 & 12 & 0 & -24 & 0 & 0 & 96
\end{bmatrix}
\begin{bmatrix} \theta_{x2} \\ \theta_{y2} \\ z_2/a \\ \theta_{x3} \\ \theta_{y3} \\ z_3/a \\ \theta_{x4} \\ \theta_{y4} \\ z_4/a \\ \theta_{x5} \\ \theta_{y5} \\ z_5/a \end{bmatrix}
$$

Since the grid is symmetrical, the following relationships are obtained:

$$\theta_{y2} = -\theta_{x2}$$
$$\theta_{y3} = -\theta_{x4}$$
$$\theta_{y4} = -\theta_{x3}$$
$$z_3 = z_4$$
$$\theta_{y5} = -\theta_{x5}$$

and the stiffness matrix may be condensed to:

$$
\begin{bmatrix} P_{\theta x2} \\ aP_{z2} \\ P_{\theta x3} \\ P_{\theta y3} \\ aP_{z3} \\ P_{\theta x5} \\ aP_{z5} \end{bmatrix}
= EI/2a
\begin{bmatrix}
18 & 0 & 4 & 1 & 12 & 0 & 0 \\
0 & 96 & -24 & 0 & -48 & 0 & 0 \\
4 & -12 & 18 & 0 & 0 & -1 & 0 \\
1 & 0 & 0 & 18 & 0 & -4 & -12 \\
12 & -24 & 0 & 0 & 96 & -12 & -24 \\
0 & 0 & -1 & -4 & -12 & 18 & 0 \\
0 & 0 & 0 & -24 & -48 & 0 & 96
\end{bmatrix}
\begin{bmatrix} \theta_{x2} \\ z_2/a \\ \theta_{x3} \\ \theta_{y3} \\ z_3/a \\ \theta_{x5} \\ z_5/a \end{bmatrix}
$$

Inverting the condensed stiffness matrix gives:

$$
\begin{bmatrix} \theta_{x2} \\ z_2/a \\ \theta_{x3} \\ \theta_{y3} \\ z_3/a \\ \theta_{x5} \\ z_5/a \end{bmatrix}
= a/500EI
\begin{bmatrix}
76.35 & -8.66 & -29.56 & -18.06 & -19.84 & -18.88 & -7.22 \\
-17.31 & 17.86 & 28.46 & 11.22 & 15.54 & 14.44 & 5.29 \\
-29.56 & 14.23 & 82.10 & 13.14 & 15.87 & 18.06 & 5.61 \\
-18.06 & 5.61 & 13.14 & 82.10 & 15.87 & 29.56 & 14.23 \\
-19.84 & 7.77 & 15.87 & 15.87 & 23.15 & 19.84 & 7.77 \\
-18.88 & 7.22 & 18.06 & 29.56 & 19.84 & 76.35 & 8.66 \\
-14.44 & 5.29 & 11.22 & 28.46 & 15.54 & 17.31 & 17.86
\end{bmatrix}
\begin{bmatrix} 0 \\ 0 \\ 0 \\ 0 \\ 0 \\ 0 \\ Wa \end{bmatrix}
$$

$$
= Wa^2/500EI
\begin{bmatrix} -7.22 \\ 5.29 \\ 5.61 \\ 14.23 \\ 7.77 \\ 8.66 \\ 17.86 \end{bmatrix}
$$

The internal moments in the members, which are identical to the actual moments since there are no fixed-end moments, are obtained by back substitution in the submatrices and are given by:

$$M_{x24} = -M_{y23} = 7.01$$
$$M_{x35} = -M_{y45} = -3.04$$
$$M_{y24} = -M_{x23} = 5.56$$
$$M_{y35} = -M_{x45} = -41.84$$
$$M_{x42} = -M_{y32} = -7.01$$
$$M_{x53} = -M_{y54} = 3.04$$
$$M_{y42} = -M_{x32} = -45.76$$
$$M_{y53} = -M_{x54} = -133.36$$
$$M_{x21} = -M_{y26} = -7.21$$
$$M_{x31} = -M_{y46} = 5.61$$
$$M_{y21} = -M_{x26} = -5.72$$
$$M_{y31} = -M_{x46} = 20.60$$
$$M_{x41} = -M_{y36} = -14.23$$
$$M_{x51} = -M_{y56} = 8.65$$
$$M_{y41} = -M_{x36} = 48.36$$
$$M_{y51} = -M_{x56} = 145.04$$

where:

$$Wa = 1000.$$

(ii) Due to the symmetry of the loading and of the grid:

$$\theta_{y2} = -\theta_{x2} = \theta_{y3} = -\theta_{x4} = -\theta_{y4} = \theta_{x3} = -\theta_{y5} = \theta_{x5}$$
$$z_3 = z_4 = z_2 = z_5$$

The stiffness matrix may be condensed to:

$$\begin{bmatrix} P_{\theta x2} \\ aP_{z2} \end{bmatrix} = EI/2a \begin{bmatrix} 13 & 12 \\ 24 & 48 \end{bmatrix} \begin{bmatrix} \theta_{x2} \\ z_2/a \end{bmatrix}$$

Inverting the condensed matrix gives:

$$\begin{bmatrix} \theta_{x2} \\ z_2/a \end{bmatrix} = a/500EI \begin{bmatrix} 142.9 & -35.71 \\ -71.43 & 38.69 \end{bmatrix} \begin{bmatrix} 0 \\ Wa \end{bmatrix}$$
$$= Wa^2/500EI \begin{bmatrix} -35.71 \\ 38.69 \end{bmatrix}$$

Back-substitution in the sub-matrices gives:

$$M_{x24} = 0$$
$$M_{y24} = 142.84$$
$$M_{x21} = -35.71$$
$$M_{y21} = -178.60$$

where:

$$Wa = 1000.$$

(c) Automation of procedure

The stiffness sub-matrices for inclined members in a structure must be referred to the main x- and y-axes before assembling the complete stiffness matrix. This is readily accomplished[12] by using an orthogonal transformation matrix that transfers displacements from the member axes to the x- and y-axes. The elements of the orthogonal transformation matrix T are the member deformations produced by unit joint displacements in the x- and y-axes, and have the property:

$$[T]^T = [T]^{-1}$$

where $[T]^T$ is the transpose of $[T]$. The use of the orthogonal transformation matrix also provides an automatic means of analyzing a structure by a digital computer.

The member displacements and internal forces are related to the joint displacements and the total internal forces referred to the x- and y-axes by the expressions:

$$\{\Delta'\} = [T]\{\Delta\}$$
$$\{P'\} = [T]\{P\}$$
$$\{P^{F'}\} = [T]\{P^F\}$$

The member internal forces and displacements referred to the member axes are given by:

$$\{P'\} = [\bar{S}]\{\Delta'\}$$

where:

$$[\bar{S}] = \begin{bmatrix} S'_{12} & 0 & 0 & 0 \\ 0 & S'_{ij} & 0 & 0 \\ 0 & 0 & \ddots & 0 \\ 0 & 0 & 0 & S'_{mn} \end{bmatrix}$$

and S'_{ij} is the stiffness sub-matrix for member ij referred to its own axis. Thus:

$$[T]\{P\} = [\bar{S}][T]\{\Delta\}$$
$$\{P\} = [T]^T[\bar{S}][T]\{\Delta\}$$
$$= \{W\} - [T]^T\{P^{F'}\}$$

Hence:

$$\{\Delta\} = [[T]^T[\bar{S}][T]]^{-1}\{\{W\} - [T]^T\{P^{F'}\}\}$$

and:

$$\{P'\} = [\bar{S}][T][[T]^T[\bar{S}][T]]^{-1}\{\{W\} - [T]^T\{P^{F'}\}\}$$

The final member forces are given by:

$$\{P'\} + \{P^{F'}\}$$

The orthogonal transformation matrix for the frame shown in Figure 10.1 is given by:

$$\begin{bmatrix} \theta_{21} \\ y'_{21} \\ \theta_{23} \\ \theta_{32} \\ \theta_{34} \\ \theta_{43} \\ y'_{34} \end{bmatrix} = \begin{bmatrix} 1 & 0 & 0 & 0 \\ 0 & 0 & 0 & 1 \\ 1 & 0 & 0 & 0 \\ 0 & 1 & 0 & 0 \\ 0 & 1 & 0 & 0 \\ 0 & 0 & 1 & 0 \\ 0 & 0 & 0 & 1 \end{bmatrix} \begin{bmatrix} \theta_2 \\ \theta_3 \\ \theta_4 \\ x_2 \end{bmatrix}$$

The matrix $[\bar{S}]$ is given by:

$$
\begin{bmatrix} M_{21} \\ lQ'_{21} \\ M_{23} \\ M_{32} \\ M_{34} \\ M_{43} \\ lQ'_{34} \end{bmatrix} = EI/l \begin{bmatrix} 4 & -6 & 0 & 0 & 0 & 0 & 0 \\ -6 & 12 & 0 & 0 & 0 & 0 & 0 \\ 0 & 0 & 4 & 2 & 0 & 0 & 0 \\ 0 & 0 & 2 & 4 & 0 & 0 & 0 \\ 0 & 0 & 0 & 0 & 4 & 2 & -6 \\ 0 & 0 & 0 & 0 & 2 & 4 & -6 \\ 0 & 0 & 0 & 0 & -6 & -6 & 12 \end{bmatrix} \begin{bmatrix} \theta_{21} \\ y'_{21}/l \\ \theta_{23} \\ \theta_{32} \\ \theta_{34} \\ \theta_{43} \\ y'_{34}/l \end{bmatrix}
$$

The joint displacements are given by:

$$
\begin{bmatrix} \theta_2 \\ \theta_3 \\ \theta_4 \\ x_{2/l} \end{bmatrix} = l/EI \begin{bmatrix} 8 & 2 & 0 & -6 \\ 2 & 8 & 2 & -6 \\ 0 & 2 & 4 & -6 \\ -6 & -6 & -6 & 24 \end{bmatrix}^{-1} \begin{bmatrix} W_{\theta 2} - P^F_{\theta 2} \\ W_{\theta 3} - P^F_{\theta 3} \\ W_{\theta 4} - P^F_{\theta 4} \\ l(W_{x2} + W_{x3} - P^F_{x2} - P^F_{x3}) \end{bmatrix}
$$

and the final member forces are readily obtained.

The stiffness sub-matrix $[S'_{12}]$ for the inclined member 12 shown in Figure 10.3, allowing for axial effects and referred to the inclined member axis, is given by:

$$
\begin{bmatrix} P'_{12} \\ Q'_{12} \\ M_{12} \\ P'_{21} \\ Q'_{21} \\ M_{21} \end{bmatrix} = \begin{bmatrix} EA/l & 0 & 0 & -EA/l & 0 & 0 \\ 0 & 12EI/l^3 & 6EI/l^2 & 0 & -12EI/l^3 & 6EI/l^2 \\ 0 & 6EI/l^2 & 4EI/l & 0 & -6EI/l^2 & 2EI/l \\ -EA/l & 0 & 0 & EA/l & 0 & 0 \\ 0 & -12EI/l^3 & -6EI/l^2 & 0 & 12EI/l^3 & -6EI/l^2 \\ 0 & 6EI/l^2 & 2EI/l & 0 & -6EI/l^2 & 4EI/l \end{bmatrix} \begin{bmatrix} x'_{12} \\ y'_{12} \\ \theta_{12} \\ x'_{21} \\ y'_{21} \\ \theta_{21} \end{bmatrix}
$$

The orthogonal transformation matrix $[T_{12}]$ for the inclined member 12 is obtained by considering unit joint displacements in the x- and y-axes, as shown in Figure 10.3 (iv), and is given by:

$$
\begin{bmatrix} x'_{12} \\ y'_{12} \\ \theta_{12} \\ x'_{21} \\ y'_{21} \\ \theta_{21} \end{bmatrix} = \begin{bmatrix} \lambda & \mu & 0 & 0 & 0 & 0 \\ -\mu & \lambda & 0 & 0 & 0 & 0 \\ 0 & 0 & 1 & 0 & 0 & 0 \\ 0 & 0 & 0 & \lambda & \mu & 0 \\ 0 & 0 & 0 & -\mu & \lambda & 0 \\ 0 & 0 & 0 & 0 & 0 & 1 \end{bmatrix} \begin{bmatrix} x_1 \\ y_1 \\ \theta_1 \\ x_2 \\ y_2 \\ \theta_2 \end{bmatrix}
$$

where $\lambda = \cos \alpha$ and $\mu = \sin \alpha$ and the angle α is measured in the positive direction of M.

The triple matrix product $[T_{12}]^T [S'_{12}][T_{12}]$, which represents the stiffness sub-matrix of member 12 referred to the x- and y-axes, is given by:

$$
\begin{bmatrix} \{P_{12}\} \\ \hline \{P_{21}\} \end{bmatrix} = \begin{bmatrix} [Z_{11}] & | & [Z_{12}] \\ \hline [Z_{21}] & | & [Z_{22}] \end{bmatrix} \begin{bmatrix} \{\Delta_1\} \\ \hline \{\Delta_2\} \end{bmatrix}
$$

where:

$$
\{P_{12}\} = \begin{bmatrix} P_{x12} \\ P_{y12} \\ M_{12} \end{bmatrix}
$$

$$
\{P_{21}\} = \begin{bmatrix} P_{x21} \\ P_{y21} \\ M_{21} \end{bmatrix}
$$

$$
\{\Delta_1\} = \begin{bmatrix} x_1 \\ y_1 \\ \theta_1 \end{bmatrix}
$$

$$
\{\Delta_2\} = \begin{bmatrix} x_2 \\ y_2 \\ \theta_2 \end{bmatrix}
$$

$$[Z_{11}] = \begin{bmatrix} \lambda^2 EA/l + \mu^2 12EI/l^3 & & \\ \lambda\mu(EA/l - 12EA/l^3) & \mu^2 EA/l + \lambda^2 12EI/l^3 & \text{symmetric} \\ -\mu 6EI/l^2 & \lambda 6EI/l^2 & 4EI/l \end{bmatrix}$$

$$[Z_{12}] = \begin{bmatrix} -\lambda^2 EA/l - \mu^2 12EI/l^3 & & \\ -\lambda\mu(EA/l - 12EI/l^3) & -\mu^2 EA/l - \lambda^2 12EI/l^3 & \text{symmetric} \\ \mu 6EI/l^2 & -\lambda 6EI/l^2 & 2EI/l \end{bmatrix}$$

$$[Z_{21}] = \begin{bmatrix} -\lambda^2 EA/l - \mu^2 12EI/l^3 & & \\ -\lambda\mu(EA/l - 12EI/l^3) & -\mu^2 EA/l - \lambda^2 12EI/l^3 & \text{symmetric} \\ -\mu 6EI/l^2 & \lambda 6EI/l^2 & 2EI/l \end{bmatrix}$$

$$[Z_{22}] = \begin{bmatrix} \lambda^2 EA/l + \mu^2 12EI/l^3 & & \\ \lambda\mu(EA/l - 12EI/l^3) & \mu^2 EA/l + \lambda^2 12EI/l^3 & \text{symmetric} \\ \mu 6EI/l^2 & -\lambda 6EI/l^2 & 4EI/l \end{bmatrix}$$

Similarly, the stiffness sub-matrices of all the members in a structure may be referred to the x- and y-axes, and these may be readily combined[13] to give the complete stiffness matrix of the structure. Thus, the triple matrix product $[T]^T[\bar{S}][T]$ has been eliminated from the analysis, and this reduces the computer capacity required and enables larger structures to be handled.

The complete stiffness matrix of the frame shown in Figure 10.12 is given by:

$$\begin{Bmatrix} \{P_1\} \\ \{P_2\} \\ \{P_3\} \\ \{P_4\} \\ \{P_5\} \\ \{P_6\} \\ \{P_7\} \\ \{P_8\} \end{Bmatrix} = \begin{bmatrix} \Sigma[Z_{11}] & [Z_{12}] & [Z_{13}] & 0 & 0 & 0 & 0 & 0 \\ [Z_{21}] & \Sigma[Z_{22}] & 0 & [Z_{24}] & 0 & 0 & 0 & 0 \\ [Z_{31}] & 0 & \Sigma[Z_{33}] & [Z_{34}] & [Z_{35}] & 0 & 0 & 0 \\ 0 & [Z_{42}] & [Z_{43}] & \Sigma[Z_{44}] & 0 & [Z_{46}] & 0 & 0 \\ 0 & 0 & [Z_{53}] & 0 & \Sigma[Z_{55}] & [Z_{56}] & [Z_{57}] & 0 \\ 0 & 0 & 0 & [Z_{64}] & [Z_{65}] & \Sigma[Z_{66}] & 0 & [Z_{68}] \\ 0 & 0 & 0 & 0 & [Z_{75}] & 0 & \Sigma[Z_{77}] & [Z_{78}] \\ 0 & 0 & 0 & 0 & 0 & [Z_{86}] & [Z_{87}] & \Sigma[Z_{88}] \end{bmatrix} \begin{Bmatrix} \{\Delta_1\} \\ \{\Delta_2\} \\ \{\Delta_3\} \\ \{\Delta_4\} \\ \{\Delta_5\} \\ \{\Delta_6\} \\ \{\Delta_7\} \\ \{\Delta_8\} \end{Bmatrix}$$

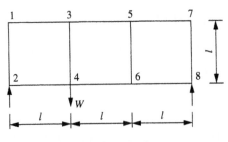

Figure 10.12

where $\{P_i\}$ and $\{\Delta_i\}$ are the three components of force and displacement at joint i and $[Z_{ij}]$ is a 3×3 matrix relating the forces at joint i to the displacements at joint j. The complete stiffness matrix is square and symmetric, and its size is $p \times q$ where p is the number of rows and q is the number of columns and:

$$p = q$$
$$= d$$

where d is the number of joint displacements. The number of joint displacements, neglecting the support restraints, is:

$$d = 3n_j$$

where n_j is the total number of joints in the structure. This provides the complete stiffness matrix of the free structure and may be utilized in the solution of structures with known support displacements. The number of joint displacements, allowing for the support restraints, is:

$$d = 3n_j - r$$

where r is the total number of support restraints.

For the frame shown in Figure 10.12, the size of the complete stiffness matrix, neglecting support restraints, is:

$$p \times q = 3n_j \times 3n_j$$
$$= (3 \times 8) \times (3 \times 8)$$
$$= 24 \times 24$$

The joint displacements are obtained by inverting the stiffness matrix, and back substituting these values in the stiffness sub-matrix $[T_{ij}]^T[S'_{ij}][T_{ij}]$ for each member gives the internal forces referred to the x- and y-axes. The shear force and axial force in each member may then be obtained by resolving forces or by using the expression:

$$\{P'_{ij}\} = [T_{ij}]\{P_{ij}\}$$

All internal forces are considered to be acting from the joint on the member.

The complete stiffness matrix of a structure is symmetric and, provided the joints are numbered in a systematic manner, banded about the leading diagonal. The band width of the stiffness matrix is:

$$b = 3j - 3i + 3$$

where member ij is that member with the maximum numerical difference between the joint numbers at its ends. Advantage may be taken of the banded form of the matrix to reduce the computer storage space required. Only the elements within the banded region need be stored in order to define the matrix, and this produces a condensed rectangular matrix with b columns and d rows where d is the number of joint displacements. The positions of the elements in each row are adjusted to bring the elements of the leading diagonal $[Z_{ij}]$ to the left-hand side of the condensed matrix.

Provided the joints are numbered systematically, the band width is dependent on the number of joints across the width of the structure. The narrower the structure, the smaller the band produced.

For the frame shown in Figure 10.12, the band width is:

$$\begin{aligned} b &= 3j - 3i + 3 \\ &= 3 \times 2 + 3 \\ &= 9 \end{aligned}$$

The number of joint displacements, neglecting the support restraints, is:

$$\begin{aligned} d &= 3n_j \\ &= 3 \times 8 \\ &= 24 \end{aligned}$$

The size of the condensed stiffness matrix, neglecting the support restraints, is:

$$\begin{aligned} p \times q &= d \times b \\ &= 24 \times 9 \end{aligned}$$

The condensed stiffness matrix of the frame shown in Figure 10.12 is:

$$\begin{bmatrix} \Sigma[Z_{11}] & [Z_{12}] & [Z_{13}] \\ \Sigma[Z_{22}] & 0 & [Z_{24}] \\ \Sigma[Z_{33}] & [Z_{34}] & [Z_{35}] \\ \Sigma[Z_{44}] & 0 & [Z_{46}] \\ \Sigma[Z_{55}] & [Z_{56}] & [Z_{57}] \\ \Sigma[Z_{66}] & 0 & [Z_{68}] \\ \Sigma[Z_{77}] & [Z_{78}] & 0 \\ \Sigma[Z_{88}] & 0 & 0 \end{bmatrix}$$

The solution of the equilibrium equations using the condensed stiffness matrix may be achieved with the Gaussian elimination process[14]. Since the band width of the stiffness matrix, rather than the number of joint displacements, determines the time required for the analysis, which is proportional to the band width squared, this provides a rapid and efficient solution procedure. In addition, as the elimination proceeds, the reduced rows overwrite the locations occupied by the original rows, thus conserving computer storage requirements. Back substitution, starting from the last row of the reduced matrix, continues in a similar manner to the elimination process to obtain the values of the displacements.

When the structure is subjected to several alternative loading conditions, these may be analyzed simultaneously by replacing the load vector with a load matrix and the displacement vector with a displacement matrix.

When the available computer capacity limits the order of the matrix that can be inverted, advantage may also be taken of the banded form of the stiffness matrix[15] by partitioning it as shown to give:

$$
\begin{bmatrix} \{P_1\} \\ \{P_C\} \\ \{P_2\} \end{bmatrix} =
\begin{bmatrix}
[S_{11}] & [S_{1C}] & 0 \\
[S_{C1}] & [S_{CC}] & [S_{C2}] \\
0 & [S_{2C}] & [S_{22}]
\end{bmatrix}
\begin{bmatrix} \{\Delta_1\} \\ \{\Delta_C\} \\ \{\Delta_2\} \end{bmatrix}
$$

This is equivalent to dividing the structure into two sub-frames with common joints 4 and 5. Sub-frame 1 consists of members 42, 21, 13, 34, and 35, while sub-frame 2 consists of members 46, 68, 87, 75, and 56.

Expanding the expression gives:

$$\{\Delta_1\} = [S_{11}]^{-1}\{\{P_1\} - [S_{1C}]\{\Delta_C\}\}$$

$$\{\Delta_2\} = [S_{22}]^{-1}\{\{P_2\} - [S_{2C}]\{\Delta_C\}\}$$

$$\{\Delta_C\} = [[S_{CC}] - [S_{C1}][S_{11}]^{-1}[S_{1C}] - [S_{C2}][S_{22}]^{-1}[S_{2C}]]^{-1}\{\{P_C\}$$
$$- [S_{C1}][S_{11}]^{-1}\{P_1\} - [S_{C2}][S_{22}]^{-1}\{P_2\}\}$$

Thus, the number of arithmetic operations is increased but the order of the matrices requiring inversion is reduced, and this reduces the computer capacity required.

Example 10.7

Determine the member forces in the frame shown in Figure 7.51, ignoring axial effects. All members have the same second moment of area.

Solution

Due to the skew symmetry, only half of the frame need be considered, with a hinge inserted at the center of both beams. The vertical deflection of both hinges is zero, and the applied load is:

$$W_{x3} = 4/2$$
$$= 2 \text{ kips}$$

The member displacement y'_{32}, produced by unit joint displacements x_3 and y_3, may be obtained from Figure 10.13 (i) and (ii). Thus:

$$y'_{32} = 0.8x_3 + 0.6y_3$$

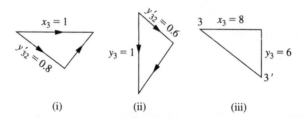

(i) (ii) (iii)

Figure 10.13

The relationship between x_3 and y_3 is obtained from the displacement diagram at (iii) as:

$$y_3 = 0.75x_3$$

Thus:

$$y'_{32} = (0.8 + 0.45)x_3$$
$$= 1.25x_3$$

Similarly:

$$y'_{23} = y'_{21}$$
$$= 1.25x_2$$

And:

$$y'_{34} = y_3$$
$$= 0.75x_3$$
$$y'_{25} = y_2$$
$$= 0.75x_2$$

The transformation matrix is given by:

$$
\begin{bmatrix} \theta_{21} \\ y'_{21} \\ \theta_{23} \\ y'_{23} \\ \theta_{32} \\ y'_{32} \\ \theta_{34} \\ y'_{34} \\ \theta_{25} \\ y'_{25} \end{bmatrix}
=
\begin{bmatrix}
1 & 0 & 0 & 0 \\
0 & 1.25 & 0 & 0 \\
1 & 0 & 0 & 0 \\
0 & 1.25 & 0 & 0 \\
0 & 0 & 1 & 0 \\
0 & 0 & 0 & 1.25 \\
0 & 0 & 1 & 0 \\
0 & 0 & 0 & 0.75 \\
1 & 0 & 0 & 0 \\
0 & 0.75 & 0 & 0
\end{bmatrix}
\begin{bmatrix} \theta_2 \\ x_2 \\ \theta_3 \\ x_3 \end{bmatrix}
$$

Taking the value of EI as 100 kip-ft units, the matrix $[\bar{S}]$ is given by:

$$
\begin{bmatrix} M_{21} \\ Q'_{21} \\ M_{23} \\ Q'_{23} \\ M_{32} \\ Q'_{32} \\ M_{34} \\ Q'_{34} \\ M_{25} \\ Q'_{25} \end{bmatrix}
= EI/l
\begin{bmatrix}
40 & -6 & 0 & 0 & 0 & 0 & 0 & 0 & 0 & 0 \\
-6 & 1.2 & 0 & 0 & 0 & 0 & 0 & 0 & 0 & 0 \\
0 & 0 & 40 & 6 & 20 & -6 & 0 & 0 & 0 & 0 \\
0 & 0 & 6 & 1.2 & 6 & -1.2 & 0 & 0 & 0 & 0 \\
0 & 0 & 20 & 6 & 40 & -6 & 0 & 0 & 0 & 0 \\
0 & 0 & -6 & -1.2 & -6 & 1.2 & 0 & 0 & 0 & 0 \\
0 & 0 & 0 & 0 & 0 & 0 & 100 & 33.3 & 0 & 0 \\
0 & 0 & 0 & 0 & 0 & 0 & 33.3 & 11.1 & 0 & 0 \\
0 & 0 & 0 & 0 & 0 & 0 & 0 & 0 & 33.3 & 3.7 \\
0 & 0 & 0 & 0 & 0 & 0 & 0 & 0 & 3.7 & 0.41
\end{bmatrix}
\begin{bmatrix} \theta_{21} \\ y'_{21} \\ \theta_{23} \\ y'_{23} \\ \theta_{32} \\ y'_{32} \\ \theta_{34} \\ y'_{34} \\ \theta_{25} \\ y'_{25} \end{bmatrix}
$$

The matrix $[T]^T[\bar{S}][T]$ is given by:

$$
\begin{bmatrix} P_{\theta 2} \\ P_{x2} \\ P_{\theta 3} \\ P_{x3} \end{bmatrix} = \begin{bmatrix} 0 \\ 0 \\ 0 \\ 2 \end{bmatrix}
$$

$$
= \begin{bmatrix} 113.33 & & & \\ 2.78 & 3.98 & \text{symmetric} & \\ 20.00 & 7.50 & 140.00 & \\ -7.50 & -1.88 & 17.50 & 8.13 \end{bmatrix} \begin{bmatrix} \theta_2 \\ x_2 \\ \theta_3 \\ x_3 \end{bmatrix}
$$

The displacements are given by:

$$
\begin{bmatrix} \theta_2 \\ x_2 \\ \theta_3 \\ x_3 \end{bmatrix} = 1/100 \begin{bmatrix} 1.12 & & & \\ 1.51 & 46.52 & \text{symmetric} & \\ -0.56 & -5.78 & 1.83 & \\ 2.59 & 24.57 & -5.79 & 32.85 \end{bmatrix} \begin{bmatrix} 0 \\ 0 \\ 0 \\ 2 \end{bmatrix}
$$

$$
= \begin{bmatrix} 0.0519 \\ 0.4195 \\ -0.1159 \\ 0.6570 \end{bmatrix}
$$

The member forces, in kip and ft units, are obtained by pre-multiplying the displacement vector by $[\bar{S}][T]$, since there are no fixed-end forces. Thus:

$$
\begin{bmatrix} M_{21} \\ Q'_{21} \\ M_{23} \\ Q'_{23} \\ M_{32} \end{bmatrix} = \begin{bmatrix} -1.61 \\ 0.426 \\ -1.48 \\ -0.632 \\ -4.84 \end{bmatrix}
$$

$$
\begin{bmatrix} Q'_{32} \\ M_{34} \\ Q'_{34} \\ M_{25} \\ Q'_{25} \end{bmatrix}
\begin{bmatrix} 0.632 \\ 4.84 \\ 1.61 \\ 3.09 \\ 0.344 \end{bmatrix}
$$

Example 10.8

Determine the member forces in the rigidly jointed truss shown in Figure 7.18. The second moment of area of members 14, 34, and 24 is 30 in^4 and of members 12 and 23 is 40 in^4. The cross-sectional area of members 14, 34, and 24 is 4 in^2 and of members 12 and 23 is 5 in^2.

Solution

Due to the symmetry, θ_2, θ_4, x_2, and x_4 are zero. Only half of the frame need be considered, with the applied loads at joints 2 and 4 and the cross-sectional area of member 24 halved in value.

For member 12, $\lambda = (3)^{0.5}/2$ and $\mu = -0.5$, and the stiffness sub-matrix referred to the x- and y-axes is given by:

$$
\begin{bmatrix} P_{x12} \\ M_{12} \\ \hline P_{y21} \end{bmatrix}
$$

$$
= E \begin{bmatrix}
\begin{array}{l} 0.75 \times 5/116 + 0.25 \times \\ \quad 12 \times 40/(116)^3 \end{array} & & & x_1 \\
0.5 \times 6 \times 40/(116)^2 & 4 \times 40/116 & \text{symmetric} & \theta_1 \\
\hline
0.433\{5/116 - 12 \times 40/(116)^3\} & -0.866 \times 6 \times 40/(116)^2 & \begin{array}{l} 0.25 \times 5/116 + 0.75 \times \\ \quad 12 \times 40/(116)^3 \end{array} & y_2
\end{bmatrix}
$$

$$
= E \begin{bmatrix}
0.03248 & & & x_1 \\
0.00895 & 1.3780 & \text{symmetric} & \theta_1 \\
\hline
0.01854 & -0.0155 & 0.011031 & y_2
\end{bmatrix}
$$

For member 14, $\lambda = 1$ and $\mu = 0$ and:

$$\begin{bmatrix} P_{x14} \\ M_{14} \\ \hline P_{y41} \end{bmatrix} = E \begin{bmatrix} 4/100 & & & \\ 0 & 4 \times 30/100 & & \text{symmetric} \\ \hline 0 & -6 \times 30/(100)^2 & 12 \times 30/(100)^3 \end{bmatrix} \begin{bmatrix} x_1 \\ \theta_1 \\ \hline y_4 \end{bmatrix}$$

$$= E \begin{bmatrix} 0.04 & & & \\ 0 & 1.2 & \text{symmetric} \\ \hline 0 & -0.018 & 0.00036 \end{bmatrix} \, | \, | \begin{bmatrix} x_1 \\ \theta_1 \\ \hline y_4 \end{bmatrix}$$

For member 24, $\lambda = 0$ and $\mu = 1$ and:

$$\begin{bmatrix} P_{y24} \\ P_{y42} \end{bmatrix} = E \begin{bmatrix} 2/58 & -2/58 \\ -2/58 & 2/58 \end{bmatrix} \begin{bmatrix} y_2 \\ y_4 \end{bmatrix}$$

$$= E \begin{bmatrix} 0.0345 & -0.0345 \\ -0.0345 & 0.0345 \end{bmatrix} \begin{bmatrix} y_2 \\ y_4 \end{bmatrix}$$

The complete stiffness matrix of the truss is given by:

$$\begin{bmatrix} P_{x1} \\ P_{\theta 1} \\ P_{y2} \\ P_{y4} \end{bmatrix} = \begin{bmatrix} 0 \\ 0 \\ 1 \\ 2.5 \end{bmatrix}$$

$$= E \begin{bmatrix} 0.07248 & & & \\ 0.00895 & 2.57800 & \text{symmetric} \\ 0.01854 & -0.01550 & 0.04553 \\ 0 & -0.01800 & -0.03450 & 0.03486 \end{bmatrix} \begin{bmatrix} x_1 \\ \theta_1 \\ y_2 \\ y_4 \end{bmatrix}$$

The joint displacements are:

$$\begin{bmatrix} x_1 \\ \theta_1 \\ y_2 \\ y_4 \end{bmatrix} = 1/E \begin{bmatrix} -146.2 \\ 8.4 \\ 567.7 \\ 638.0 \end{bmatrix}$$

The member forces, referred to the x- and y-axes and acting from the joint on the member, are obtained by back substitution in the sub-matrices for each member. Thus:

$$
\begin{bmatrix} P_{x12} \\ M_{12} \\ P_{y21} \\ P_{x14} \\ M_{14} \\ P_{y41} \\ P_{y24} \end{bmatrix} = \begin{bmatrix} 5.84 \\ 1.39 \\ 3.42 \\ -5.84 \\ -1.39 \\ 0.08 \\ -2.42 \end{bmatrix}
$$

The axial forces and remaining moments are given by:

$$P'_{12} = 0.86603P_{x12} - 0.5P_{y12}$$
$$= 6.82 \text{ kips compression}$$
$$P'_{14} = P_{x14}$$
$$= -5.84 \text{ kips tension}$$
$$P'_{24} = 2P_{y24}$$
$$= -4.84 \text{ kips tension}$$
$$M_{21} = 100P_{y12} + 58P_{x12} - M_{12}$$
$$= -4.39 \text{ kip-in}$$
$$M_{41} = 100P_{y14} - M_{14}$$
$$= -6.61 \text{ kip-in}$$

Example 10.9

Determine the member forces in the Vierendeel girder shown in Figure 10.12 for values of $l = 12$ ft and $W = 10$ kips. All the members are of uniform cross-section.

Solution

Due to the skew symmetry, only the lower half of the frame need be considered, with a hinge inserted at the center of each post and the applied load at joint 4 halved in value.

For members 21, 43, 65, and 87, $\lambda = 0$ and $\mu = -1$, and the stiffness sub-matrix referred to the x- and y-axes is given by:

$$M_{21} = [3EI/6]\theta_2$$
$$= 72\theta_2$$

where axial effects are ignored and the value of EI is taken as $(12)^2$ kip-ft^2 units.

For member 46, $\lambda = 1$ and $\mu = 0$, and the stiffness sub-matrix is given by:

$$
\begin{bmatrix} P_{y46} \\ M_{46} \\ \hline P_{y64} \\ M_{64} \end{bmatrix} = EI \begin{bmatrix} 12/l^3 & & & \\ 6/l^2 & 4/l & \text{symmetric} \\ \hline -12/l^3 & -6/l^2 & 12/l^3 & \\ 6/l^2 & 2/l & -6/l^2 & 4/l \end{bmatrix} \begin{bmatrix} y_4 \\ \theta_4 \\ y_6 \\ \theta_6 \end{bmatrix}
$$

$$
= \begin{bmatrix} 1 & & & \\ 6 & 48 & \text{symmetric} \\ \hline -1 & -6 & 1 & \\ 6 & 24 & -6 & 48 \end{bmatrix} \begin{bmatrix} y_4 \\ \theta_4 \\ y_6 \\ \theta_6 \end{bmatrix}
$$

For member 24, $y_2 = 0$, $\lambda = 1$, and $\mu = 0$ and:

$$
\begin{bmatrix} M_{24} \\ P_{y42} \\ M_{42} \end{bmatrix} = \begin{bmatrix} 48 & \text{symmetric} \\ -6 & 1 \\ 24 & -6 & 48 \end{bmatrix} \begin{bmatrix} \theta_2 \\ y_4 \\ \theta_4 \end{bmatrix}
$$

For member 68, $y_8 = 0$, $\lambda = 1$, and $\mu = 0$ and

$$
\begin{bmatrix} P_{y68} \\ M_{68} \\ M_{86} \end{bmatrix} = \begin{bmatrix} 1 & & \\ 6 & 48 & \text{symm} \\ 6 & 24 & 48 \end{bmatrix} \begin{bmatrix} y_6 \\ \theta_6 \\ \theta_8 \end{bmatrix}
$$

The complete stiffness matrix of the frame is:

$$
\begin{bmatrix} P_{\theta2} \\ P_{y4} \\ P_{\theta4} \\ \hline P_{y6} \\ P_{\theta6} \\ \hline P_{\theta8} \end{bmatrix} = \begin{bmatrix} 0 \\ 5 \\ 0 \\ \hline 0 \\ 0 \\ \hline 0 \end{bmatrix}
$$

$$
= \begin{bmatrix}
120 & & & & & \\
-6 & 2 & & \text{symmetric} & & \\
24 & 0 & 168 & & & \\
\hline
0 & -1 & -6 & 2 & & \\
0 & 6 & 24 & 0 & 168 & \\
\hline
0 & 0 & 0 & 6 & 24 & 120
\end{bmatrix}
\begin{bmatrix}
\theta_2 \\
y_4 \\
\theta_4 \\
\hline
y_6 \\
\theta_6 \\
\hline
\theta_8
\end{bmatrix}
$$

Using the partitioning shown:

$$
\begin{bmatrix} y_6 \\ \theta_6 \end{bmatrix} = \begin{bmatrix} \begin{bmatrix} 2 & 0 \\ 0 & 168 \end{bmatrix} - \begin{bmatrix} 0.76 & -4.18 \\ -4.18 & 23.62 \end{bmatrix} \end{bmatrix} \begin{bmatrix} 0.3 & 1.2 \\ 1.2 & 4.8 \end{bmatrix}^{-1} \begin{bmatrix} 2.83 \\ -17.22 \end{bmatrix}
$$

$$
= \begin{bmatrix} 1.1410 & -0.0244 \\ -0.0244 & 0.0077 \end{bmatrix} \begin{bmatrix} 2.83 \\ -17.22 \end{bmatrix} = \begin{bmatrix} 3.65 \\ -0.20 \end{bmatrix}
$$

$$
\begin{bmatrix} \theta_2 \\ y_4 \\ \theta_4 \end{bmatrix} = \begin{bmatrix} 0.0101 & \text{symmetric} & \\ 0.0305 & 0.5913 & \\ -0.0014 & -0.0043 & 0.0062 \end{bmatrix} \begin{bmatrix} 0 \\ 9.85 \\ 26.70 \end{bmatrix} = \begin{bmatrix} 0.26 \\ 5.71 \\ 0.12 \end{bmatrix}
$$

$M_{21} = 18.7$ kip-ft

$M_{43} = 8.6$ kip-ft

$M_{65} = -14.4$ kip-ft

$M_{87} = -10.1$ kip-ft

$M_{42} = 24 \times 0.26 - 6 \times 5.71 + 48 \times 0.12$

$\quad = -22.3$ kip-ft

$M_{68} = 6 \times 3.65 + 48 \times -0.20 + 24 \times -0.14$

$\quad = 8.9$ kip-ft

10.3 Flexibility matrix method

(a) Introduction

The structure shown in Figure 10.14 is two degrees redundant, and these redundants may be considered to be the internal moments M_2 and M_3 at joints 2 and 3. The structure is cut back to a statically determinate condition by

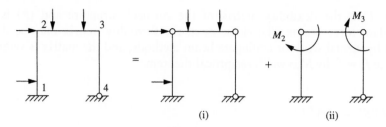

Figure 10.14

introducing releases corresponding to the redundants. The external loads are applied in system (i) and the redundant forces applied in system (ii). The actual structure may be considered as the sum of system (i) and system (ii). The loads applied in system (i) produce discontinuities $\{\Delta^W\}$ at the releases, and these discontinuities may be calculated by virtual work or conjugate beam methods. The redundant forces applied in system (ii) produce discontinuities $\{\Delta^R\}$ at the releases. Since there are no discontinuities in the original structure at the positions of the releases:

$$\{\Delta^R\} = -\{\Delta^W\}$$

The discontinuities produced in system (ii) are given by:

$$\delta_1^R = f_{11}R_1 + f_{12}R_2 + \ldots + f_{1n}R_n$$
$$\delta_2^R = f_{21}R_1 + f_{22}R_2 + \ldots + f_{2n}R_n$$
$$\vdots \qquad \vdots \qquad \vdots \qquad \vdots$$
$$\delta_n^R = f_{n1}R_1 + f_{n2}R_2 + \ldots + f_{nn}R_n$$

where the flexibility coefficient f_{ij} is the displacement produced at point i by a unit force replacing the redundant at j, and R_j is the redundant at j. Thus:

$$
\begin{bmatrix} \delta_1^R \\ \delta_2^R \\ \cdot \\ \cdot \\ \cdot \\ \delta_n^R \end{bmatrix}
=
\begin{bmatrix} f_{11} & f_{12} & & f_{1n} \\ f_{21} & f_{22} & & f_{2n} \\ \cdot & \cdot & & \cdot \\ \cdot & \cdot & & \cdot \\ \cdot & \cdot & & \cdot \\ f_{n1} & f_{n2} & & f_{nn} \end{bmatrix}
\begin{bmatrix} R_1 \\ R_2 \\ \cdot \\ \cdot \\ \cdot \\ R_n \end{bmatrix}
$$

or:

$$\{\Delta^R\} = [F]\{R\}$$

where $[F]$ is the flexibility matrix of the cut-back structure and $\{R\}$ is the redundant force vector. The elements in the flexibility matrix may be determined by virtual work or conjugate beam methods, and the matrix is symmetric since $f_{ij} = f_{ji}$ by Maxwell's reciprocal theorem.
Thus:

$$[F]\{R\} = -\{\Delta^W\}$$

and $\{R\}$ may be determined. The internal forces in any member may now be obtained by summing the internal forces in system (i) and system (ii).

To ensure that the equations $[F]\{R\} = -\{\Delta^W\}$ are well conditioned, the elements on the main diagonal should predominate. This is achieved by choosing a cut-back structure such that the unit value of each redundant produces its maximum displacement at its own release.

Lack of fit in the members of a structure is equivalent to initial discontinuities $\{\Delta^I\}$ at the releases.
Thus:

$$[F]\{R\} = -\{\{\Delta^I\} + \{\Delta^W\}\}$$

The relation between deflections and redundant forces in the cut-back structure is:

$$\{\Delta^R\} = [F]\{R\}$$

Thus:

$$\{R\} = [F]^{-1}\{\Delta^R\}$$

and the stiffness matrix of the cut-back structure is the inverse of the flexibility matrix of the cut-back structure.

Example 10.10

Determine the bending moment at the support 1 of the frame shown in Figure 10.15.

Solution

The cut-back structure with the three redundants H_1, V_1, M_1 is shown at (i), and the displacements corresponding to the redundants are shown at (ii).

The external load W applied to the cut-back structure produces the bending moment diagram shown at (iii). The resulting displacements x_1^W, y_1^W, θ_1^W may be determined from the conjugate frame shown at (iv), where the elastic

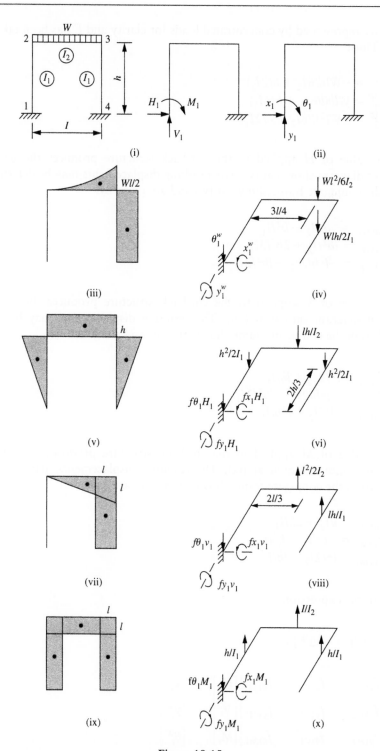

Figure 10.15

loads are represented by concentrated loads for clarity and E has been taken as unity. Thus:

$$\theta_1^W = -Wl(l/6I_2 + h/2I_1)$$
$$x_1^W = Wlh(l/6I_2 + h/4I_1)$$
$$y_1^W = -Wl^2(l/8I_2 + h/2I_1)$$

Unit value of H applied to the cut-back structure produces the bending moment diagram shown at (v). The resulting displacements may be determined from the conjugate frame shown at (vi) and are given by:

$$f_{\theta 1H1} = -lh/I_2 - h^2/I_1$$
$$f_{x1H1} = lh^2/I_2 + 2h^3/3I_1$$
$$fy_{1H1} = -l^2h/2I_2 - lh^2/2I_1$$

Unit value of V applied to the cut-back structure produces the bending moment diagram shown at (vii). The resulting displacements may be determined from the conjugate frame shown at (viii) and are given by:

$$f_{\theta 1V1} = l^2/2I_2 + lh/I_1$$
$$f_{x1V1} = -l^2h/2I_2 - lh^2/2I_1$$
$$f_{y1V1} = l^3/3I_2 + l^2h/I_1$$

Unit value of M applied to the cut-back structure produces the bending moment diagram shown at (ix). The resulting displacements may be determined from the conjugate frame shown at (x) and are given by:

$$f_{\theta 1M1} = l/I_2 + 2h/I_1$$
$$f_{x1M1} = -lh/I_2 - h^2/I_1$$
$$f_{y1M1} = l^2/2I_2 + lh/I_1$$

Then, the expression:

$$[F]\{R\} = -\{\Delta^W\} \text{ is}$$

$$
\begin{bmatrix} f_{x1H1} & f_{x1V1} & f_{x1M1} \\ f_{y1H1} & f_{y1V1} & f_{y1M1} \\ f_{\theta 1H1} & f_{\theta 1V1} & f_{\theta 1M1} \end{bmatrix}
\begin{bmatrix} H_1 \\ V_1 \\ M_1 \end{bmatrix}
= -
\begin{bmatrix} x_1^W \\ y_1^W \\ \theta_1^W \end{bmatrix}
$$

Substituting $b = hI_2/lI_1$ this becomes:

$$\begin{bmatrix} (2b+3)h^2/3 & -(b+1)lh/2 & -(b+1)h \\ -(b+1)lh/2 & (3b+1)l^2/3 & (2b+3)l/2 \\ -(b+1)h & (2b+1)l/2 & (2b+1) \end{bmatrix} \begin{bmatrix} H_1 \\ V_1 \\ M_1 \end{bmatrix} = Wl \begin{bmatrix} -(3b+2)h/12 \\ (4b+1)l/8 \\ (3b+1)l/6 \end{bmatrix}$$

Multiplying the first row by $3l(b + 1)/2h(2b + 3)$ and adding to the second row give:

$$\begin{bmatrix} (2b+3)h^2/3 & -(b+1)lh/2 & -(b+1)h \\ 0 & l^2(15b^2+26b+3)/12 & (b^2+2b)l/2 \\ -(b+1)h & (2b+1)l/2 & (2b+1) \end{bmatrix} \begin{bmatrix} H_1 \\ V_1 \\ M_1 \end{bmatrix} = Wl \begin{bmatrix} -(3b+2)h/12 \\ (5b^2+9b+1)l/8 \\ (3b+1)l/6 \end{bmatrix}$$

Thus:

$$V_1 l^2 (15b^2 + 26b + 3)/12 + M_1 l(b^2 + 2b)/2 = Wl^2(5b^2 + 9b + 1)/8$$

From symmetry,

$$V_1 = W/2$$

hence:

$$M_1 = Wl/12(6b + 2)$$

(b) Automation of procedure

The flexibility matrix [F] of the cut-back structure may be assembled automatically from the flexibility sub-matrices of the individual members. The sub-matrices for the members shown in Figure 10.16 may be determined by the conjugate beam method or by inversion of the corresponding stiffness matrix. For the member 12, fixed-ended at 2 as shown at (i), the sub-matrix is given by:

$$\begin{bmatrix} x_1 \\ y_1 \\ \theta_1 \end{bmatrix} = \begin{bmatrix} l/EA & 0 & 0 \\ 0 & l^3/3EI & -l^2/2EI \\ 0 & -l^2/2EI & l/EI \end{bmatrix} \begin{bmatrix} P_{12} \\ Q_{12} \\ M_{12} \end{bmatrix}$$

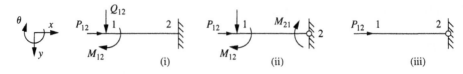

Figure 10.16

For the member 12, pinned at each end as shown at (ii), the sub-matrix is given by:

$$
\begin{bmatrix} x_1 \\ \theta_1 \\ \theta_2 \end{bmatrix} = \begin{bmatrix} l/EA & 0 & 0 \\ 0 & l/3EI & -l/6EI \\ 0 & -l/6EI & l/3EI \end{bmatrix} \begin{bmatrix} P_{12} \\ M_{12} \\ M_{21} \end{bmatrix}
$$

For the member 12 of a pin-jointed frame subjected to an axial force only as shown at (iii), the sub-matrix is given by:

$$x_1 = [l/EA]P_{12}$$

A structure containing n members, subjected to m applied loads and indeterminate to the degree D, may be cut back to a determinate condition by removing D redundants. The force produced in member i by the application of the redundants to the cut-back structure is:

$$P_i^R = \sum_{j=1}^{j=D} u_{ij} R_j$$

In general,

$$\{P^R\} = [U]\{R\}$$

where the element u_{ij}, of the matrix $[U]$ is the force in member i due to a unit value of the redundant R_j, $\{R\}$ is the column vector of redundants, and $\{P^R\}$ is the column vector of the member forces produced by the redundants acting on the cut-back structure. The discontinuity produced at release j by the application of all the redundants to the cut-back structure is:

$$\delta_j^R = \sum_{i=1}^{i=n} u_{ij} f_i P_i^R$$

where f_i is the flexibility of member i. In general:

$$\{\Delta^R\} = [U]^T[\bar{F}]\{P^R\}$$
$$= [U]^T[\bar{F}][U]\{R\}$$

where:

$$[\bar{F}] = \begin{bmatrix} [F_1] & 0 & 0 & 0 \\ 0 & [F_i] & 0 & 0 \\ 0 & 0 & \ddots & 0 \\ 0 & 0 & 0 & [F_n] \end{bmatrix}$$

and $[F_i]$ is the flexibility sub-matrix for member i. The triple matrix product $[U]^T[\bar{F}][U]$ is equivalent to the flexibility matrix $[F]$ of the cut-back structure.

The force produced in member i by the application of the external loads to the cut-back structure is:

$$P_i^W = \sum_{k=1}^{k=m} v_{ik} W_k$$

In general:

$$\{P^W\} = [V]\{W\}$$

where the element v_{ik} of the matrix $[V]$ is the force in member i due to a unit value of the applied load W_k, $\{W\}$ is the column vector of applied loads, and $\{P^W\}$ is the column vector of the member forces produced by the applied loads acting on the cut-back structure. The discontinuity produced at release j by the application of all the external loads to the cut-back structure is:

$$\delta_j^W = \sum_{i=1}^{i=n} u_{ij} f_i P_i^W$$

In general:

$$\{\Delta^W\} = [U]^T[\bar{F}]\{P^W\}$$
$$= [U]^T[\bar{F}][V]\{W\}$$

For the case of zero initial discontinuities:

$$\{\Delta^R\} = -\{\Delta^W\}$$

and: $\{R\} = -[[U]^{\mathrm{T}}[\bar{F}][U]]^{-1}[U]^{\mathrm{T}}[\bar{F}][V]\{W\}$

The final member forces in the structure are given by:

$$\{P^W\} + \{P^R\} = [[V] - [U][[U]^{\mathrm{T}}[\bar{F}][U]]^{-1}[U]^{\mathrm{T}}[\bar{F}][V]]\{W\}$$

The deflection produced at the point of application of load W_k by all the external loads acting on the real structure is:

$$\delta_k = \sum_{i=1}^{i=n} v_{ik} f_i (P_i^W + P_i^R)$$

or, in general:

$$\{\Delta\} = [V]^{\mathrm{T}}[\bar{F}]\{\{P^W\} + \{P^R\}\}$$

The discontinuity produced at release j by the final member forces in the structure is

$$\delta_j^W + \delta_j^R = \sum_{i=1}^{i=n} u_{ij} f_i (P_i^W + P_i^R)$$
$$= 0$$

In general:

$$[U]^{\mathrm{T}}[\bar{F}]\{\{P^W\} + \{P^R\}\} = 0$$

and this may be used as a check on the computation.

The discontinuity produced at release j by lack of fit h of the members, due to thermal changes or manufacturing errors, is:

$$\delta_j^I = \sum_{i=1}^{i=n} u_{ij} h_i$$

or: $\{\Delta^I\} = [U]^T\{H\}$
$= -\{\Delta^R\}$
$= -[U]^T[\bar{F}][U]\{R\}$

The final member forces in the structure are given by:

$\{P^R\} = [U]\{R\}$
$= -[U][U]^T[\bar{F}][U]^{-1}[U]^T\{H\}$

The analysis of structures with loads applied between the joints may be obtained in a similar manner[16]. The displacements at the joints in the cut-back structure due to the loads applied between the joints are used to form the matrix $\{H\}$.

Example 10.11

Determine the forces in the members of the pin-jointed frame shown in Figure 10.17. All members have the same value for EA/l.

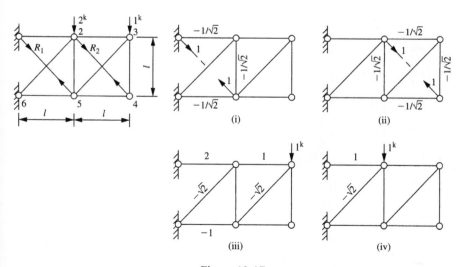

Figure 10.17

Solution

The two redundants may be considered to be the tensile forces R_1 and R_2 in members 15 and 24. The matrix $[U]$ is obtained from (i) and (ii) and the matrix $[V]$ from (iii) and (iv), and these are given by:

$$
\begin{bmatrix} P_{12}^R \\ P_{23}^R \\ P_{34}^R \\ P_{45}^R \\ P_{56}^R \\ P_{62}^R \\ P_{15}^R \\ P_{53}^R \\ P_{24}^R \\ P_{25}^R \end{bmatrix}
=
\begin{bmatrix} -1/a & 0 \\ 0 & -1/a \\ 0 & -1/a \\ 0 & -1/a \\ -1/a & 0 \\ 1 & 0 \\ 1 & 0 \\ 0 & 1 \\ 0 & 1 \\ -1/a & -1/a \end{bmatrix}
\begin{bmatrix} R_1 \\ R_2 \end{bmatrix}
$$

$$
\begin{bmatrix} P_{12}^W \\ P_{23}^W \\ P_{34}^W \\ P_{45}^W \\ P_{56}^W \\ P_{62}^W \\ P_{15}^W \\ P_{53}^W \\ P_{24}^W \\ P_{25}^W \end{bmatrix}
=
\begin{bmatrix} 2 & 1 \\ 1 & 0 \\ 0 & 0 \\ 0 & 0 \\ -1 & 0 \\ -a & -a \\ 0 & 0 \\ -a & 0 \\ 0 & 0 \\ 1 & 0 \end{bmatrix}
\begin{bmatrix} W_{y3} \\ W_{y2} \end{bmatrix}
$$

where $a = (2)^{0.5}$ and tensile forces have been considered positive.

The matrix $[\bar{F}]$ is:

$$
\begin{bmatrix}
1 & & & & & & & & & \\
0 & 1 & & & & & & & & \\
0 & 0 & 1 & & & & & & & \\
0 & 0 & 0 & 1 & & \text{symmetric} & & & & \\
0 & 0 & 0 & 0 & 1 & & & & & \\
0 & 0 & 0 & 0 & 0 & 1 & & & & \\
0 & 0 & 0 & 0 & 0 & 0 & 1 & & & \\
0 & 0 & 0 & 0 & 0 & 0 & 0 & 1 & & \\
0 & 0 & 0 & 0 & 0 & 0 & 0 & 0 & 1 & \\
0 & 0 & 0 & 0 & 0 & 0 & 0 & 0 & 0 & 1
\end{bmatrix}
$$

and:

$$
[U]^T[\bar{F}] = l/aEA
\begin{bmatrix}
-1 & 0 & 0 & 0 & -1 & a & a & 0 & 0 & -1 \\
0 & -1 & -1 & -1 & 0 & 0 & 0 & a & a & -1
\end{bmatrix}
$$

$$
[U]^T[\bar{F}][U] = l/2EA
\begin{bmatrix}
7 & 1 \\
1 & 8
\end{bmatrix}
$$

$$
[[U]^T[\bar{F}][U]]^{-1} = 2EA/55l
\begin{bmatrix}
8 & -1 \\
-1 & 7
\end{bmatrix}
$$

$$
[U]^T[\bar{F}][V] = l/aEA
\begin{bmatrix}
-4 & -3 \\
-4 & 0
\end{bmatrix}
$$

$$
[[U]^T[\bar{F}][U]]^{-1}[U]^T[\bar{F}][V] = a/55
\begin{bmatrix}
-28 & -24 \\
-24 & 3
\end{bmatrix}
$$

$$[U][[U]^T[\bar{F}][U]]^{-1}[U]^T[\bar{F}][V] = 1/55 \begin{bmatrix} 28 & 24 \\ 24 & -3 \\ 24 & -3 \\ 24 & -3 \\ 28 & 24 \\ -28a & -24a \\ -28a & -24a \\ -24a & 3a \\ -24a & 3a \\ 52 & 21 \end{bmatrix}$$

The final member forces are given by:

$$\begin{bmatrix} P_{12}^W \\ P_{23}^W \\ P_{34}^W \\ P_{45}^W \\ P_{56}^W \\ P_{62}^W \\ P_{15}^W \\ P_{53}^W \\ P_{24}^W \\ P_{25}^W \end{bmatrix} + \begin{bmatrix} P_{12}^R \\ P_{23}^R \\ P_{34}^R \\ P_{45}^R \\ P_{56}^R \\ P_{62}^R \\ P_{15}^R \\ P_{53}^R \\ P_{24}^R \\ P_{25}^R \end{bmatrix} = 1/55 \begin{bmatrix} 82 & 31 \\ 31 & 3 \\ -24 & 3 \\ -24 & 3 \\ -83 & -24 \\ -27a & -31a \\ 28a & 24a \\ -31a & -3a \\ 24a & -3a \\ 3 & -21 \end{bmatrix} \begin{bmatrix} 1 \\ 2 \end{bmatrix}$$

$$= 1/55 \begin{bmatrix} 144 \\ 37 \\ -18 \\ -18 \\ -131 \\ -89a \\ 76a \\ -37a \\ 18a \\ -39 \end{bmatrix}$$

Example 10.12

Determine the support moments of the continuous beam shown in Figure 7.8. The relative EI/l values are shown ringed.

Solution

The support moments M_2 and M_3 may be considered to be the redundants, and matrix $[U]$ is obtained from Figure 10.18 (i) and (ii) as:

$$\begin{bmatrix} M_{21}^R \\ M_{23}^R \\ M_{32}^R \\ M_{34}^R \end{bmatrix} = \begin{bmatrix} 1 & 0 \\ -1 & 0 \\ 0 & 1 \\ 0 & -1 \end{bmatrix} \begin{bmatrix} M_2 \\ M_3 \end{bmatrix}$$

The flexibility sub-matrices for members 21 and 23 are:

$$\theta_{21} = [l/3EI]M_{21}$$

$$\begin{bmatrix} \theta_{23} \\ \theta_{32} \end{bmatrix} = l/EI \begin{bmatrix} 1/3 & -1/6 \\ -1/6 & 1/3 \end{bmatrix} \begin{bmatrix} M_{23} \\ M_{32} \end{bmatrix}$$

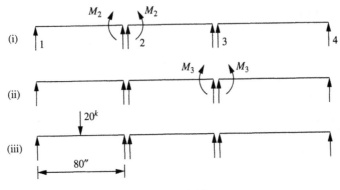

Figure 10.18

The matrix $[\bar{F}]$ is:

$$[\bar{F}] = 1/72 \begin{bmatrix} 12 & 0 & 0 & 0 \\ 0 & 8 & -4 & 0 \\ 0 & -4 & 8 & 0 \\ 0 & 0 & 0 & 6 \end{bmatrix}$$

The discontinuities produced by the application of the redundants to the cut-back structure are given by $[U]^T[\bar{F}][U]\{R\}$ and are:

$$\begin{bmatrix} \theta_2^R \\ \theta_3^R \end{bmatrix} = 1/72 \begin{bmatrix} 20 & 4 \\ 4 & 14 \end{bmatrix} \begin{bmatrix} M_2 \\ M_3 \end{bmatrix}$$

The displacements produced in the cut-back structure by the applied load are obtained from (iii) as:

$$\{H\} = \begin{bmatrix} \theta_{21}^H \\ \theta_{23}^H \\ \theta_{32}^H \\ \theta_{34}^H \end{bmatrix} = \begin{bmatrix} -Wl^2/16EI \\ 0 \\ 0 \\ 0 \end{bmatrix} = \begin{bmatrix} -50 \\ 0 \\ 0 \\ 0 \end{bmatrix}$$

The discontinuities produced by the applied load are given by $[U]^T\{H\}$ and are:

$$\begin{bmatrix} \theta_2^H \\ \theta_3^H \end{bmatrix} = \begin{bmatrix} -50 \\ 0 \end{bmatrix}$$

The support moments are given by $-[[U]^T[\bar{F}][U]]^{-1}[U]^T\{H\}$ and are:

$$\begin{bmatrix} M_2 \\ M_3 \end{bmatrix} = 12/55 \begin{bmatrix} 17.5 & -5 \\ -5 & 25 \end{bmatrix} \begin{bmatrix} 50 \\ 0 \end{bmatrix} = \begin{bmatrix} 191 \\ -54 \end{bmatrix}$$

Example 10.13

Determine the forces in the members of the frame shown in Figure 10.19. All the members are of uniform section.

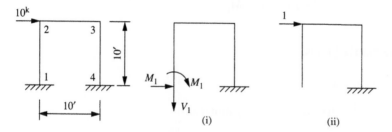

Figure 10.19

Solution

The reactions H_1, V_1, and M_1 may be considered to be the redundants, and the matrices $[U]$ and $[V]$ are obtained from (i) and (ii) as:

$$\begin{bmatrix} Q_{12}^R \\ M_{12}^R \\ Q_{23}^R \\ M_{23}^R \\ Q_{34}^R \\ M_{34}^R \end{bmatrix} = \begin{bmatrix} 1 & 0 & 0 \\ 0 & 0 & 1 \\ 0 & 1 & 0 \\ -10 & 0 & 1 \\ -1 & 0 & 0 \\ -10 & -10 & 1 \end{bmatrix} \begin{bmatrix} H_1 \\ V_1 \\ M_1 \end{bmatrix}$$

$$\begin{bmatrix} Q^W_{12} \\ M^W_{12} \\ Q^W_{23} \\ M^W_{23} \\ Q^W_{34} \\ M^W_{34} \end{bmatrix} = \begin{bmatrix} 0 \\ 0 \\ 0 \\ 0 \\ -1 \\ 0 \end{bmatrix} W_2$$

Ignoring axial effects, the flexibility sub-matrix for member 12 is:

$$\begin{bmatrix} y_{12} \\ \theta_{12} \end{bmatrix} = l/6EI \begin{bmatrix} 2l^2 & -3l \\ -3l & 6 \end{bmatrix} \begin{bmatrix} Q_{12} \\ M_{12} \end{bmatrix}$$

$$= l/6EI \begin{bmatrix} 200 & -30 \\ -30 & 6 \end{bmatrix} \begin{bmatrix} Q_{12} \\ M_{12} \end{bmatrix}$$

The matrix $[\bar{F}]$ is:

$$[\bar{F}] = \begin{bmatrix} 200 \\ -30 & 6 & & & & \text{symmetric} \\ 0 & 0 & 200 \\ 0 & 0 & -30 & 6 \\ 0 & 0 & 0 & 0 & 200 \\ 0 & 0 & 0 & 0 & -30 & 6 \end{bmatrix}$$

$$[U]^T[\bar{F}][U] = \begin{bmatrix} 1000 & 600 & -120 \\ 600 & 800 & -90 \\ -120 & -90 & 18 \end{bmatrix}$$

$$[U]^T[\bar{F}][V] = \begin{bmatrix} -100 \\ -300 \\ 30 \end{bmatrix}$$

$$[U][[U]^T[\bar{F}][U]]^{-1}[U]^T[\bar{F}][V] = \begin{bmatrix} 0.5 \\ 2.86 \\ -0.429 \\ -2.14 \\ -0.50 \\ 2.14 \end{bmatrix}$$

The final member forces, in kips and kip-ft units, are given by:

$$\begin{bmatrix} Q^W_{12} \\ M^W_{12} \\ Q^W_{23} \\ M^W_{23} \\ Q^W_{34} \\ M^W_{34} \end{bmatrix} + \begin{bmatrix} Q^R_{12} \\ M^R_{12} \\ Q^R_{23} \\ M^R_{23} \\ Q^R_{34} \\ M^R_{34} \end{bmatrix} = \begin{bmatrix} -5.00 \\ -28.60 \\ 4.29 \\ 21.40 \\ -5.00 \\ -21.40 \end{bmatrix}$$

(c) Nonprismatic members

When the value of EI varies along the length of a member, the integrals involved in the application of the flexibility matrix method may be evaluated by Simpson's rule. The member is divided into an even number of segments of equal length, and the integral of the function shown in Figure10.20 is given by:

$$\int_1^n e\,ds = \delta s(e_1 + 4e_2 + 2e_3 + 4e_4 + \cdots + 2e_{n-2} + 4e_{n-1} + e_n)/3$$

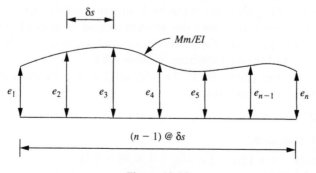

Figure 10.20

Example 10.14

Determine the support reactions of the symmetrical, continuous beam shown in Figure 10.21. The relative EI values are given in Table 10.1.

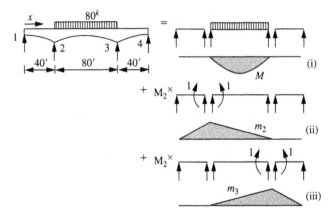

Figure 10.21

Table 10.1 Properties of continuous beam in Example 10.14

x ft	0	20	40	60	80	100	120
EI	1	3	10	3	1	3	10
M kip-ft	0	0	0	-600	-800	-600	0
m_2 kip-ft	0	1/2	1	3/4	1/2	1/4	0
m_3 kip-ft	0	0	0	1/4	1/2	3/4	1
Mm_2/EI	0	0	0	-150	-400	-50	0
m_2^2/EI	0	1/12	1/10	3/16	1/4	1/48	0
m_2m_3/EI	0	0	0	1/16	1/4	1/16	0

Solution

Because of the symmetry of the structure and loading, the support reactions at 2 and 3 are equal. The moment M_2 at these supports is taken as the redundant and releases introduced at 2 and 3 to produce the cut-back structure.

The external load applied to the cut-back structure produces the distribution of moment M shown at (i). Unit value of each redundant applied in turn to the cut-back structure produces the moments m_2 and m_3 shown at (ii) and (iii). Values of M, m_2 and m_3 are tabulated in Table 10.1.

The discontinuities produced at the releases 2 and 3 by the external load applied to the cut-back structure are:

$$\theta_2^W = \theta_3^W$$
$$= \int Mm_2 \, ds/EI$$
$$= 20(0 - 4 \times 150 - 2 \times 400 - 4 \times 50 - 0)/3$$
$$= -32{,}000/3$$

The discontinuity produced at release 2 by the application of unit value of M_2 in (ii) is:

$$f_{22} = \int m_2^2 \, ds/EI$$
$$= 20(0 + 4/12 + 2/10 + 12/16 + 2/4 + 4/48 + 0)/3$$
$$= 112/9$$
$$= f_{33}$$

where f_{33} is the discontinuity produced at release 3 by the application of unit value of M_2 in (iii).

The discontinuity produced at release 2 by the application of unit value of M_2 in (iii) is:

$$f_{23} = \int m_2 m_3 \, ds/EI$$
$$= 20(0 + 4/16 + 2/4 + 4/16 + 0)$$
$$= 20/3$$
$$= f_{32}$$

where f_{32} is the discontinuity produced at release 3 by the application of unit value of M_2 in (ii).

Since there are no discontinuities in the original structure at the positions of the releases:

$$\begin{bmatrix} f_{22} & f_{23} \\ f_{32} & f_{33} \end{bmatrix} \begin{bmatrix} M_2 \\ M_2 \end{bmatrix} = - \begin{bmatrix} \theta_2^W \\ \theta_3^W \end{bmatrix}$$

$$\begin{bmatrix} 112/9 & 20/3 \\ 20/3 & 112/9 \end{bmatrix} \begin{bmatrix} M_2 \\ M_2 \end{bmatrix} = - \begin{bmatrix} -32,000/3 \\ -32,000/3 \end{bmatrix}$$

and:

$$M_2 = 32,000/(37.3 + 20)$$
$$= 558 \text{ kip-ft}$$
$$V_1 = 558/40$$
$$= 14 \text{ kips} \cdots \text{ downward}$$
$$V_2 = 54 \text{ kips} \cdots \text{ upward}$$

(d) Concordant cable profile

The application of the pre-stressing force P, with an eccentricity e, to the continuous beam shown in Figure 10.22 results in the production of indeterminate reactions at the supports. The pre-stressing force tends to deflect the beam,

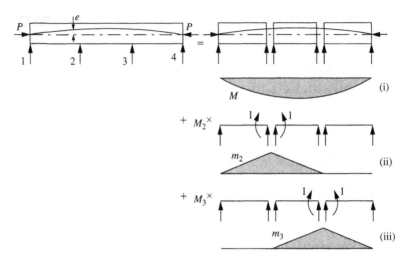

Figure 10.22

which is restrained against lateral displacement by the supports. This causes secondary moments in the beam, and the resultant line of thrust no longer coincides with the cable profile, as is the case with a statically determinate beam. The indeterminate moments M_2 and M_3 at the supports may be considered the redundants and releases introduced at 2 and 3 to produce the cut-back structure.

The application of the pre-stressing force to the cut-back structure produces the distribution of moment $M = Pe$ shown at (i). Unit value of each redundant applied in turn to the cut-back structure produces the moments m_2 and m_3 shown at (ii) and (iii).

The discontinuities produced at the releases 2 and 3 by the pre-stressing force applied to the cut-back structure are:

$$\theta_2^P = \int Pe m_2 \, ds/EI$$
$$\theta_3^P = \int Pe m_3 \, ds/EI$$

The discontinuities produced at release 2 by unit values of M_2 and M_3 applied in turn to the cut-back structure are:

$$f_{22} = \int m_2^2 \, ds/EI$$
$$f_{23} = \int m_2 m_3 \, ds/EI$$

The discontinuities produced at release 3 by unit values of M_2 and M_3 applied in turn to the cut-back structure are:

$$f_{32} = \int m_2 m_3 \, ds/EI$$
$$f_{33} = \int m_3^2 \, ds/EI$$

Since there are no discontinuities in the original structure at the positions of the releases:

$$\begin{bmatrix} \int m_2^2 \, ds/EI & \int m_2 m_3 \, ds/EI \\ \int m_2 m_3 \, ds/EI & \int m_3^2 \, ds/EI \end{bmatrix} \begin{bmatrix} M_2 \\ M_3 \end{bmatrix} = - \begin{bmatrix} \int Pem_2 \, ds/EI \\ \int Pem_3 \, ds/EI \end{bmatrix}$$

Expanding this expression gives:

$$\int (Pe + M_2 m_2 + M_3 m_3) m_2 \, ds/EI = 0$$
$$\int (Pe + M_2 m_2 + M_3 m_3) m_3 \, ds/EI = 0$$

The final distribution of moment in the beam due to the pre-stressing force and secondary effects is:

$$Pe' = Pe + M_2 m_2 + M_3 m_3$$

where the effective cable eccentricity is:

$$e' = e + M_2 m_2 / P + M_3 m_3 / P$$

and the line of thrust has been displaced by an amount:

$$M_2 m_2 / P + M_2 m_3 / P$$

A cable with an initial eccentricity e' produces discontinuities at releases 2 and 3 of:

$$\theta_2 = \int Pe' m_2 \, ds/EI$$
$$= \int (Pe + M_2 m_2 + M_3 m_3) m_2 \, ds/EI$$
$$= 0$$

$$\theta_3 = \int Pe' m_3 \, ds/EI$$
$$= \int (Pe + M_2 m_2 + M_3 m_3) m_3 \, ds/EI$$
$$= 0$$

Thus, there are no secondary moments produced on tensioning this concordant cable, and the resultant line of thrust coincides with the cable profile.

Example 10.15

Determine the distribution of moment produced in the frame shown in Figure 10.23 by the pre-stressing force and secondary effects. The magnitude of the pre-stressing force in the beam is 500 kips and in each column is 100 kips. The pre-stressing cable has zero eccentricity in the left-hand column, and the eccentricities in the beam and right-hand column are tabulated in Table 10.2 together with the EI values. The effects of axial compression in the frame may be neglected.

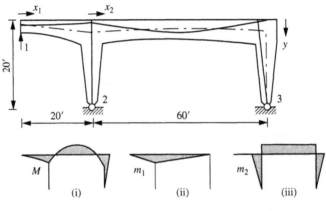

Figure 10.23

Solution

The vertical reaction V at support 1 and the horizontal reaction H at support 2 may be considered the redundants and releases introduced at 1 and 2.

The application of the pre-stressing force to the cut-back structure produces the distribution of moment M shown at (i). Unit values of V and H applied in turn to the cut-back structure produce the moments m_1 and m_2 shown at (ii) and (iii). These values are tabulated in Table 10.2 together with values of Mm_1/EI, Mm_2/EI, m_1^2/EI, m_2^2/EI, and m_1m_2/EI.

The discontinuities produced at the releases 1 and 2 by the pre-stressing force are:

$$y_1^W = \int Mm_1 \, ds/EI$$
$$= 10(0 + 832 + 667)/3 + 15(667 + 0 - 2500 + 208 + 0)/3$$
$$= -3128$$

$$x_2^W = \int Mm_2 \, ds/EI$$
$$= 15(-667 + 0 + 5000 - 832 - 667)/3 + 10(111 + 252 + 0)/3$$
$$= 15,380$$

Table 10.2 Properties of prestressed beam in Example 10.15

x_1 ft	0	10	20	—	—	—	—	—	—	—	—
x_2 ft	—	—	—	0	15	30	45	60	—	—	—
y ft	—	—	—	—	—	—	—	—	0	10	20
e ft	0	0.25	1.0	1.0	0	-1.0	0.25	2.0	0.5	0.25	0
EI	2	6	15	15	9	4	12	30	9	4	1
M	0	125	500	500	0	-500	125	1000	50	25	0
m_1	0	10	20	20	15	10	5	0	0	0	0
m_2	0	0	0	-20	-20	-20	-20	-20	20	10	0
Mm_1/EI	0	208	667	667	0	-1250	52	0	0	0	0
Mm_2/EI	0	0	0	-667	0	2500	-208	-667	111	63	0
m_1^2/EI	0	17	27	27	25	25	2	0	0	0	0
m_2^2/EI	0	0	0	27	44.4	100	33.3	13.3	44.4	25	0
m_1m_2/EI	0	0	0	-27	-33.3	-50	-8.3	0	0	0	0
Pe	0	125	500	500	0	-500	125	1000	50	25	0
Vm_1	0	-44	-88	-88	-66	-44	-22	0	0	0	0
Mm_2	0	0	0	118	118	118	118	118	-118	-59	0
Pe'	0	81	412	530	52	-426	221	1118	-68	-34	0

The discontinuities produced at releases 1 and 2 by the application of unit value of V are:

$$f_{11} = \int m_1^2 \, ds/EI$$
$$= 10(0 + 68 + 27)/3 + 15(27 + 100 + 50 + 8 + 0)/3$$
$$= 1242$$

$$f_{12} = \int m_1 m_2 \, ds/EI$$
$$= 15(-27 - 132 - 100 - 32 + 0)/3$$
$$= 1460$$

The discontinuity produced at release 2 by the application of unit value of H is:

$$f_{22} = \int m_2^2 \, ds/EI$$
$$= 15(27 + 176 + 200 + 132 + 13)/3 + 2 \times 10(44 + 100 + 0)/3$$
$$= 3700$$

Since there are no discontinuities in the original structure at the positions of the releases:

$$\begin{bmatrix} 1242 & -1460 \\ -1460 & 3700 \end{bmatrix} \begin{bmatrix} V \\ M \end{bmatrix} = - \begin{bmatrix} -3128 \\ 15380 \end{bmatrix}$$

and:

$$\begin{bmatrix} V \\ M \end{bmatrix} = - \begin{bmatrix} 4.42 \\ 5.90 \end{bmatrix}$$

The final distribution of moment in the frame due to the pre-stressing force and secondary effects is:

$$Pe' = Pe + Vm_1 + Mm_2$$

and values of Pe' are tabulated in Table 10.2.

(e) Interchanging cut-back systems

The vertical reaction V at the support 1 of the propped cantilever shown in Figure 10.24 is taken as the redundant. The external load applied to the cut-back structure at (i) produces the distribution of moment M, and unit value of

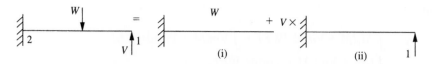

Figure 10.24

V applied to the cut-back structure at (ii) produces the moment m. The actual moment in the original structure is $(M + Vm)$.

The deflection under the applied load is obtained by applying unit virtual load, in place of W, to the structure, as shown in Figure 10.25. The moment in the structure is then $(M + Vm)/W$. The deflection under the load is given by:

$$y = \int (M + Vm)(M + Vm)\,ds/WEI$$
$$= \int M(M + Vm)\,ds/WEI + \int Vm(M + Vm)\,ds/WEI$$

Figure 10.25

However, to obtain the deflection y, it was shown in Section 3.2 that the unit virtual load may be applied to any cut-back structure that will support it. Applying the virtual load to the cut-back structures shown in Figure 10.26 (i) and (ii) produces the moments M/W and M_1/W, where M_1 is the moment produced in the cut-back structure at (ii) by the applied load W. Thus, the required deflection y is also given by:

$$y = \int M(M + Vm)\,ds/WEI$$

Figure 10.26

and:

$$y = \int M_1(M + Vm)\,ds/WEI$$

and:

$$y = \int M_1(M + Vm)\,ds/WEI + \int Vm(M + Vm)\,ds/WEI$$
$$= \int (M_1 + Vm)(M + Vm)\,ds/WEI$$

Thus, the actual moment in the original structure is given by:

$$(M + Vm) = (M_1 + Vm)$$

and this is equivalent to employing the two cut-back systems shown in Figure 10.27.

This interchanging of the cut-back systems may result in considerable simplification in the flexibility matrix method.

Example 10.16

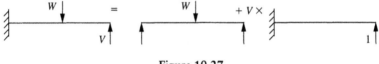

Figure 10.27

Determine the bending moment at the support 1 of the frame shown in Figure 10.28.

Using the cut-back systems shown, the redundants H_5, V_5, and M_5 are given by:

$$\begin{bmatrix} f_{11} & f_{12} & f_{13} \\ f_{21} & f_{22} & f_{23} \\ f_{31} & f_{32} & f_{33} \end{bmatrix} \begin{bmatrix} H_5 \\ V_5 \\ M_5 \end{bmatrix} = - \begin{bmatrix} x_5^W \\ y_5^W \\ \theta_5^W \end{bmatrix}$$

$$f_{11} = \int m_1^2 \, ds/EI$$
$$= 2h^3/3EI_1$$
$$f_{12} = \int m_1 m_2 \, ds/EI$$
$$= 0$$
$$= f_{21}$$
$$f_{13} = \int m_1 m_3 \, ds/EI$$
$$= -2h^2/2EI_1$$
$$= f_{31}$$

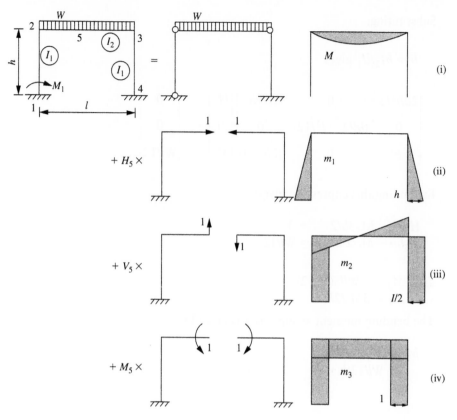

Figure 10.28

$$f_{22} = \int m_2^2 \, ds/EI$$
$$= hl^2/2EI_1 + l^3/12EI_2$$
$$f_{23} = \int m_2 m_3 \, ds/EI$$
$$= 0$$
$$= f_{32}$$
$$f_{33} = \int m_3^2 \, ds/EI$$
$$= 2h/I_1 + l/I_2$$
$$x_5^W = \int M m_1 \, ds/EI$$
$$= 0$$
$$y_5^W = \int M m_2 \, ds/EI$$
$$= 0$$
$$\theta_5^W = \int M m_3 \, ds/EI$$
$$= Wl^2/12EI_2$$

Substituting:

$b = hI_2/lI_1$ gives

$$\begin{bmatrix} 2bh^2/3 & 0 & -bh \\ 0 & (bl/2 + l^2/12) & 0 \\ -bh & 0 & (2b + 1) \end{bmatrix} \begin{bmatrix} H_5 \\ 0 \\ M_5 \end{bmatrix} = - \begin{bmatrix} 0 \\ 0 \\ Wl/12 \end{bmatrix}$$

Expanding this expression gives:

$H_5 bh - M_5 3b/2 = 0$
$H_5 bh - M_5 (2b + 1) = Wl/12$

and: $M_5 = -Wl/6(b + 2)$
$H_5 = 3M_5/2h$

The bending moment at support 1 is given by:

$M_1 = M_5 - hH_5$
$\quad = Wl/12(b + 2)$

Supplementary problems

Use the stiffness matrix technique to solve the following problems.
S10.1 The continuous beam shown in Figure S10.1 has the relative EI/l values shown ringed alongside the members. Set up the stiffness matrix for the beam and determine the bending moments at the supports due to the applied loading.

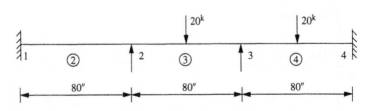

Figure S10.1

S10.2 Figure S10.2 shows a propped frame with members of uniform section and with a value of 16 kip-ft for Wl. Set up the stiffness matrix for the frame and determine the bending moments at the ends of each member.

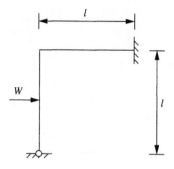

Figure S10.2

S10.3 Set up the stiffness matrix for the rigid frame shown in Figure S10.3 (i), in which all the members are of uniform section and there is a fixed support at 1 and a hinged support at 4. The inverse of this matrix is:

$$
\begin{bmatrix} \theta_2 \\ \theta_3 \\ \theta_4 \\ x_2/l \end{bmatrix} = l/EI \begin{bmatrix} 0.182 & & & \\ -0.023 & 0.159 & \text{symmetric} & \\ 0.114 & -0.046 & 0.477 & \\ 0.068 & 0.023 & 0.136 & 0.099 \end{bmatrix} \begin{bmatrix} P_{\theta2} \\ P_{\theta3} \\ P_{\theta4} \\ lP_{x2} \end{bmatrix}
$$

Determine the horizontal displacement of joint 2 and the clockwise rotation of joint 4 due to a clockwise unit moment applied at support 4. Hence, for the frame shown in Figure S10.3 (ii), determine the bending moment M_4 due to the indicated load H.

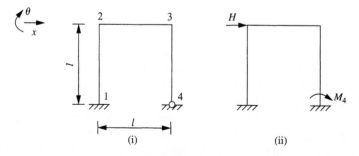

Figure S10.3

S10.4 Set up the stiffness matrix for the structure shown in Figure S10.4 and determine the forces in the members. The relative *EA* values are shown ringed alongside the members.

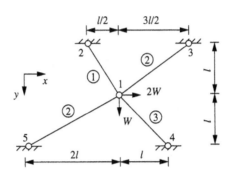

Figure S10.4

S10.5 Set up the stiffness matrix for the pile group shown in Figure S10.5 and determine the force in each pile. The piles are end-bearing and may be considered pin-ended. The pile cap may be assumed to be rigid, and the soil under the cap carries no load. The value of *AE/l* is constant for all piles, and it may be assumed that $\sin \alpha = 1.0$ and $\cos \alpha = 1/4$, where α = angle of inclination of a pile.

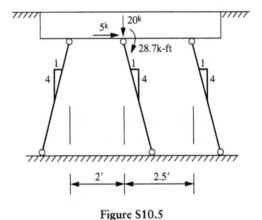

Figure S10.5

Use the flexibility matrix method to solve the following problems.
S10.6 The cable-stayed bridge shown in Figure S10.6 consists of a continuous main girder 12, supported on rollers where it crosses the rigid piers, and two cables, which are continuous over frictionless saddles at the tops of the towers. The modulus of elasticity, cross-sectional areas, and second moments of area of the members are given in the table. Consider the force R in the cable as the redundant and set up the flexibility matrix for the cut-back structure.

Determine the force R in the cables produced by a uniform load of 1 kip/ft over the girder.

Member	E kips/in²	I in⁴	A in²
Cable	30,000	-	20
Tower	3000	-	100
Girder	3000	100,000	1000

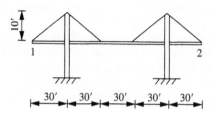

Figure S10.6

S10.7 The symmetrical, two-hinged arch shown in Figure S10.7 has a constant second moment of area, and the arch axis is defined by the coordinates given in the table. Determine the value of the horizontal thrust at the supports.

x ft	0	7.5	16.8	27.4	39.0	50.0
y ft	0	9.4	17.0	22.6	26.0	27.2
s ft	0	12.0	24.0	36.0	48.0	60.0

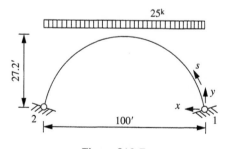

Figure S10.7

References

1. Felton, L. P. and Nelson, R. B. Matrix structural analysis. New York. John Wiley and Sons. 1996.
2. Sack, L. S. Matrix structural analysis. Prospect Heights, IL. Waveland Press. 1994.

3. Sennett, R. E. Matrix analysis of structures. Upper Saddle River, NJ. Prentice Hall. 1994.
4. McGuire, W. et al. Matrix structural analysis. New York. John Wiley and Sons. 2000.
5. Hoit, M. I. Computer assisted structural analysis and modeling. Upper Saddle River, NJ. Prentice Hall. 1995.
6. Balfour, J. A. D. Computer analysis of structural frameworks. Reisterstown, MD. Nichols Publishing Company. 1986.
7. Fleming, J. F. Computer analysis of structural systems. New York. McGraw Hill. 1989.
8. Mehringer, V. Computer-aided analysis and design of steel frames. AISC Engineering Journal. July 1985. pp. 143–149.
9. Wright, P. M. Simplified formulation of stiffness matrices. Proc. Am. Soc. Civil Eng. 89 (ST2). April 1963. pp. 33–48, and 89(ST6). December 1963. pp. 45–49.
10. Lightfoot, E. The analysis of grid frameworks and floor systems by the electronic computer. Struct Eng. 38. December 1960. pp. 388–402.
11. Tezcan, S. S. Stiffness analysis of grid frameworks. Trans. Eng. Inst. Canada. 8. July 1965. pp. 1–15.
12. Tezcan, S. S. Direct stiffness analysis of large frames by computers. Trans. Eng. Inst. Canada. 8. December 1965. pp. 1–10.
13. Livesley, R. K. The analysis of rigid frames by an electronic digital computer. Engineering. 176. August 1953. pp. 230–233, and 277–278.
14. Tocher, J. L. Selective inversion of stiffness matrices. Proc. Am. Soc. Civil Eng. 92 (ST1). February 1966. pp. 75–88.
15. Allwood, R. J. The solution of large elastic structures. Proceedings of the symposium on the use of computers in civil engineering, Lisbon. October 1962. Vol. 1. Lisbon, Laboratorio Nacional de Engenharia Civil. 1962. pp. 7.1–7.16.
16. Filho, F. V. Matrix analysis of plane rigid frames. Proc. Am. Soc. Civil Eng. 86 (ST7). July 1960. pp. 95–108.

11 Elastic instability

Notation

c_{12}	carry-over factor for a member 12 from end 1 to end 2
E	Young's modulus
I	second moment of area of a member
j	stability function $= 2(q + r) - \rho\pi^2$
j'	stability function $= q' - \rho\pi^2$
l	length of a member
m	stability function $= 2(q + r)/j$
m'	stability function $= q'/j'$
M_{12}	moment acting at end 1 of a member 12
M_{12}^F	fixed-end moment at end 1 of a member 12
n	stability function $= q - m(q + r)/2$
n'	stability function $= n\{1 - (o/n)^2\}$
o	stability function $= r - m(q + r)/2$
P	axial force in a member
P_E	Euler load
P_{x1}	total internal force, acting in the x-direction, produced at joint 1 by the joint displacements
$P_{\theta 1}$	total internal moment produced at joint 1 by the joint displacements
P_{12}'	axial force produced at end 1 of a member 12 by the joint displacements, referred to the member axis
q	stability function $= sl/EI = \alpha l(\sin\alpha l - \alpha l\cos\alpha l)/(2 - 2\cos\alpha l - \alpha l\sin\alpha l)$
q'	stability function $= q(1 - c^2)$
Q_{12}	shear force produced at end 1 of a member 12 by the joint displacements
Q_{12}'	shear force produced at end 1 of a member 12 by the joint displacements, referred to the member axis
r	stability function $= qc = \alpha l(\alpha l - \sin\alpha l)/(2 - 2\cos\alpha l - \alpha l\sin\alpha l)$
s_{12}	restrained stiffness at end 1 of a member 12
$[S]$	stiffness matrix for the whole structure
t	stability function $= 6/(q + r)$
W	applied load
W_c	critical load
$\{W\}$	vector of external loads applied at the joints
x	horizontal displacement

y vertical displacement
α $= (P/EI)^{0.5}$
θ rotation
λ $= \cos \phi$
μ $= \sin \phi$
ρ $= P/P_E$
τ $= \lambda + \mu \tan \phi$
ϕ angle of rotation due to sway
$\{\Delta\}$ vector of joint displacements

11.1   Introduction

If the applied loads on a structure are continuously increased, the axial forces and bending moments in the members increase until collapse eventually occurs either by plastic yielding, by buckling of the members, or by a combination of both. When the axial forces are appreciable, the dominant factor is buckling. In this chapter, attention will be confined to structures in which the stresses in the members are entirely within the elastic range at collapse and buckling occurs in the plane of the structure. The load at which the structure collapses is known as the critical load.

The effect of an axial compression on a member is to reduce its stiffness and of an axial tension is to increase its stiffness. As the applied loads on a structure increase, the member forces increase and the overall resistance of the structure to any random disturbance decreases. At the critical load, the structure offers no resistance to the disturbance, the configuration of the structure is not unique, and any displaced position may be maintained without additional load. An alternative definition of the elastic failure load for structures subjected to primary bending moments is the load at the transition from stable to unstable equilibrium. The members of such a structure are subjected to deformations associated with the buckling mode before failure occurs. The load-displacement curve reaches a peak at the failure load, with increased displacement being produced by decreasing load.

Several methods are available for determining the critical load. The stiffness matrix of the structure may be formed and, by trial and error, the critical load is obtained as the load that produces a singular matrix[1,2]. Alternatively, the largest latent root of a matrix, derived from the stiffness matrix, may be used as a criterion[3]. Moment-distribution methods may be applied to determine the stiffness offered by the structure to a random disturbance[4,5,6]. Alternatively, the rate of convergence of the moment-distribution technique may be used as a criterion[7]. An estimate of the critical load may be determined from a mathematical approximation to the applied load-stiffness relationship of the structure[8]. The first method is readily applied to simple structures; for complex structures, the determinant of the stiffness matrix may be evaluated by computer.

The analysis here is confined to structures consisting of straight prismatic members. The critical loads of structures consisting of non-prismatic members may be obtained similarly with the aid of additional tabulated functions[9].

The stability of structures in both the elastic and partially plastic ranges has been covered by Horne and Merchant[10].

11.2 Effect of axial loading on rigid frames without sway

(a) Modified moment-distribution procedure

The restrained stiffness s of the straight, prismatic member 12 subjected to an axial force P, as shown in Figure 11.1, is the bending moment required to produce unit rotation at end 1, end 2 being fixed. The carry-over factor is the ratio of the moment induced at 2 to the moment required at 1. The bending moment at any section at distance x from 1 is given by:

$$-EI\ d^2y/dx^2 = s - sx(1 + c)/l + Py$$

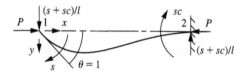

Figure 11.1

The general solution of this differential equation is:

$$y = A \cos \alpha x + B \sin \alpha x - s/P + sx(1 + c)/Pl$$

where $\alpha = (P/EI)^{0.5}$. The constants A and B are determined from the end conditions, $y = 0$ at $x = 0$ and at $x = l$ and:

$$y = s\{\cos \alpha x - (\cot \alpha l + c \operatorname{cosec} \alpha l) \sin \alpha x - 1 + x(1 + c)/l\}/P$$

The value of c is determined from the end condition, $dy/dx = 0$ at $x = l$ and:

$$c = (\alpha l - \sin \alpha l)/(\sin \alpha l - \alpha l \cos \alpha l)$$

The value of s is determined from the end condition, $dy/dx = 1$ at $x = 0$ and:

$$s = \alpha EI(1 - \alpha l \cot \alpha l)/2(\tan \alpha l/2 - \alpha l/2)$$

Values of s and c have been tabulated[11,12] and graphed[13,14,15,16]. In Table 11.1 the functions $q = sl/EI$ and $r = qc$ have been tabulated in terms of the variable $\rho = P/P_E = \alpha^2 l^2/\pi^2$ where P_E is the Euler buckling load for a column hinged at

Table 11.1 Stability functions for frames without sway

ρ	q	r	q'	t	ρ	q	r	q'	t
4.0	$-\infty$	∞	0.00	∞	1.3	1.89	2.69	-1.94	1.310
3.9	-78.33	78.57	0.49	24.770	1.2	2.09	2.61	-1.17	1.277
3.8	-38.17	38.65	0.96	12.610	1.1	2.28	2.54	-0.53	1.245
3.7	-24.68	25.39	1.42	8.555	1.0	2.47	2.47	0.00	1.216
3.6	-17.87	18.79	1.89	6.523	0.9	2.65	2.41	0.46	1.188
3.5	-13.72	14.85	2.35	5.309	0.8	2.82	2.35	0.86	1.162
3.4	-10.91	12.24	2.83	4.497	0.7	2.98	2.29	1.22	1.138
3.3	-8.86	10.40	3.33	3.916	0.6	3.14	2.24	1.54	1.115
3.2	-7.30	9.02	3.86	3.480	0.5	3.29	2.19	1.83	1.093
3.1	-6.05	7.96	4.42	3.141	0.4	3.44	2.15	2.10	1.073
3.0	-5.03	7.12	5.05	2.868	0.3	3.59	2.11	2.35	1.053
2.9	-4.18	6.44	5.77	2.646	0.2	3.73	2.07	2.58	1.035
2.8	-3.44	5.88	6.61	2.460	0.1	3.87	2.03	2.80	1.127
2.7	-2.81	5.42	7.63	2.302	0.0	4.00	2.00	3.00	1.000
2.6	-2.25	5.02	8.95	2.167	-0.2	4.26	1.94	3.37	0.969
2.5	-1.75	4.68	10.75	2.049	-0.4	4.50	1.88	3.71	0.940
2.4	-1.30	4.38	13.47	1.946	-0.6	4.74	1.83	4.03	0.913
2.3	-0.89	4.13	18.19	1.855	-0.8	4.96	1.79	4.31	0.889
2.2	-0.52	3.90	28.78	1.774	-1.0	5.18	1.75	4.58	0.867
2.1	-0.18	3.70	77.83	1.702	-1.5	5.68	1.67	5.19	0.817
2.0	0.14	3.53	-86.86	1.636	-2.0	6.15	1.60	5.73	0.775
1.9	0.44	3.37	-25.35	1.577	-2.5	6.58	1.54	6.22	0.738
1.8	0.72	3.22	-13.78	1.522	-3.0	6.99	1.50	6.67	0.707
1.7	0.98	3.10	-8.83	1.473	-3.5	7.37	1.46	7.08	0.679
1.6	1.22	2.98	-6.03	1.427	-4.0	7.74	1.43	7.47	0.655
1.5	1.46	2.87	-4.22	1.385	-5.0	8.42	1.38	8.18	0.612
1.4	1.68	2.78	-2.92	1.346	-7.0	9.62	1.30	9.45	0.549

both ends and is given by $P_E = \pi^2 EI/l^2$. Positive values of ρ indicate that P is compressive and negative values indicate that P is tensile. For negative values of ρ, α is imaginary, and the trigonometrical functions in the expressions for c and s are replaced by their hyperbolic equivalents.

The modified stiffness of the member 12 when the end 2 is hinged is given by:

$$s' = s(1 - c^2)$$

and values of $q' = s'l/EI$ are tabulated in Table 11.1.

Axial loading in a member also modifies the fixed-end moments in the member. The bending moment at any section a distance x from 1, for the uniformly loaded member shown in Figure 11.2, is given by:

$$-EI\,d^2y/dx^2 = -M^F + Wx/2 - Wx^2/2l + Py$$

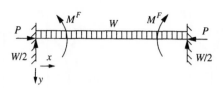

Figure 11.2

The general solution of this differential equation is:

$$y = A\cos \alpha x + B\sin \alpha x + M^F/P + W(x^2 - xl - 2/\alpha^2)/2Pl$$

where $\alpha = (P/EI)^{0.5}$. The constants A and B are determined from the end conditions, $y = 0$ at $x = 0$ and $dy/dx = 0$ at $x = l/2$ and:

$$dy/dx = (W/l - \alpha^2 M^F)(-\sin \alpha x + \cos \alpha x \tan \alpha l/2)/P\alpha + Wx/Pl - Wl/2P$$

The value of M^F is determined from the end condition, $dy/dx = 0$ at $x = 0$ and:

$$M^F = W(2 - \alpha l \cot \alpha l/2)/2\alpha^2 l$$
$$= tWl/12$$

where $t = 6/(q + r)$ is a magnification factor to allow for the effect of the axial load and is tabulated in Table 11.1. Similarly, the fixed-end moments for members subjected to concentrated loads may be obtained, and a wide range of values have been presented[13,14,15]. The notation adopted in the stability analysis of frames without sway is shown in Figure 11.3.

A graphical representation of the functions is shown in Figure 11.4.

Example 11.1

Determine the moments in the frame shown in Figure 11.5 for a value of $W = 0.625P_E$. All the members are of uniform section.

Solution

Neglecting the axial force in the beam, the stiffness (allowing for symmetry) and fixed-end moments for the beam are:

$$s_{23} = 2EI/l \quad \text{and} \quad M^F_{23} = -0.25Wl$$

For a value of $W = 0.625P_E$ and $\rho = 1.25$, the stiffness and carry-over factors for the columns are obtained from Table 11.1 as:

$$s_{21} = qEI/l = 2EI/l$$
$$c_{21} = r/q = 2.65/2 = 1.325$$

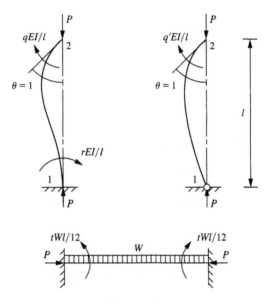

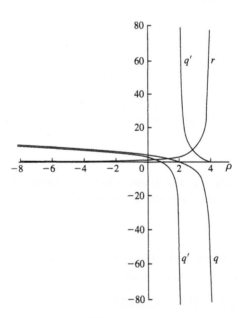

Figure 11.3

Figure 11.4

The distribution factors at joint 2 are 0.5 and the final moments are:

$M_{21} = 0.25\,Wl \times 0.5 = 0.125Wl$

$M_{12} = 0.125Wl \times 1.325 = 0.166Wl$

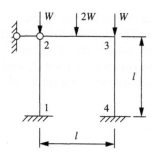

Figure 11.5

The corresponding values obtained when the axial forces in the columns are neglected are:

$$M_{21} = 0.25Wl \times 0.67 = 0.167Wl$$
$$M_{12} = 0.167Wl \times 0.5 = 0.083Wl$$

(b) Determination of the critical load

Consider the small clockwise moment M applied to the rigid joint o of Figure 11.6, causing the joint to rotate through an angle θ. The stiffness of joint o is:

$$s_o = \Sigma s_{on}$$

and

$$\theta = M/s_o$$

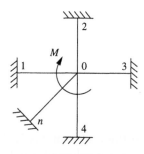

Figure 11.6

If axial compressive forces are applied to the n members, their stiffnesses decrease, s_o decreases, and θ increases. As the axial forces are continuously increased, s_o continues to decrease until eventually, at the critical load, elastic instability occurs at the joint and θ becomes infinite. In general, it is unnecessary to apply the exciting moment M since initial imperfections in the members and unavoidable eccentricity of the axial forces produce sufficient disturbance.

Instability of the propped frame shown in Figure 11.7 occurs in the symmetrical mode shown at (i). The applied loads are transferred directly to the foundations, and there is no axial force in beam 23. The rotation of joint 3 is equal and opposite to the rotation of joint 2, and the critical load may be determined by considering the stiffness of joint 2 only. Thus:

$$s_2 = s_{23} + s_{21}$$
$$= 2EI/l + q'_{21}EI/l$$

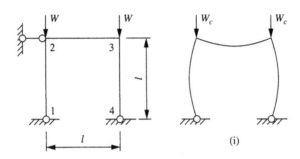

Figure 11.7

At the critical load:

$$s_2 = 0$$

and

$$q'_{21} = -2$$

From Table 11.1:

$$\rho = W_c/P_E$$
$$= 1.31$$

and

$$W_c = 1.31\pi^2 EI/l^2$$

For zero applied load, the stiffness of joint 2 is $5EI/l$. The variation of the stiffness may be determined for different values of ρ and is plotted in Figure 11.8.

When several joints in a structure are involved, it is necessary to set up the stiffness matrix of the structure. In section 10.2 it was established that this is given by:

$$\{W\} = [S]\{\Delta\}$$

where $\{W\}$ is the vector of external loads applied at the joints, $\{\Delta\}$ is the vector of joint displacements, and $[S]$ is the stiffness matrix of the whole structure.

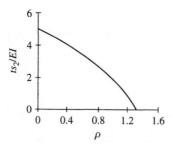

Figure 11.8

As the external loads are continuously increased, the axial forces in the members increase and the elements in the stiffness matrix change. At any particular loading stage, the joint displacements are given by:

$$\{\Delta\} = [S]^{-1}\{W\}$$

and

$$[S]^{-1} = \text{adj } [S]/|S|$$

where adj $[S]$ is the transpose of the matrix of the cofactors of $[S]$ and $|S|$ is the determinant of $[S]$. At the critical load, all joint displacements become infinite and $|S| = 0$, i.e., the stiffness matrix is singular.

To determine the critical load for a particular structure, a trial-and-error procedure is adopted. A trial value is chosen for the load factor, and the axial forces in the members are estimated. From Table 11.1 the values of the stability functions are obtained and substituted in the stiffness matrix. The determinant of the stiffness matrix is now evaluated. Several values of the load factor are tried until the value producing a singular stiffness matrix is obtained. A lower bound and an upper bound on the critical load may be obtained by considering the member with the largest axial force. If this member is pin-ended, instability will occur at a value of $\rho = 1$. If this member is fixed-ended, instability will occur at a value of $\rho = 4$. It is unnecessary to evaluate the determinant for all trial values, as the condition of the structure may be deduced by inspection of the stiffness matrix. Thus, if the matrix is dominated by its leading diagonal, the structure is stable; and if any element on the leading diagonal is zero or negative, the structure is unstable and the critical load has been exceeded.

Example 11.2

Determine the value of W at which elastic instability occurs in the rigid frame shown in Figure 11.9. All the members are of uniform section.

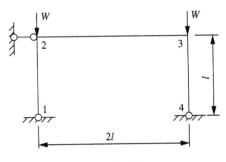

Figure 11.9

Solution

The stiffness of joint 2 is:

$$s_2 = s_{23} + s_{21}$$
$$= 2EI/2l + q'_{21}EI/l$$

At the critical load:

$$s_2 = 0$$

and

$$q'_{21} = -1$$
$$\rho_{21} = 1.18$$

Hence, instability occurs at a value of the applied load of:

$$W_c = 1.18 P_E$$

where P_E is the Euler load for a column.

Example 11.3

Determine the value of W at which elastic instability occurs in the rigid frame shown in Figure 11.10. All the members are of uniform section.

Solution

Instability occurs in the mode shown at (i), and the rotation of joint 5 is equal and of the same sense as the rotation of joint 2. Thus, instability at joints 2 and 3 only need be considered.

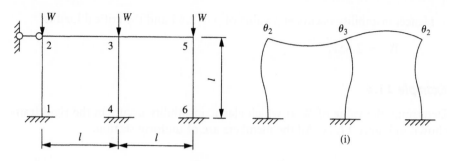

Figure 11.10

The axial strains in the columns produce no flexural deformations, and the stiffness sub-matrices of the members are given by:

$$\begin{bmatrix} M_{23} \\ M_{32} \end{bmatrix} = EI/l \begin{bmatrix} 4 & 2 \\ 2 & 4 \end{bmatrix} \begin{bmatrix} \theta_2 \\ \theta_3 \end{bmatrix}$$

$$\begin{bmatrix} M_{35} \\ M_{53} \end{bmatrix} = EI/l \begin{bmatrix} 4 & 2 \\ 2 & 4 \end{bmatrix} \begin{bmatrix} \theta_3 \\ \theta_2 \end{bmatrix}$$

$$M_{21} = qEI\theta_2/l$$
$$M_{34} = qEI\theta_3/l$$

Collecting the relevant terms gives:

$$\begin{bmatrix} P_{\theta2} \\ P_{\theta3} \end{bmatrix} = EI/l \begin{bmatrix} (4+q) & 2 \\ 2 & (8+q) \end{bmatrix} \begin{bmatrix} \theta_2 \\ \theta_3 \end{bmatrix}$$

For a value of $\rho = 2.6$, $q = -2.25$ and the determinant of the stiffness matrix is:

$$\begin{vmatrix} 1.75 & 2 \\ 4 & 5.75 \end{vmatrix} = 2.10$$

For a value of $\rho = 2.7$, $q = -2.81$ and the determinant is:

$$\begin{vmatrix} 1.19 & 2 \\ 4 & 5.19 \end{vmatrix} = -1.82$$

Hence, instability occurs at a value of $\rho = 2.65$ and the critical load is:

$$W_c = 2.65P_E$$

Example 11.4

Determine the value of W at which elastic instability occurs in the rigid frame shown in Figure 11.11. All the members are of uniform section.

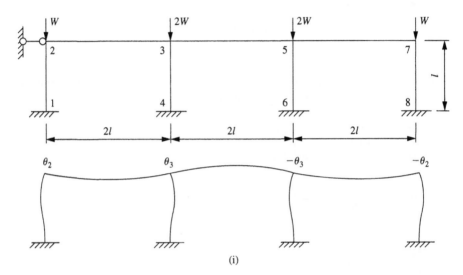

(i)

Figure 11.11

Solution

Instability occurs in the mode shown at (i), and only joints 2 and 3 need be considered.

Neglecting the effect of axial strains in the columns, the stiffness sub-matrices of the members are given by:

$$\begin{bmatrix} M_{23} \\ M_{32} \end{bmatrix} = EI/l \begin{bmatrix} 2 & 1 \\ 1 & 2 \end{bmatrix} \begin{bmatrix} \theta_2 \\ \theta_3 \end{bmatrix}$$

$$M_{35} = EI\theta_3/l$$
$$M_{21} = q_{21}EI\theta_2/l$$
$$M_{34} = q_{34}EI\theta_3/l$$

Collecting the relevant terms gives:

$$\begin{bmatrix} P_{\theta 2} \\ P_{\theta 3} \end{bmatrix} = EI/l \begin{bmatrix} (2 + q_{21}) & 1 \\ 1 & (3 + q_{34}) \end{bmatrix} \begin{bmatrix} \theta_2 \\ \theta_3 \end{bmatrix}$$

For values of $\rho_{21} = 1.3$ and $\rho_{34} = 2.6$, $q_{21} = 1.89$ and $q_{34} = -2.25$ and the determinant of the stiffness matrix is:

$$\begin{vmatrix} 3.89 & 1 \\ 1 & 0.75 \end{vmatrix} = 1.92$$

For values of $\rho_{21} = 1.4$ and $\rho_{34} = 2.8$, $q_{21} = 1.68$ and $q_{34} = -3.44$ and the determinant of the stiffness matrix is:

$$\begin{vmatrix} 3.68 & 1 \\ 1 & -0.44 \end{vmatrix} = -2.62$$

Hence, instability occurs at a value of $\rho_{21} = 1.34$ and the critical load is:

$$W_c = 1.34 P_E$$

Example 11.5

Determine the value of W at which elastic instability occurs in the rigid truss shown in Figure 11.12. All the members are the same length and are of uniform section.

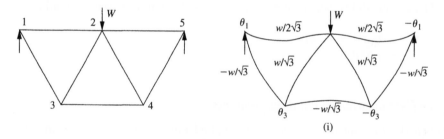

Figure 11.12

Instability occurs in the mode shown at (i), there is zero rotation at joint 2, and only joints 1 and 3 need be considered. The axial forces in the members, assuming all joints in the truss are pinned, are shown at (i), and the relationships between the P/P_E values are $\rho_{23} = -\rho_{13} = -\rho_{34} = 2\rho_{12}$.

The stiffness sub-matrices of the members are given by:

$$\begin{bmatrix} M_{13} \\ M_{31} \end{bmatrix} = EI/l \begin{bmatrix} q_{13} & r_{13} \\ r_{13} & q_{13} \end{bmatrix} \begin{bmatrix} \theta_1 \\ \theta_3 \end{bmatrix}$$

$$\begin{bmatrix} M_{34} \\ M_{43} \end{bmatrix} = EI/l \begin{bmatrix} q_{34} & r_{34} \\ r_{34} & q_{34} \end{bmatrix} \begin{bmatrix} \theta_3 \\ -\theta_3 \end{bmatrix}$$

$$M_{12} = q_{12}EI\theta_1/l$$
$$M_{32} = q_{23}EI\theta_3/l$$

Collecting the relevant terms gives:

$$\begin{bmatrix} P_{\theta 1} \\ P_{\theta 3} \end{bmatrix} = EI/l \begin{bmatrix} (q_{12} + q_{13}) & r_{13} \\ r_{13} & (q_{13} + q_{23} + q_{34} - r_{34}) \end{bmatrix} \begin{bmatrix} \theta_1 \\ \theta_3 \end{bmatrix}$$

For a value of $\rho_{12} = 1.7$, $q_{12} = 0.98$, $q_{13} = 7.30$, $q_{23} = -10.91$, $q_{34} = 7.30$, and $r_{34} = r_{13} = 1.47$ and the determinant of the stiffness matrix is:

$$\begin{vmatrix} 8.28 & 1.47 \\ 1.47 & 2.22 \end{vmatrix} = 16.22$$

For a value of $\rho_{12} = 1.75$, $q_{12} = 0.85$, $q_{13} = 7.37$, $q_{23} = -13.72$, $q_{34} = 7.37$, and $r_{34} = r_{13} = 1.46$ and the determinant of the stiffness matrix is:

$$\begin{vmatrix} 8.22 & 1.46 \\ 1.46 & -0.44 \end{vmatrix} = -5.75$$

Hence, instability occurs at a value of $\rho_{12} = 1.74$ and the critical load is:

$$W_c = 2(3)^{0.5} \times 1.74 P_E$$
$$= 6.03 P_E$$

(c) Effect of primary bending moments

When the loading on a structure is applied only at the joints, the members remain straight prior to buckling and no bending moments are produced by the loads. When the loading on a structure is applied within the lengths of the members, primary bending moments are produced that in general cause deformations associated with the buckling mode. In the case of a single-bay frame with loading applied to the beam, the buckling load is lower than the critical load of an identical frame with the loading system replaced by statically equivalent loads at the joints[17,18,19]. The buckling load, associated with the symmetrical mode of instability, is considerably decreased by primary bending moments, owing to the fact that the axial force in the beam reduces its stiffness. The buckling load, associated with the side-sway mode of instability, is only slightly decreased by primary bending moments, and the actual loading system may be replaced by statically equivalent loads at the joints with very small error.

Example 11.6

Determine the value of W at which elastic instability occurs in the rigid frame shown in Figure 11.13. All the members are of uniform section.

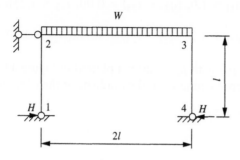

Figure 11.13

Solution

The primary moments produce an axial thrust in the beam that is equal to the horizontal reaction at the base of the columns. Allowing for this axial thrust, the stiffness of joint 2 and the fixed-end moment at joint 2 due to the distributed load are:

$$s_2 = s_{21} + s_{23}$$
$$= \{q'_{21} + (q_{23} - r_{23})/2\}EI/l$$
$$M^F_{23} = -t_{23}Wl/6$$

Equating internal and external moments at joint 2:

$$\{q'_{21} + (q_{23} - r_{23})/2\}EI\theta_2/l - t_{23}Wl/6 = 0$$

The axial thrust in the beam is:

$$P_{23} = H$$
$$= EI\theta_2 q'_{21}/l^2$$

Eliminating θ_2 from these two equations gives:

$$P_{23}\{q'_{21} + (q_{23} - r_{23})/2\} - q'_{21}t_{23}W/6 = 0$$

Dividing by $\pi^2 EI/l^2$ gives:

$$\rho_{23}\{q'_{21} + (q_{23} - r_{23})/2\} - 4q'_{21}t_{23}\rho_{21}/3 = 0$$

Values of ρ_{23} are assumed, and the corresponding values of ρ_{21} are obtained from this expression by trial and error.

For a value of $\rho_{23} = 0.4$, $(q_{23} - r_{23}) = 1.29$, $t_{23} = 1.073$ and $\rho_{21} = 0.36$, $q'_{21} = 2.20$.

For a value of $\rho_{23} = 0.8$, $(q_{23} - r_{23}) = 0.47$, $t_{23} = 1.162$ and $\rho_{21} = 0.59$, $q'_{21} = 1.58$.

For a value of $\rho_{23} = 1.0$, $(q_{23} - r_{23}) = 0.00$, $t_{23} = 1.216$ and $\rho_{21} = 0.617$, $q'_{21} = 1.49$.

For a value of $\rho_{23} = 1.2$, $(q_{23} - r_{23}) = -0.52$, $t_{23} = 1.277$ and $\rho_{21} = 0.59$, $q'_{21} = 1.58$.

The variation of ρ_{21} with ρ_{23} is shown plotted in Figure 11.14. At the origin, with ρ_{21} and ρ_{23} approaching zero, the gradient of the curve is given by:

$$\rho_{21}/\rho_{23} = 3(3 + 2/2)/12$$
$$= 1$$

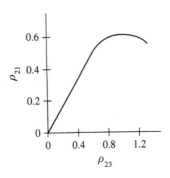

Figure 11.14

The maximum value of W that the frame can sustain is the critical value and is given by the maximum value of ρ_{21}. Hence, instability occurs at a value of $\rho_{21} = 0.617$, and the critical load is

$$W_c = 2 \times 0.617 P_E$$
$$= 1.234 P_E$$

where P_E is the Euler load for a column.

This result may be compared with the value of $\rho_{21} = 1.18$ obtained in Example 11.2 for an identical frame with the loading applied at the joints.

11.3 Effect of axial loading on rigid frames subjected to sway

(a) Modified moment-distribution procedure

The methods of Sections 7.8 and 7.9 may be readily modified and applied to the analysis of structures subjected to axial loads. An initial estimate is required of the axial forces in the members and the corresponding stiffness and carry-over factors obtained. In addition, in setting up the sway equations allowance must be made for the magnification effect of the axial loads. A subsequent analysis may be performed if necessary, using revised values for the axial forces.

The sway moments produced in the straight, prismatic member 12, shown in Figure 11.15(i), are:

$$M_{12}^F = M_{21}^F = Ql/2$$
$$= s(1 + c)x/l$$

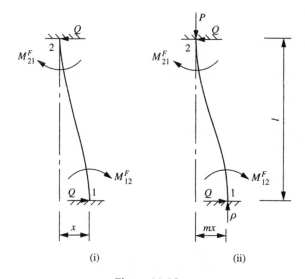

Figure 11.15

The application of an axial force P to the member, as shown at (ii), results in the increased lateral displacement mx, and the sway moments are:

$$M_{12}^F = M_{21}^F = mQl/2$$
$$= s(1 + c)mx/l$$

where m is the magnification factor. Thus, by establishing a relationship between P and m, the fixed-end moments in an axially loaded member subjected to sway may be obtained as m times the moments in the member with the axial load neglected. In addition, the lateral displacement of an axially loaded member subjected to a shear force Q is given by:

$$\delta = mx$$
$$= mQl^2/2s(1 + c)$$

Taking moments about end 1 for the axially loaded member shown at (ii):

$$M_{12}^F + M_{21}^F = Ql + Pmx$$
$$2s(1 + c)mx/l = Ql + Pmx$$

Neglecting the magnification effect and the moment due to P gives:

$$2s(1 + c)x/l = Ql$$

Thus:

$$2s(1 + c) \times (m - 1) = Pmxl = \rho\pi^2 EImx/l$$

and:

$$m = 2(q + r)/(2q + 2r - \rho\pi^2)$$

Similarly, when end 1 is hinged, the magnification factor is given by:

$$m' = q'/(q' - \rho\pi^2)$$

The sway equation for the second story of the symmetrical single-bay frame shown in Figure 11.16 is:

$$-2(M_{12} + M_{21}) = mh_2(W_3 + W_2)$$

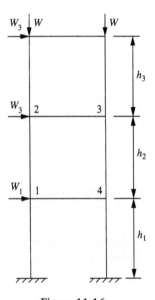

Figure 11.16

The initial fixed-end moments required to satisfy the sway equation are:

$$M_{12}^F = M_{21}^F$$
$$= -mh_2(W_3 + W_2)/4$$

The out-of-balance moment at joint 1 is distributed while allowing joint 2 to translate laterally without rotation so that the sway equation remains satisfied.

From Figure 11.17 where s' and c' refer to the modified stiffness of 12 and the modified carry-over factor from 1 to 2:

$$s' = s - mxs(1 + c)/l$$
$$s'c' = sc - mxs(1 + c)/l$$

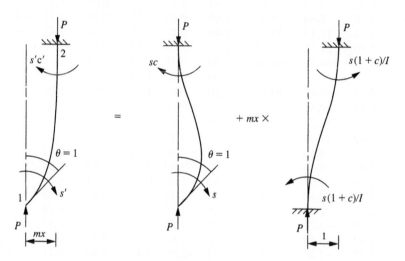

Figure 11.17

Neglecting the magnification effect, it is shown in Section 7.9 that:

$$s' = s - s(1 + c)/2$$
$$s'c' = sc - s(1 + c)/2$$

Allowing for the magnification effect:

$$s' = s - ms(1 + c)/2$$
$$= nEI/l$$
$$s'c' = sc - ms(1 + c)/2$$
$$= oEI/l$$

The modified carry–over factor is:

$$c' = o/n$$

The modified stiffness of the member 12 when end 2 is hinged is given by:

$$s'' = nEI\{1 - (o/n)^2\}/l$$
$$= n'EI/l$$

Values of m, n, n', and o have been tabulated[11] and graphed[13] and are presented in Table 11.2. The notation adopted in the stability analysis of frames with sway is shown in Figure 11.18.

Table 11.2 Stability functions for frames with sway

ρ	m	n	o	n'	ρ	m	n	o	n'
1.50	-1.41	4.51	5.93	-3.28	0.27	1.30	-0.10	-1.64	26.46
1.40	-1.82	5.73	6.83	-2.41	0.25	1.27	0.00	-1.57	∞
1.30	-2.50	7.60	8.40	-1.69	0.23	1.24	0.10	-1.51	-23.45
1.20	-3.85	11.13	11.65	-1.06	0.20	1.21	0.24	-1.43	-8.4
1.10	-7.90	21.32	21.57	-0.51	0.15	1.15	0.45	-1.30	-3.29
1.02	-40.33	101.47	101.51	-0.20	0.10	1.09	0.65	-1.19	-1.53
1.00	∞	∞	∞	0.00	0.05	1.04	0.83	-1.09	-0.59
0.98	40.73	-98.47	-98.51	0.20	0.00	1.00	1.00	-1.00	0.00
0.95	16.41	-38.41	-38.53	0.24	-0.10	0.93	1.31	-0.85	0.75
0.90	8.31	-18.33	-18.57	0.48	-0.20	0.86	1.59	-0.73	1.25
0.85	5.60	-11.58	-11.93	0.72	-0.40	0.76	2.06	-0.56	1.91
0.80	4.25	-8.16	-8.63	0.97	-0.60	0.69	2.47	-0.43	2.40
0.75	3.44	-6.08	-6.66	1.22	-0.80	0.63	2.83	-0.34	2,79
0.70	2.90	-4.67	-5.35	1.48	-1.00	0.58	3.15	-0.27	3.13
0.65	2.51	-3.63	-4.43	1.77	-1.20	0.55	3.45	-0.22	3.43
0.60	2.22	-2.84	-3.74	2.08	-1.40	0.51	3.72	-0.18	3.71
0.55	2.00	-2.21	-3.21	2.46	-1.60	0.49	3.98	-0.15	3.97
0.50	1.82	-1.69	-2.79	2.92	-1.80	0.46	4.22	-0.13	4.21
0.45	1.67	-1.25	-2.45	3.54	-2.00	0.44	4.44	-0.11	4.44
0.40	1.55	-0.88	-2.17	4.50	-2.50	0.40	4.97	-0.07	4.97
0.35	1.44	-0.55	-1.94	6.28	-3.00	0.36	5.44	-0.05	5.44
0.30	1.35	-0.26	-1.74	11.39	-3.50	0.34	5.88	-0.03	5.88

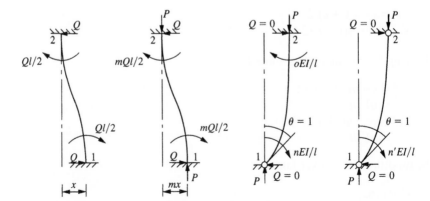

Figure 11.18

A graphical representation of the functions is shown in Figure 11.19.

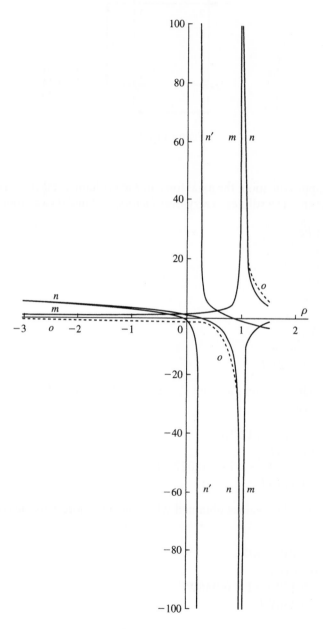

Figure 11.19

Example 11.7

Determine the moments in the frame shown in Figure 11.20 for a value of $W = 0.01375 P_E$. All the members are of uniform section.

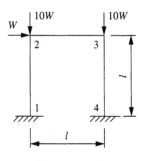

Figure 11.20

As a first approximation, the axial force in each column is taken as $10W$ and in the beam as zero. The stiffness, carry-over factors, and initial sway moments are:

$$s_{23} = 6EI/l$$
$$s_{21} = nEI/l$$
$$= 0.5EI/l$$
$$c_{21} = o/n$$
$$= -1.27/0.5$$
$$= -2.54$$
$$M_{12}^F = M_{21}^F$$
$$= -mWl/4$$
$$= -0.282Wl$$

The distribution factors at joint 2 are $d_{23} = 12/13$ and $d_{21} = 1/13$, and the final moments are:

$$M_{23} = 0.282Wl \times 12/13$$
$$= 0.260Wl$$
$$M_{12} = -0.282Wl - 2.54 \times 0.022Wl$$
$$= -0.338Wl$$

The corresponding values obtained when the axial forces in the columns are neglected are:

$$M_{23} = 0.25Wl \times 6/7$$
$$= 0.214Wl$$
$$M_{12} = -0.25Wl - 1 \times 0.036Wl$$
$$= -0.286Wl$$

Example 11.8

Determine the moments in the frame shown in Figure 11.21. For the columns, $EI = 3 \times 10^9$ lb/in^2 and for the beams, $EI = 6 \times 10^9$ lb/in^2.

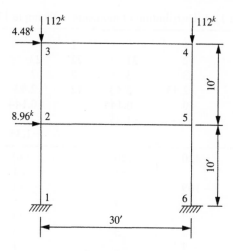

Figure 11.21

Solution

The Euler load for each column is:

$$P_E = \pi^2 \times 3 \times 10^9 / (120)^2$$
$$= 2,055,000 \text{ lb}$$

As a first approximation, the axial force in each column is taken as 112 kips and in each beam as zero. The ρ, m, n, and o factors are:

$$\rho = 112,000/2,055,000$$
$$= 0.055$$

and:

$$m = 1.045, \quad n = 0.81, \quad o = -1.1$$

Because of the skew symmetry, only the left half of the frame needs to be considered, with a modified stiffness of $6EI/l$ applied to the beams and no carry-over between the two halves.

The initial sway moments are:

$$M_{23}^F = M_{32}^F$$
$$= -4.48 \times 10 \times 12 \times 1.045/4$$
$$= -140 \text{ kip-in}$$
$$M_{12}^F = M_{21}^F$$
$$= -13.44 \times 10 \times 12 \times 1.045/4$$
$$= -420 \text{ kip-in}$$

The distribution procedure and the final moments are given in Table 11.3.

Table 11.3 Distribution of moments in Example 11.8

Joint	1		2		3	
Member	12	21	22'	23	32	33'
Relative EI/l	3	3	2	3	3	2
Modified stiffness	2.43	2.43	12	2.43	2.43	12
Distribution factor	0	0.144		0.144	0.168	
Carry-over factor	←	−1.35		←	−1.35	
				−1.35	→	
M^F sway	−420	−420		−140	−140	
Distribution		81		81	24	
Carry-over	−109			−32	−109	
Distribution		5		5	18	
Carry-over	−7			−24	−7	
Distribution		3		3	1	
Carry-over	−4			−1	−4	
Final moments, kip-in	−540	−331	438	−107	−217	217

(b) Modified matrix methods

The methods of Section 10.2 may be readily modified and applied to the analysis of structures subjected to axial loads. The stiffness matrix for a structure must be expressed in terms of the m, q, and r functions. An initial estimate is required of the axial force in each member in order to determine initial values of the stability functions, and the analysis is then carried out. A subsequent analysis may be required with revised values of the stability functions if the axial forces derived in the first analysis differ appreciably from the initial estimates.

The stiffness matrix for a straight prismatic member 12, allowing for axial strains and referred to the member axis, is given by:

$$
\begin{bmatrix} P'_{12} \\ lQ'_{12} \\ M_{12} \\ P'_{21} \\ lQ'_{21} \\ M_{21} \end{bmatrix} = EI/l
\begin{bmatrix}
A/I & & & & & \\
0 & 2(q+r)/m & & & \text{symmetric} & \\
0 & (q+r) & q & & & \\
-A/I & 0 & 0 & A/I & & \\
0 & -2(q+r)/m & -(q+r) & 0 & 2(q+r)/m & \\
0 & (q+r) & r & 0 & -(q+r) & q
\end{bmatrix}
\begin{bmatrix} x'_{12} \\ y'_{12}/l \\ \theta_{12} \\ x'_{21} \\ y'_{21}/l \\ \theta_{21} \end{bmatrix}
$$

Example 11.9

Determine the moments in the frame shown in Figure 11.20 for a value of $W = 1.01375P_E$. All the members are of uniform section.

Solution

The axial forces in the members may be estimated from the results of Example 11.7, neglecting the sway displacement, as:

$$P_{23} = 0.260W + 0.338W$$
$$= 0.598W$$
$$P_{34} = 10W + W - 0.676W$$
$$= 10.324W$$
$$P_{21} = 20W - 10.324W$$
$$= 9.676W$$

The corresponding stability functions are:

$$\rho_{23} = 0.0082, \ q_{23} = 3.995, \ r_{23} = 2.003$$
$$\rho_{34} = 0.1420, \ q_{34} = 3.810, \ r_{34} = 2.049, \ m_{34} = 1.136$$
$$\rho_{21} = 0.1330, \ q_{21} = 3.822, \ r_{21} = 2.048, \ m_{21} = 1.126$$

Neglecting axial strains, the stiffness submatrices for the members are given by:

$$\begin{bmatrix} M_{21} \\ IQ_{21} \end{bmatrix} = EI/l \begin{bmatrix} 3.822 & -5.870 \\ -5.870 & 10.426 \end{bmatrix} \begin{bmatrix} \theta_2 \\ x_2/l \end{bmatrix}$$

$$\begin{bmatrix} M_{23} \\ M_{32} \end{bmatrix} = EI/l \begin{bmatrix} 3.995 & 2.003 \\ 2.003 & 3.995 \end{bmatrix} \begin{bmatrix} \theta_2 \\ \theta_3 \end{bmatrix}$$

$$\begin{bmatrix} M_{34} \\ IQ_{34} \end{bmatrix} = EI/l \begin{bmatrix} 3.810 & -5.859 \\ -5.859 & 10.314 \end{bmatrix} \begin{bmatrix} \theta_3 \\ x_2/l \end{bmatrix}$$

Collecting the relevant terms gives:

$$\begin{bmatrix} P_{\theta 2} \\ P_{\theta 3} \\ IP_{x2} + IP_{x3} \end{bmatrix} = \begin{bmatrix} 0 \\ 0 \\ Wl \end{bmatrix} = EI/l \begin{bmatrix} 7.817 & 2.003 & -5.870 \\ 2.003 & 7.805 & -5.859 \\ -5.870 & -5.859 & 20.740 \end{bmatrix} \begin{bmatrix} \theta_2 \\ \theta_3 \\ x_2/l \end{bmatrix}$$

The joint displacements are:

$$\theta_2 = 0.04354Wl^2/EI$$
$$\theta_3 = 0.04349Wl^2/EI$$
$$x_2 = 0.07283Wl^3/EI$$

The moments and axial forces are:

$$M_{34} = 3.810\,\theta_3 - 5.859x_2/l = -0.2610Wl$$
$$M_{43} = 2.049\,\theta_3 - 5.859x_2/l = -0.3376Wl$$
$$M_{21} = 3.822\,\theta_2 - 5.870x_2/l = -0.2611Wl$$
$$M_{12} = 2.048\,\theta_2 - 5.870x_2/l = -0.3383Wl$$
$$P_{23} = -5.859\,\theta_3 + 10.314x_2/l = 0.4964W$$
$$P_{34} = 10W + (M_{23} + M_{32})/l = 10.5221W$$
$$P_{21} = 20W - P_{34} = 9.4779W$$

An additional analysis is unnecessary, as the revised values of the axial forces differ only slightly from the initial estimate.

(c) Determination of the critical load

Instability of frames that are liable to sway occurs in a skew-symmetrical mode, and the critical load is invariably lower than that of an identical frame in which sway is prevented. In the case of a symmetrical single-bay frame, the stiffness of the columns is readily expressed in terms of the n and o functions, and the critical load is the one that produces a singular stiffness matrix.

Instability of the frame, shown in Figure 11.22, occurs in the skew-symmetrical mode shown at (i). The rotation of joint 3 is equal and of the same sense as the rotation of joint 2, and the critical load may be determined by considering the stiffness of joint 2 only. Thus:

$$s_2 = s_{23} + s_{21}$$
$$= 6EI/l + n'_{21}EI/l$$

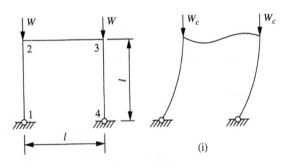

Figure 11.22

At the critical load:

$$s_2 = 0$$

and

$$n'_{21} = -6$$

From Table 11.2:

$$\rho = W_c / P_E$$
$$= 0.177$$

and

$$W_c = 0.177\pi^2 EI/l^2$$

This result may be compared with the value of $\rho = 1.31$ obtained in Section 11.2(b) for an identical frame with sway prevented.

In the case of multi-bay frames, a procedure similar to that used in Section 7.9 may be employed to reduce the structure to an equivalent single-bay symmetrical frame. However, for the determination of the critical load, it is unnecessary for the original structure to satisfy exactly the principle of multiples[20]. The frame, shown in Figure 11.23, may be reduced with sufficient accuracy to the single-bay frame shown at (i). The EI/l value for the beam of the equivalent frame is given by Σk_i^b, where k_i^b is the EI/l for beam i of the original frame. The EI/l value for each column of the equivalent frame is given by $\Sigma k_j^c/2$, where k_j^c is the EI/l value for column j of the original frame. The ρ value for each column of the equivalent frame is given by $\Sigma W_n/\Sigma P_{En}$, where ΣW_n is the sum of the applied loads and ΣP_{En} is the sum of the column Euler loads of the original frame.

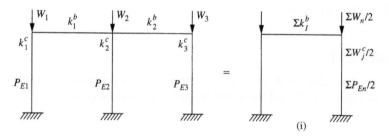

Figure 11.23

Example 11.10

Determine the value of W at which elastic instability occurs in the rigid frame shown in Figure 11.24. All the members are of uniform section.

Solution

The stiffness of joint 2 is:

$$s_2 = s_{23} + s_{21}$$
$$= 6EI/2l + n'_{21}EI/l$$

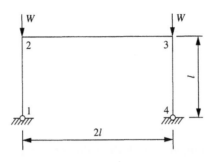

Figure 11.24

At the critical load:

$$s_2 = 0$$

and:

$$n'_{21} = -3$$
$$\rho_{21} = 0.142$$

Hence, instability occurs at a value of the applied load:

$$W_c = 0.142\pi^2 EI/l^2$$

Example 11.11

Determine the value of W at which elastic instability occurs in the rigid frame shown in Figure 11.25. All the members are of uniform section.

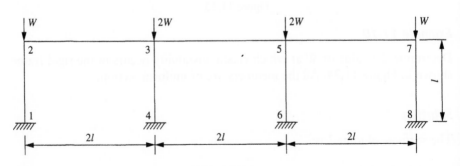

Figure 11.25

Solution

The frame may be replaced by an equivalent single-bay frame with a beam stiffness of:

$$6E(3I)/2l = 9EI/l$$

and a column stiffness of:

$$nE(4I)/2l = 2nEI/l$$

At the critical load:

$$n = -4.5$$
$$\rho = 0.69$$
$$= 6W/4P_E$$

Hence, the critical load is:

$$W_c = 0.46P_E$$

A more accurate value may be obtained by setting up the stiffness matrix for the original frame. Instability occurs in a skew-symmetrical mode with $\theta_3 = \theta_5$ and $\theta_2 = \theta_7$. Neglecting axial strains, the stiffness sub-matrices for the members are given by:

$$\begin{bmatrix} M_{23} \\ M_{32} \end{bmatrix} = EI/l \begin{bmatrix} 2 & 1 \\ 1 & 2 \end{bmatrix} \begin{bmatrix} \theta_2 \\ \theta_3 \end{bmatrix}$$

$$M_{35} = 3EI\theta_3/l$$

$$\begin{bmatrix} M_{21} \\ lQ_{21} \end{bmatrix} = EI/l \begin{bmatrix} q_{21} & -(q_{21} + r_{21}) \\ -(q_{21} + r_{21}) & 2(q_{21} + r_{21})/m_{21} \end{bmatrix} \begin{bmatrix} \theta_2 \\ x_2/l \end{bmatrix}$$

$$\begin{bmatrix} M_{34} \\ lQ_{34} \end{bmatrix} = EI/l \begin{bmatrix} q_{34} & -(q_{34} + r_{34}) \\ -(q_{34} + r_{34}) & 2(q_{34} + r_{34})/m_{34} \end{bmatrix} \begin{bmatrix} \theta_3 \\ x_2/l \end{bmatrix}$$

Collecting the relevant terms gives:

$$\begin{bmatrix} P_{\theta 2} \\ P_{\theta 3} \\ lP_{x2} + lP_{x3} \end{bmatrix} = EI/l \begin{bmatrix} (2 + q_{21}) & 1 & -(q_{21} + r_{21}) \\ 1 & (5 + q_{34}) & -(q_{34} + r_{34}) \\ -(q_{21} + r_{21}) & -(q_{34} + r_{34}) & (j_{21} + j_{34}) \end{bmatrix} \begin{bmatrix} \theta_2 \\ \theta_3 \\ x_2/l \end{bmatrix}$$

where:

$$j_{21} = 2(q_{21} + r_{21})/m_{21}$$

and:

$$j_{34} = 2(q_{34} + r_{34})/m_{34}$$

For values of $\rho_{21} = 0.44$ and $\rho_{34} = 0.88$ the determinant of the stiffness matrix is:

$$\begin{vmatrix} 5.39 & 1.00 & -5.55 \\ 1.00 & 7.68 & -5.07 \\ -5.55 & -5.07 & 8.22 \end{vmatrix} = 10.8$$

For values of $\rho_{21} = 0.46$ and $\rho_{34} = 0.92$ the determinant of the stiffness matrix is:

$$\begin{vmatrix} 5.34 & 1.00 & -5.53 \\ 1.00 & 7.61 & -5.03 \\ -5.53 & -5.03 & 7.49 \end{vmatrix} = -17.0$$

Hence, the critical load is:

$$W_c = 0.448 P_E$$

Example 11.12

Determine the value of W at which elastic instability occurs in the rigid frame shown in Figure 11.26. All the members are of uniform section.

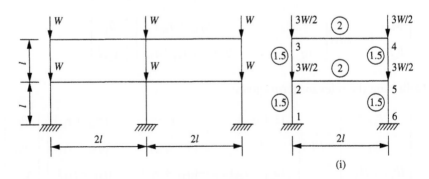

(i)

Figure 11.26

Solution

The frame may be replaced by the equivalent single-bay frame shown at (i), with the relative second moment of area values ringed. The stiffness sub-matrices for the members are given by:

$$\begin{bmatrix} M_{23} \\ M_{32} \end{bmatrix} = 3EI/2l \begin{bmatrix} n_{23} & o_{23} \\ o_{23} & n_{23} \end{bmatrix} \begin{bmatrix} \theta_2 \\ \theta_3 \end{bmatrix}$$

$$M_{21} = 3n_{21}EI\theta_2/2l$$
$$M_{34} = 6EI\theta_3/l$$
$$M_{25} = 6EI\theta_2/l$$

Collecting the relevant terms gives:

$$\begin{bmatrix} P_{\theta 2} \\ P_{\theta 3} \end{bmatrix} = 3EI/2l \begin{bmatrix} (4 + n_{23} + n_{21}) & o_{23} \\ o_{23} & (4 + n_{23}) \end{bmatrix} \begin{bmatrix} \theta_2 \\ \theta_3 \end{bmatrix}$$

For values of $\rho_{23} = 0.30$ and $\rho_{21} = 0.60$ the determinant of the stiffness matrix is:

$$\begin{vmatrix} 0.90 & -1.74 \\ -1.74 & 3.74 \end{vmatrix} = 0.34$$

For values of $\rho_{23} = 0.31$ and $\rho_{21} = 0.62$ the determinant of the stiffness matrix is:

$$\begin{vmatrix} 0.54 & -1.78 \\ -1.78 & 3.68 \end{vmatrix} = -1.18$$

Hence instability occurs at a value of $\rho_{23} = 0.302$ and the critical load is:

$$W_c = 0.302 P_E$$

(d) Effect of primary bending moments

The buckling load of a structure liable to side sway is only slightly reduced by primary moments caused by vertical loads. Immediately prior to buckling, the expressions derived in Section 11.2(c) for the no-sway case are applicable. As buckling occurs, an additional expression involving the n and o functions must be satisfied.

A structure subjected to lateral loads, as in Figure 11.20, has no defined buckling load[21]. The frame displacements approach infinity as the vertical loads approach the elastic critical load of an identical frame subjected only to vertical loads.

Example 11.13

Determine the value of W at which elastic instability occurs in the rigid frame shown in Figure 11.27. All the members are of uniform section.

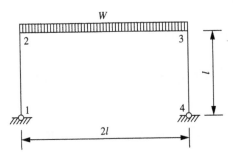

Figure 11.27

Solution

There is no sway of the frame until buckling occurs, and the relationship between ρ_{21} and ρ_{23} established in Example 11.6 and plotted in Figure 11.14 is applicable.

As buckling occurs, the stiffness of joint 2 is:

$$s_2 = 0$$
$$= s_{23} + s_{21}$$
$$= (q_{23} + r_{23})EI/2l + n'_{21}EI/l$$
$$= q_{23} + r_{23} + 2n'_{21}$$

At the skew-symmetrical buckling load, the relationship between ρ_{21} and ρ_{23} is almost linear with: $\rho_{21} \approx \rho_{23}$.

For a value of: $\rho_{21} = \rho_{23} = 0.12$

$$q_{23} + r_{23} + 2n'_{21} = 1.42$$

For a value of: $\rho_{21} = \rho_{23} = 0.14$

$$q_{23} + r_{23} + 2n'_{21} = -0.02$$

Hence, instability occurs at a value of $\rho_{21} = 0.14$ and the critical load is:

$$W_c = 2 \times 0.14 P_E$$
$$= 0.28 P_E$$

where P_E is the Euler load for a column.

This result may be compared with the value of $\rho_{21} = 0.142$ obtained in Example 11.10 for an identical frame with the loading applied at the joints.

Example 11.14

Determine the value of W at which elastic instability occurs in the rigid frame shown in Figure 11.28. All the members are of uniform section.

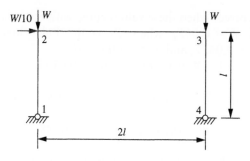

Figure 11.28

Solution

The stiffness sub-matrices for the members are given by:

$$\begin{bmatrix} M_{23} \\ M_{32} \end{bmatrix} = EI/2l \begin{bmatrix} q_{23} & r_{23} \\ r_{23} & q_{23} \end{bmatrix} \begin{bmatrix} \theta_2 \\ \theta_3 \end{bmatrix}$$

$$\begin{bmatrix} M_{21} \\ lQ_{21} \end{bmatrix} = EI/l \begin{bmatrix} q'_{21} & -q'_{21} \\ -q'_{21} & q'_{21}/m'_{21} \end{bmatrix} \begin{bmatrix} \theta_2 \\ x_2/l \end{bmatrix}$$

$$\begin{bmatrix} M_{34} \\ lQ_{34} \end{bmatrix} = EI/l \begin{bmatrix} q'_{34} & -q'_{34} \\ -q'_{34} & q'_{34}/m'_{34} \end{bmatrix} \begin{bmatrix} \theta_3 \\ x_2/l \end{bmatrix}$$

Collecting the relevant terms gives:

$$\begin{bmatrix} P_{\theta 2} \\ P_{\theta 3} \\ lP_{x2} + lP_{x3} \end{bmatrix} = \begin{bmatrix} 0 \\ 0 \\ Wl/10 \end{bmatrix} = EI/l \begin{bmatrix} (q'_{21} + q_{23}/2) & r_{23}/2 & -q'_{21} \\ r_{23}/2 & (q'_{34} + q_{23}/2) & -q'_{34} \\ -q'_{21} & -q'_{34} & (j'_{21} + j'_{34}) \end{bmatrix} \begin{bmatrix} \theta_2 \\ \theta_3 \\ x_2/l \end{bmatrix}$$

where

$$j'_{21} = q'_{21}/m'_{21} \text{ and } j'_{34} = q'_{34}/m'_{34}$$

Values of P_{23} and P_{34} are estimated and the corresponding stability functions inserted in the stiffness matrix. Inverting this matrix gives the displacements, and thus Q_{34}, M_{23}, and M_{32} may be obtained. The axial forces in the members are given by:

$$P_{23} = Q_{34}$$

and:

$$P_{34} = W + (M_{23} + M_{32})/2l$$

The analysis is correct when these values agree with the initial estimates.
For a value of $W = 0.04P_E$ where P_E is the Euler load for a column, $\rho_{23} = 0.0078$, $\rho_{34} = 0.0427$, and $\theta_2 = 0.0090$ rad.
For a value of $W = 0.08P_E$, $\rho_{23} = 0.0139$, $\rho_{34} = 0.887$, and $\theta_2 = 0.0290$ rad.
For a value of $W = 0.12P_E$, $\rho_{23} = -0.0089$, $\rho_{34} = 0.1541$, and $\theta_2 = 1.145$ rad.
For a value of $W = 0.13P_E$, $\rho_{23} = -0.0788$, $\rho_{34} = 0.1908$, and $\theta_2 = 2.047$ rad.
In Figure 11.29, values of W/P_E are plotted against θ_2 and it may be seen that θ_2 approaches infinity as W/P_E approaches 0.142. In the same figure values

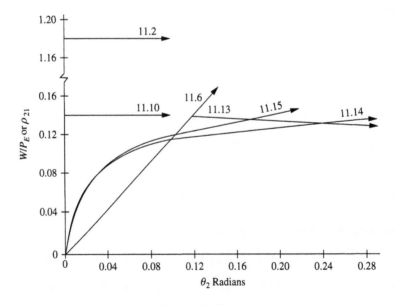

Figure 11.29

of W/P_E are plotted against θ_2 for Examples 11.2 and 11.10, and values of ρ_{21} are plotted against θ_2 for Examples 11.6 and 11.13.

(e) Effect of finite displacements

The previous analyses have been based on small deflection theory, and the stiffness matrix of the structure when instability occurs has been assumed identical to the stiffness matrix of the unloaded structure. In practice, the finite deflections produced before instability may significantly change the geometry of the structure. The columns of the frame shown in Figure 11.30 (i) are initially vertical. On the application of a lateral load, as shown at (ii), the geometry of the frame changes, and the columns are inclined at a sway angle $\phi = x_2/l$. The elements in the stiffness matrix of the frame depend on the inclination of the columns, and thus the stiffness matrix is continuously changing as the lateral load increases. In structures that undergo a severe change in geometry, elastic instability may not occur[22].

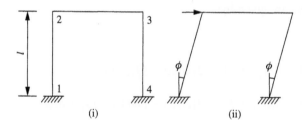

(i)   (ii)

Figure 11.30

Example 11.15

Determine the relationship between W and θ_2 for the frame shown in Figure 11.28. All the members are of uniform section.

Solution

At any given loading stage, the columns are inclined at an angle ϕ to the vertical where $\phi = x_2/l$. The relationship between horizontal and vertical displacement of a column top is $y_2 = x_2 \tan \phi$, and the orthogonal transformation matrix $[T_{21}]$ of column 21 is given by:

$$
\begin{bmatrix} \theta_{21} \\ y'_{21} \end{bmatrix} = \begin{bmatrix} 1 & 0 \\ 0 & (\lambda + \mu \tan \phi) \end{bmatrix} \begin{bmatrix} \theta_2 \\ x_2 \end{bmatrix}
$$

where $\lambda = \cos \phi$, $\mu = \sin \phi$, and y'_{21} is the displacement of 2 perpendicular to member 21.

The stiffness sub-matrix for column 21, referred to the x- and y-axes, is given by $[T_{21}]^T[S_{21}][T_{21}]$ and:

$$\begin{bmatrix} M_{21} \\ lQ_{21} \end{bmatrix} = EI/l \begin{bmatrix} q'_{21} & -q'_{21}\tau \\ -q'_{21}\tau & j'_{21}\tau^2 \end{bmatrix} \begin{bmatrix} \theta_2 \\ x_2/l \end{bmatrix}$$

where $\tau = (\lambda + \mu \tan \phi)$.

The complete stiffness matrix of the frame is given by:

$$\begin{bmatrix} P_{\theta 2} \\ P_{\theta 3} \\ lP_{x2} + lP_{x3} \end{bmatrix} = \begin{bmatrix} 0 \\ 0 \\ Wl/10 \end{bmatrix} = EI/l \begin{bmatrix} (q'_{21} + q_{23}/2) & r_{23}/2 & -q'_{21}\tau \\ r_{23}/2 & (q'_{34} + q_{23}/2) & -q'_{34}\tau \\ -q'_{21}\tau & -q'_{34}\tau & (j'_{21} + j'_{34}\tau^2) \end{bmatrix} \begin{bmatrix} \theta_2 \\ \theta_3 \\ x_2/l \end{bmatrix}$$

Values of P_{21}, P_{23}, P_{34}, and ϕ are estimated and the corresponding stability functions inserted in the stiffness matrix. Inverting this matrix gives the displacements, and thus Q_{34}, M_{23}, and M_{32} may be determined. The axial forces in the members are given by:

$$P_{23} = Q_{34}$$

$$P_{34} = \{W + (M_{23} + M_{32})/2\}\cos\phi - Q_{34}\sin\phi$$

$$P_{21} = \{W - (M_{23} + M_{32})/2\}\cos\phi - (W/10 - Q_{34})\sin\phi$$

The analysis is correct when these values and the value of ϕ agree with the initial estimates.

For a value of $W = 0.04P_E$, $\rho_{23} = 0.0078$, $\rho_{34} = 0.0427$, $\rho_{21} = 0.0372$, $\phi = 0.0182$, and $\theta_2 = 0.00898\,\text{rad}$.

For a value of $W = 0.08P_E$, $\rho_{23} = 0.0139$, $\rho_{34} = 0.0883$, $\rho_{21} = 0.0710$, $\phi = 0.0586$, and $\theta_2 = 0.02869\,\text{rad}$.

For a value of $W = 0.12P_E$, $\rho_{23} = -0.0019$, $\rho_{34} = 0.1467$, $\rho_{21} = 0.0862$, $\phi = 0.1995$, and $\theta_2 = 0.0995\,\text{rad}$.

For a value of $W = 0.13P_E$, $\rho_{23} = -0.0266$, $\rho_{34} = 0.1668$, $\rho_{21} = 0.0801$, $\phi = 0.2736$, and $\theta_2 = 0.1392\,\text{rad}$.

Values of W/P_E are plotted against θ_2 in Figure 11.29, and it may be seen that θ_2 continuously increases as W increases and that elastic instability does not occur.

11.4   Stability coefficient matrix method

A relatively direct determination of the critical load is possible if member forces are related to joint displacements by the normal stiffness matrix plus a stability coefficient matrix to allow for the axial load. The stability coefficient matrix is derived from the assumption that the elastic curve of a member may be defined by a cubic polynomial. For a member 12, subjected to the joint displacements and member forces shown in Figure 11.31, the lateral displacement is given by:

$$y = a_1 + a_2x + a_3x^2 + a_4x^3$$

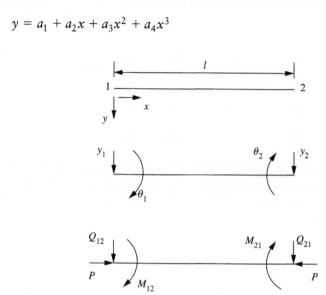

Figure 11.31

Thus:

$$\theta = dy/dx$$
$$= a_2 + 2a_3x + 3a_4x^2$$

and:

$$M/EI = d^2y/dx^2$$
$$= 2a_3 + 6a_4x$$

where M is the bending moment in the member due to the member end forces.

The joint displacements are obtained by substituting the coordinates of joints 1 and 2 into the displacement functions. Thus:

$$
\begin{bmatrix} y_1 \\ \theta_1 \\ y_2 \\ \theta_2 \end{bmatrix}
=
\begin{bmatrix}
1 & 0 & 0 & 0 \\
0 & 1 & 0 & 0 \\
1 & l & l^2 & l^3 \\
0 & 1 & 2l & 3l^2
\end{bmatrix}
\begin{bmatrix} a_1 \\ a_2 \\ a_3 \\ a_4 \end{bmatrix}
$$

or:

$$\{\Delta\} = [B]\{A\}$$

The coefficients of the cubic polynomial are given by:

$$\{A\} = [B]^{-1}\{\Delta\}$$

$$= \begin{bmatrix} 1 & 0 & 0 & 0 \\ 0 & 1/l^2 & 0 & 0 \\ -3/l^2 & -2/l^3 & 3/l^2 & -1/l^3 \\ 2/l^3 & 1/l^4 & -2/l^3 & 1/l^4 \end{bmatrix} \begin{bmatrix} y_1 \\ \theta_1 \\ y_2 \\ \theta_2 \end{bmatrix}$$

where $\{A\}$ is the vector of undetermined constants in the displacement functions. The total moment in the member is the sum of M due to the member end forces and $\bar{M}$ due to the axial load P. The moment due to the member end forces is given by:

$$M = EId^2y/dx^2$$

$$= EI\begin{bmatrix} 0 & 0 & 2 & 6x \end{bmatrix} \begin{bmatrix} a_1 \\ a_2 \\ a_3 \\ a_4 \end{bmatrix}$$

$$= EI[C]\{A\}$$
$$= EI[C][B]^{-1}\{\Delta\}$$

The member deformations are given by:

$$d\theta = M/EI$$
$$= [C][B]^{-1}\{\Delta\}$$

The moment due to the axial load is given by:

$$\bar{M} = Py$$

$$= P\begin{bmatrix} 1 & x & x^2 & x^3 \end{bmatrix} \begin{bmatrix} a_1 \\ a_2 \\ a_3 \\ a_4 \end{bmatrix}$$

$$= P[D]\{A\}$$
$$= P[D][B]^{-1}\{\Delta\}$$

The member end forces $\{P\}$ are statically compatible with the total moment in the member $(M + \bar{M})$. Hence, during any imposed virtual displacements, the internal work done and the external work done sum to zero. The external work done is the sum of each member end force multiplied by the corresponding virtual displacement. The internal work done is the product of the virtual deformation in the member and the total moment, including the moment due to the axial load[23]. Applying unit virtual joint displacements successively while the remaining displacements are prevented produces the virtual deformations $[C][B]^{-1}[I]$. Then, equating external and internal work:

$$[I]\{P\} = \int_0^l ([C][B]^{-1}[I])^T (M + \bar{M})\, dx$$

Expanding each term separately:

$$\int_0^l ([C][B]^{-1}[I])^T M\, dx = EI \int_0^l ([C][B]^{-1}[I])^T ([C][B]^{-1}\{\Delta\})\, dx$$
$$= EI([B]^{-1})^T \int_0^l [C]^T[C]\, dx\, [B]^{-1}\{\Delta\}$$

where:

$$[C]^T[C] = \begin{bmatrix} 0 & 0 & 0 & 0 \\ 0 & 0 & 0 & 0 \\ 0 & 0 & 4 & 12x \\ 0 & 0 & 12x & 36x^2 \end{bmatrix}$$

$$\int_0^l [C]^T[C]\, dx = \begin{bmatrix} 0 & 0 & 0 & 0 \\ 0 & 0 & 0 & 0 \\ 0 & 0 & 4l & 6l^2 \\ 0 & 0 & 6l^2 & 12l^3 \end{bmatrix}$$

$$([B]^{-1})^T \int_0^l [C]^T[C]\, dx[B]^{-1} = \begin{bmatrix} 12/l^3 & 6/l^2 & -12/l^3 & 6/l^2 \\ 6/l^2 & 4/l & -6/l^2 & 2/l \\ -12/l^3 & -6/l^2 & 12/l^3 & -6/l^2 \\ 6/l^2 & 2/l & -6/l^2 & 4/l \end{bmatrix}$$

and this is the normal stiffness matrix for a member,

$$\int_0^l ([C][B]^{-1}[I])^T M\, dx = \int_0^l ([C][B]^{-1}[I])^T (P[D][B]^{-1}\{\Delta\})\, dx$$
$$= P([B]^{-1})^T \int_0^l [C]^T[D]\, dx\, [B]^{-1}\{\Delta\}$$

where:

$$[C]^T[D] = \begin{bmatrix} 0 & 0 & 0 & 0 \\ 0 & 0 & 0 & 0 \\ 2 & 2x & 2x^2 & 2x^3 \\ 6x & 6x^2 & 6x^3 & 6x^4 \end{bmatrix}$$

$$\int_0^l [C]^T[D]\,dx = \begin{bmatrix} 0 & 0 & 0 & 0 \\ 0 & 0 & 0 & 0 \\ 2l & l^2 & 2l^3/3 & l^4/2 \\ 3l^2 & 2l^3 & 3l^4/2 & 6l^5/5 \end{bmatrix}$$

$$([B]^{-1})\int_0^l [C]^T[D]\,dx[B]^{-1} = \begin{bmatrix} -6/5l & -1/10 & 6/5l & -1/10 \\ -1/10 & -2l/15 & 1/10 & l/30 \\ 6/5l & 1/10 & -6/5l & 1/10 \\ -1/10 & l/30 & 1/10 & -2l/15 \end{bmatrix}$$

and this is the stability coefficient matrix. The individual elements in the matrix do not involve the axial load as a parameter. The matrix may be computed explicitly for each member and remains independent of the axial load.

Thus the relationship between member forces and joint displacement is:

$$\begin{bmatrix} lQ_{12} \\ M_{12} \\ lQ_{21} \\ M_{21} \end{bmatrix} = EI/l \begin{bmatrix} 12 & 6 & -12 & 6 \\ 6 & 4 & -6 & 2 \\ -12 & -6 & 12 & -6 \\ 6 & 2 & -6 & 4 \end{bmatrix} + \rho\pi^2/30 \begin{bmatrix} -36 & -3 & 36 & -3 \\ -3 & -4 & 3 & 1 \\ 36 & 3 & -36 & 3 \\ -3 & 1 & 3 & -4 \end{bmatrix} \begin{bmatrix} x_1/l \\ \theta_1 \\ x_2/l \\ \theta_2 \end{bmatrix}$$

and this is valid for a tensile load with a change of sign.

The displacement function used in the above derivation is exact so far as member forces are concerned and provides an approximate expression for the displacements produced by an axial load. Provided that failure of the structure occurs in the sway mode, the axial loads are small and the deformations of the members agree closely with the assumed displacement function. Thus, a good estimate of the critical load is obtained, and this is an upper bound because of the constraints imposed by the assumed displacement function. The accuracy may be further improved by dividing each member into a number of smaller segments[24], thus increasing the number of degrees of freedom. Alternatively, additional terms may be introduced into the sub-matrix for each member, corresponding to the first and second buckling modes of a built-in strut[25].

For the column shown in Figure 11.32 the relationship between member forces and joint displacements is similarly obtained as:

$$\begin{bmatrix} lQ_{12} \\ M_{12} \end{bmatrix} = EI/l \left[\begin{bmatrix} 3 & -3 \\ -3 & 3 \end{bmatrix} + \rho\pi^2/5 \begin{bmatrix} -6 & 1 \\ 1 & -1 \end{bmatrix} \right] \begin{bmatrix} x_1/l \\ \theta_1 \end{bmatrix}$$

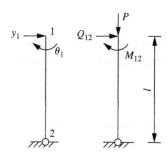

Figure 11.32

Example 11.16

Determine the value of W at which elastic instability occurs in the rigid frame shown in Figure 11.22. All the members are of uniform section.

Solution

The stiffness sub-matrices for the members are given by:

$$M_{23} = 6EI\theta_2/l$$

and

$$\begin{bmatrix} lQ_{21} \\ M_{21} \end{bmatrix} = EI/l \left[\begin{bmatrix} 3 & -3 \\ -3 & 3 \end{bmatrix} + \rho\pi^2/5 \begin{bmatrix} -6 & 1 \\ 1 & -1 \end{bmatrix} \right] \begin{bmatrix} x_2/l \\ \theta_2 \end{bmatrix}$$

Collecting the relevant terms gives:

$$\begin{bmatrix} lP_{x2} \\ P_{\theta2} \end{bmatrix} = EI/5l \left[\begin{bmatrix} 15 & -15 \\ -15 & 45 \end{bmatrix} + \rho\pi^2 \begin{bmatrix} -6 & 1 \\ 1 & -1 \end{bmatrix} \right] \begin{bmatrix} x_2/l \\ \theta_2 \end{bmatrix}$$

The determinant of the stiffness matrix is zero for a value of:

$$\rho = 0.185$$

and:

$$W_c = 0.185\pi^2 EI/l^2$$

Example 11.17

Determine the value of W at which elastic instability occurs in the rigid frame shown in Figure 11.25. All the members are of uniform section.

Solution

The stiffness sub-matrices for the members are given by:

$$M_{35} = 3EI\theta_3/l$$

$$\begin{bmatrix} M_{23} \\ M_{32} \end{bmatrix} = EI/l \begin{bmatrix} 2 & 1 \\ 1 & 2 \end{bmatrix} \begin{bmatrix} \theta_2 \\ \theta_3 \end{bmatrix}$$

$$\begin{bmatrix} M_{21} \\ lQ_{21} \end{bmatrix} = EI/l \left(\begin{bmatrix} 4 & -6 \\ -6 & 12 \end{bmatrix} + \rho\pi^2/30 \begin{bmatrix} -4 & 3 \\ 3 & -36 \end{bmatrix} \right) \begin{bmatrix} \theta_2 \\ x_2/l \end{bmatrix}$$

$$\begin{bmatrix} M_{34} \\ lQ_{34} \end{bmatrix} = EI/l \left(\begin{bmatrix} 4 & -6 \\ -6 & 12 \end{bmatrix} + 2\rho\pi^2/30 \begin{bmatrix} -4 & 3 \\ 3 & -36 \end{bmatrix} \right) \begin{bmatrix} \theta_3 \\ x_2/l \end{bmatrix}$$

Collecting the relevant terms gives:

$$\begin{bmatrix} P_{\theta2} \\ P_{\theta3} \\ lP_{x_2} + lP_{x_3} \end{bmatrix} = EI/30l \left(\begin{bmatrix} 180 & 30 & -180 \\ 30 & 270 & -180 \\ -180 & -180 & 720 \end{bmatrix} + \rho\pi^2 \begin{bmatrix} -4 & 0 & 3 \\ 0 & -8 & 6 \\ 3 & 6 & -108 \end{bmatrix} \right) \begin{bmatrix} \theta_2 \\ \theta_3 \\ x_2/l \end{bmatrix}$$

The determinant of the stiffness matrix is zero for a value of:

$$\rho = 0.453$$

and:

$$W_c = 0.453\pi^2 EI/l^2$$

Supplementary problems

S11.1 Determine the value of W at which elastic instability occurs in the rigid frame shown in Figure S11.1. The frame is braced against lateral translation, and all the members are of uniform section.

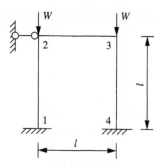

Figure S11.1

S11.2 All the members of the symmetrical rigid frame shown in Figure S11.2 are of uniform section. Determine the value of W at which elastic instability of the frame occurs in its own plane.

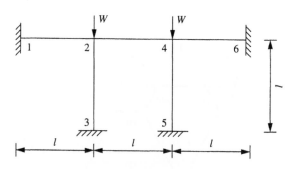

Figure S11.2

S11.3 All the members of the braced frame shown in Figure S11.3 are of uniform section. Set up the stiffness matrix for the frame and hence determine the value of W at which elastic instability occurs. The loading on the members may be replaced by equivalent static loading at the joints.

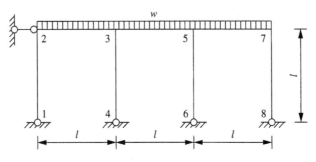

Figure S11.3

S11.4 All the members of the rigid frame shown in Figure S11.4 are of uniform section, and the frame is hinged at supports 1 and 5. The dimensions and loading are indicated on the figure, and the Euler load for member 12 is 8 W. Determine the load factor against collapse by elastic instability of the frame in its own plane.

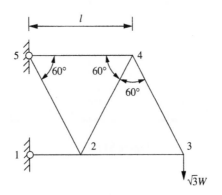

Figure S11.4

S11.5 Set up the stiffness matrix and determine the value of W in terms of the Euler load at which elastic instability occurs in the symmetrical rigid frame shown in Figure S11.5. All the members are of uniform section.

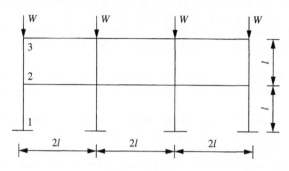

Figure S11.5

References

1. Bleich, F. Buckling strength of metal structures. New York. McGraw Hill Book Company, Inc. 1952. pp. 196–213.
2. Masur, E. F. On the lateral stability of multi storey bents. Proc. Am. Soc. Civil Eng. 81 (EM2). April 1955.
3. McMinn, S. J. The determination of the critical loads of plane frames. The Structural Engineer. 39. July 1961. pp. 221–227.
4. Merchant, W. Critical loads of tall building frames. The Structural Engineer. 33. March 1955. pp. 84–89.
5. Smith, R. B. L. and Merchant, W. Critical loads of tall building frames. The Structural Engineer. 34. August 1956. pp. 284–292.
6. Bolton, A. A quick approximation to the critical load of rigidly jointed trusses. The Structural Engineer. 33. March 1955. pp. 90–95.
7. Bolton, A. A convergence technique for determining the elastic critical load of rigidly jointed trusses. The Structural Engineer. 37. August 1959. pp. 233–242.
8. Waters, H. A direct approximation to the critical loads of rigidly jointed plane structures. Civil Eng. and Pub. Works Rev. 59. February 1964. pp. 193–196, March 1964. pp. 355-357.
9. Krynicki, E. J. and Mazurkiewicz, Z. E. Frames of solid bars of varying cross sections. Proc. Am. Soc. Civil Eng. 90 (ST4). August 1964. pp. 145–174, and 91(ST1). January 1965. pp. 318–327.
10. Horne, M. R. and Merchant, W. The stability of frames. Oxford. Pergamon Press Ltd. 1965.
11. Livesley, R. K. and Chandler, D. B. Stability functions for structural frameworks. Manchester University Press. 1956.
12. Goldberg, J. E. Buckling of one storey frames and buildings. Proc. Am. Soc. Civil Eng. 86 (ST10). October 1960. pp. 53–85.
13. Lightfoot, E. Moment distribution. London. E. & F. N. Spon Ltd. 1961. pp. 61–74, 226–251.

14. Gere, J. M. Moment distribution. Princeton. N. J. D. Van Nostrand Company, Inc. 1963. pp. 166–211.
15. Hoff, N. J. The analysis of structures. New York. John Wiley & Sons, Inc. 1956. pp. 162–168.
16. Merchant, W. The failure load of rigid jointed frameworks as influenced by stability. The Structural Engineer. 32. July 1954. pp. 185–190.
17. Masur, E. F., Chang, I. C., Donnell, L. H. Stability of frames in the presence of primary bending moments. Proc. Am. Soc. Civil Eng. 87 (EM4). August 1961. pp. 19–34.
18. Horne, M. R. The effect of finite deformations in the elastic stability of plane frames. Proceedings, Royal Society of London. Series A. 266. February 1962. pp. 47–67.
19. Lu, L. W. Stability of frames under primary bending moments. Proc. Am. Soc. Civil. Eng. 89 (ST3). June 1963. pp. 35–62, 89(ST6). December 1963. pp. 455–458 and 90(ST1). February 1964. pp. 234–239.
20. Bowles, R. E. and Merchant, W. Critical loads of tall building frames. The Structural Engineer. 34. June 1956. pp. 324–329, and 36. June 1958. pp. 187–190.
21. Chu, K. H. and Pabarcius, A. Elastic and inelastic buckling of portal frames. Proc. Am. Soc. Civil Eng. 90 (EM5). October 1964. pp. 221–249.
22. Saafan, S. A. Nonlinear behavior of structural plane frames. Proc. Am Soc. Civil Eng. 89 (ST4). August 1963. pp. 557–579, and 91(STI). February 1965. pp. 279–281.
23. Stevens, L. K. Correction of virtual work relationships for the effect of axial load. The Structural Engineer. 42. May 1964. pp. 153–159.
24. Hartz, B. J. Matrix formulation of structural stability problems. Proc. Am Soc. Civil Eng. 91 (ST6). December 1965. pp. 141–157, and 92(ST3). June 1966. pp. 287–291.
25. Jennings, A. The elastic stability of rigidly jointed frames. Int. J. Mech. Sciences. 5. 1963. pp. 99–113.

12 Elastic-plastic analysis

Notation

c	carry-over factor for a member
$\{C\}$	plastic force vector
E	modulus of elasticity
I	second moment of area of a member
j	stability function = $2(q + r) - \rho\pi^2$
j'	stability function = $q' - \rho\pi^2$
l	length of a member
m	stability function = $2(q + r)/j$
M_{12}	moment acting at end 1 of a member 12
n	stability function = $q - m(q + r)/2$
N	load factor
N_c	load factor against elastic instability
N_d	load factor against elastic instability of the deteriorated structure
N_f	load factor against elastic-plastic failure
N_R	Rankine-Merchant load factor
N_u	load factor against rigid-plastic collapse
o	stability function = $r - m(q + r)/2$
P	axial force in a member
P_E	Euler load
q	stability function = sl/EI = $\alpha l(\sin \alpha l - \alpha l \cos \alpha l)/(2 - 2 \cos \alpha l - \alpha l \sin \alpha l)$
q'	stability function = equals $q(1 - c^2)$
r	stability function = $qc = \alpha l(\alpha l - \sin \alpha l)/(2 - 2 \cos \alpha l - \alpha l \sin \alpha l)$
$[S]$	stiffness matrix for the whole structure
W	applied load
W_c	critical load
W_d	critical load of the deteriorated structure
W_f	load producing elastic-plastic failure
W_R	Rankine-Merchant load
W_u	load producing rigid-plastic collapse
$\{W\}$	vector of external loads applied at the joints
x	horizontal displacement
y	vertical displacement
α	equals $(P/EI)^{0.5}$
θ	rotation
ρ	equals P/P_E
$\{\Delta\}$	vector of joint displacement

12.1 Introduction

The collapse of a normal structure is influenced by both elastic instability effects and by plastic yielding in the members. In chapter 9 plastic theory is used to determine the collapse load of an ideal rigid-plastic structure in which bending effects predominate and axial effects may be neglected. Chapter 11

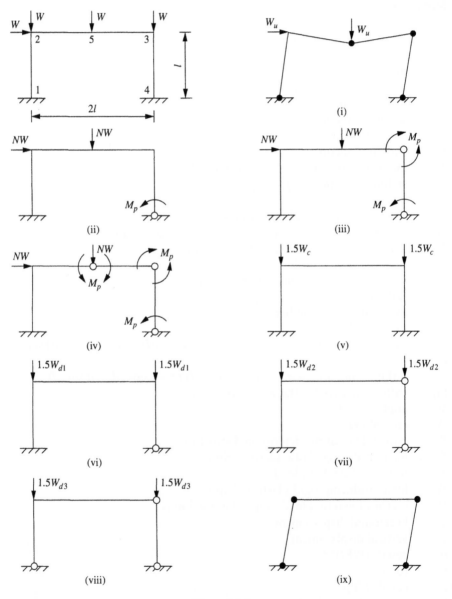

Figure 12.1

considers the ideal elastic structure in which axial forces are appreciable and the stresses in the members are entirely within the elastic range at collapse. Thus, failure occurs due to buckling effects at the elastic critical load of the structure. The present chapter deals with the determination of the failure load of a structure, taking account of both plastic yielding and instability effects.

(a) Linear elastic response

The uniform frame shown in Figure 12.1 is designed with a load factor N_u against rigid-plastic collapse (ignoring instability effects) and with a load factor N_c against elastic instability (assuming that the members exhibit indefinite elastic behavior). Disregarding both instability effects and plastic yielding, the linear elastic response of the frame to a proportional increase in the applied loads is linear, as shown in Figure 12.2. A linear relationship exists between the horizontal deflection, x_2, of joint 2 and the applied load W.

(b) Rigid-plastic response

The rigid-plastic collapse mode of the frame is shown in Figure 12.1 (i), and the ultimate load for proportional loading is:

$$W_u = 3M_p/l$$

where M_p is the plastic moment of resistance of the member.

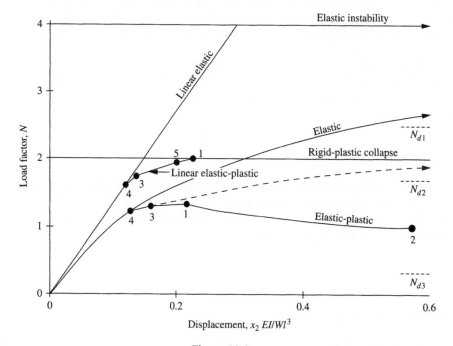

Figure 12.2

For a value of the plastic moment of resistance of $0.667Wl$, the load factor against rigid-plastic collapse is:

$$N_u = W_u/W$$
$$= 2$$

Ignoring the change in the geometry of the frame as loading increases, the rigid-plastic response curve is shown in Figure 12.2 as a horizontal line, with no displacements being produced until the load W_u is applied.

(c) Linear elastic-plastic response

The linear elastic-plastic response curve is determined by tracing the formation of plastic hinges as loading increases. Allowing for the linear elastic effects at loads less than W_u, the linear response curve is followed as the loads are increased until the first plastic hinge forms in the frame at joint 4 at a load factor of 1.60. The equivalent loading applied to the frame subsequent to the formation of this plastic hinge is shown in Figure 12.1 (ii). The formation of the hinge may be simulated by the insertion of a frictionless hinge at 4 and the application of a moment of magnitude M_p. The stiffness of the frame is reduced, and displacements, for a given increase in applied load, are greater than in the original frame. The linear elastic-plastic response is thus a series of straight lines with the rate of growth of displacements increasing as the stiffness deteriorates with the formation of each plastic hinge. The sequence of formation of the hinges is shown in Figure 12.2, and the equivalent frame, on the formation of each additional hinge, is shown in Figure 12.1 at (iii) and (iv). The second hinge forms at joint 3 at a load factor of 1.74, and the third hinge forms at 5 at a load factor of 1.94. The collapse mechanism is produced when the last hinge forms at joint 1 on the application of the load W_u at a load factor of 2.00.

(d) Elastic instability

Assuming indefinite elastic behavior, the elastic critical load may be obtained from the equivalent frame and loading shown in Figure 12.1 (v) by equating the stiffness of joint 2 to zero. Thus, allowing for the skew symmetry and ignoring the axial force in the beam:

$$n_{12}EI/l_{12} + 6EI/l_{23} = 0$$
$$n_{12} + 3 = 0$$

and from Table 11.2:

$$1.5W_c/P_E = 0.61$$

where P_E is the Euler load of column 12. For a value of P_E of 9.82 W the load factor against elastic instability is:

$$N_c = W_c/W$$
$$= 4$$

No displacement occurs in the frame until the critical load is applied when infinite displacements are produced, resulting in the horizontal line shown in Figure 12.2.

(e) Elastic response

The elastic response of the actual frame to the actual loading, allowing for instability effects and assuming indefinite elastic behavior, is non-linear, and displacements approach infinity as the elastic critical load is approached. A close approximation to the elastic response is obtained by multiplying the linear elastic displacement at a given load factor N by the amplification factor $1/(1 - N/N_c)$. This follows since the deflected shape of the structure is largely controlled by the sway deflections, and the displacement components of the lowest critical mode predominate as the first critical load is approached[1,2].

(f) Elastic-plastic response

In the actual frame, allowing for both instability effects and plastic yielding of the members, the elastic response is followed until the first plastic hinge forms at joint 4 at a load factor of 1.21. Displacements now increase more rapidly as the response follows a new elastic curve corresponding to the reduced, or deteriorated, structure. This deteriorated structure is obtained by taking only that part of the actual structure that is still behaving elastically. The deteriorated structure, on the formation of the first hinge, is shown in Figure 12.1 (vi), and its deteriorated critical load, W_{d1}, is the value of the applied load that causes the stiffness matrix of the deteriorated structure to become singular. This value is given by:

$$1.5 W_{d1}/P_E = 0.38$$

Thus, the first deteriorated structure has a load factor against instability of:

$$N_{d1} = W_{d1}/W$$
$$= 2.5$$

After the formation of the first hinge, the displacements follow the curve shown by the broken line in Figure 12.2, and this is asymptotic to the load factor N_{d1}. The first plastic hinge forms at a lower load than in the linear elastic-plastic case due to the reduction in the stiffness of the frame by the instability effects.

The second plastic hinge forms at joint 3 at a load factor of 1.30, and the deteriorated critical load, for the second deteriorated structure shown in Figure 12.1 (vii), is $N_{d2} = 1.7$. This is lower than the rigid-plastic load factor, and it is clear that this load factor cannot be attained. The load-displacement curve again exhibits a discontinuity of slope on the formation of the plastic hinge at joint 3, and displacements follow a curve that is asymptotic to the load factor N_{d2}.

The third plastic hinge forms at joint 1 at a load factor of 1.33, and the critical load for the deteriorated structure shown in Figure 12.1 (viii) is $N_{d3} = 0.35$. The structure is now unstable; loading must be decreased to maintain equilibrium; and the load factor N_f at which the third hinge forms represents the load factor against elastic-plastic failure. The increasing displacements produced by a reduction in the loading follow a curve that approaches the load factor N_{d3} from above.

The last plastic hinge forms at joint 2 at a load factor of 0.98, and this produces the collapse mechanism shown in Figure 12.1 (ix). The elastic-plastic collapse mechanism is thus quite different from the rigid-plastic collapse mechanism. In addition, the sequence of hinge formation differs from that obtained in a linear elastic-plastic analysis, and this is a general characteristic[3].

Several methods have been proposed for determining the elastic-plastic failure load, and three of these methods will be considered here.

12.2 The Rankine-Merchant load

The Rankine-Merchant method[4] is a means of estimating the failure load of a structure from values of the rigid-plastic collapse load and the elastic instability collapse load. A good approximation to the elastic-plastic load factor is provided by the Rankine load factor, N_R, which is given by:

$$1/N_R = 1/N_c + 1/N_u$$

When yield effects predominate, this empirical expression equates N_R to N_u. Similarly, when instability effects predominate, N_R is equated to N_c. Hence, for these limiting states, N_R provides an exact estimate of N_f. For intermediate conditions the Rankine expression also gives good results[5,6].

The linear elastic load-displacement relationship of a typical frame is shown in Figure 12.3 (i). The load factor against elastic instability is N_c, the linear displacement at this load factor is δ_c, and the linear displacement at an arbitrary load factor, N, is δ_l. Then:

$$N\delta_c = N_c\delta_l$$

The point P is obtained from the point of intersection of the horizontal and vertical lines RP and QP. The linear displacement of the frame at the load factor N' is δ'. Then:

$$N'\delta_c = N_c\delta'$$

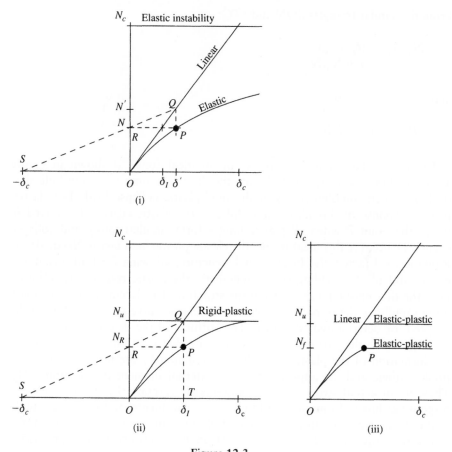

Figure 12.3

From the similar triangles RPQ and SON:

$$\delta' = \delta_c(N' - N)/N$$
$$= N_c(\delta' - \delta_l)/N$$
$$= \delta_l/(1 - N/N_c)$$

where $1/(1 - N/N_c)$ is the amplification factor discussed in section 12.1. Thus, δ' is the elastic displacement of the frame at the load factor N, and the point P lies on the elastic response curve.

The load factor against rigid-plastic collapse of a typical frame is N_u, and the linear displacement at this load factor is δ_u. Then from Figure 12.3 (ii):

$$\delta_u/\delta_c = N_u/N_c$$

From the similar triangles SON and STQ:

$$N_u/N_R = (\delta_c + \delta_u)/\delta_c$$
$$= 1 + N_u/N_c$$

and:

$$1/N_R = 1/N_c + 1/N_u$$

The linear elastic-plastic response of an ideal frame is shown in Figure 12.3 (iii). Frame displacements follow the linear response curve until all plastic hinges form simultaneously at the rigid-plastic collapse load. The elastic-plastic response of this ideal frame follows the elastic curve OP, derived in (ii) to the point P when all plastic hinges form simultaneously and collapse occurs. Hence, for this ideal frame, the elastic-plastic load factor N_f equals the Rankine load factor N_R. In an actual structure, allowing for both instability effects and plastic yielding of the members, the elastic response is followed until the first plastic hinge forms in the structure. Displacements now increase more rapidly as the response follows a new elastic curve corresponding to the reduced, or deteriorated, structure. The stiffness of the deteriorated structure corresponds to the part of the actual structure that is still behaving elastically, the formation of the plastic hinge being simulated by the insertion of a frictionless hinge and the application of an external moment of magnitude M_p. Thus, in an actual structure, the elastic-plastic response lies below the ideal response as hinges form one at a time and displacements increase more rapidly as the stiffness deteriorates. The value of N_f thus appears to be lower than N_R. In practice, it is found that N_R gives a reasonable lower bound to N_f, provided that side loads are not excessive[7].

The Rankine load factor for the frame shown in Figure 12.1 is, given by:

$$N_R = 4/(1 + 2)$$
$$= 1.333$$

The actual elastic-plastic load factor is:

$$N_f = 1.336$$

Example 12.1

Determine the load factors against elastic instability and rigid-plastic collapse of the frame shown in Figure 12.4. Determine the Rankine load factor for a ratio $M_p/I = 2$ kip in^{-3}. The values of the plastic moments of resistance and the second moments of area of the members are shown in the figure, and the modulus of elasticity is 29,000 kip/in^2.

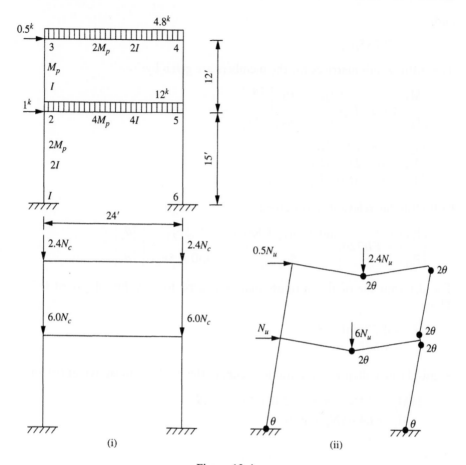

Figure 12.4

The Euler load of column 21 is:

$$P_E = \pi^2 E(2I)/(180)^2$$
$$= 17.67I \text{ kips}$$

The Euler load of column 23 is:

$$P_E = \pi^2 EI/(144)^2$$
$$= 13.81I \text{ kips}$$

At the elastic critical load, as shown at (i), the P/P_E ratios in the columns are:

$$\rho_{21} = 8.4N_c/17.67I$$
$$= 0.475N_c/I$$
$$\rho_{23} = 2.4N_c/13.81I$$
$$= 0.174N_c/I$$

and:

$$\rho_{21} = 2.735\rho_{23}$$

The stiffness sub-matrices for the members are given by:

$$\begin{bmatrix} M_{23} \\ M_{32} \end{bmatrix} = EI/l_{23} \begin{bmatrix} n_{23} & o_{23} \\ o_{23} & n_{23} \end{bmatrix} \begin{bmatrix} \theta_2 \\ \theta_3 \end{bmatrix}$$

$$M_{21} = 2n_{21}EI\theta_2/l_{21}$$
$$M_{34} = 6 \times 2EI\theta_3/l_{34}$$
$$M_{25} = 6 \times 4EI\theta_2/l_{25}$$

Collecting the relevant terms gives:

$$\begin{bmatrix} P_{\theta 2} \\ P_{\theta 3} \end{bmatrix} = EI/720 \begin{bmatrix} (60 + 5n_{23} + 8n_{21}) & 5o_{23} \\ 5o_{23} & (30 + 5n_{23}) \end{bmatrix} \begin{bmatrix} \theta_2 \\ \theta_3 \end{bmatrix}$$

The determinant of the stiffness matrix is zero for a value of ρ_{21} of 0.775. Thus:

$$N_c = 0.775I/0.475$$
$$= 1.63I$$

Rigid-plastic collapse of the frame occurs in the mechanism shown at (ii) and:

$$24M_p = 12N_u(28.8 + 72.0 + 6.0 + 22.5)$$
$$M_p = 64.65N_u \text{ kip-in}$$

thus:

$$N_u = 0.0155 M_p$$
$$= 0.031I$$

The Rankine load factor is given by:

$$1/N_R = 1/N_c + 1/N_u$$

and:

$$N_R = 0.030I$$

12.3 The deteriorated critical load

The formation of plastic hinges in a frame is equivalent to the insertion of frictionless hinges in the frame. The stability of the frame now depends on the stiffness of the remaining elastic structure, and in general the elastic critical load of the deteriorated frame is lower than the critical load of the original frame.

The frame shown in Figure 12.5 collapses as the combined mechanism shown at (i) at a load factor N_u. The load factor against elastic instability N_c is assumed to be higher than N_u. As the loading is progressively increased, plastic hinges are produced, and their formation at the center of each beam results in the deteriorated structure shown at (ii). This deteriorated structure has a critical load identical with that of the original frame, since the hinges are at points of contraflexure in the beams. The formation of plastic hinges at the right-hand end of each beam results in the deteriorated structure shown at (iii). This deteriorated structure has a critical load lower than that of the original frame, since the beam stiffnesses are reduced to approximately 75 percent of their original values. The formation of plastic hinges at both the center and right-hand end of each beam results in the deteriorated structure shown at (iv). The beams have now been reduced to zero stiffness, and the remaining elastic structure consists of two freestanding cantilevers, three stories in height, which have a much lower critical load than that of the original frame. If, at any stage in the formation of plastic hinges, the critical load of the deteriorated frame is lower than the applied load producing the hinges, the failure load has been attained.

The procedure for determining the failure load consists of tracing the development of plastic hinges as the applied loading is increased and determining

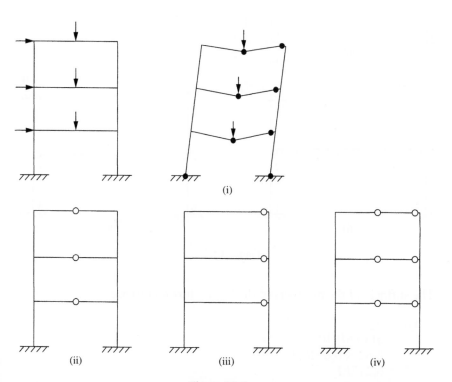

Figure 12.5

the critical load for each deteriorated frame[8]. For simple structures, this may be effected by moment distribution methods, making an initial estimate of the axial forces in the members and allowing for the modified stiffness and carry-over values. For more complex structures, the use of a digital computer is necessary, and this procedure is presented in section 12.4.

Example 12.2

The bottom two stories of a multistory frame are shown in Figure 12.6, where the second-story columns, at 3 and 4, are free to sway but are fixed against rotation. Beam 25 has a second moment of area of I in^4 and a plastic moment of resistance of 180 kip-in. All columns have a second moment of area of $0.6I$ and a plastic moment of resistance of 100 kip-in. Determine the load factor against elastic-plastic failure if the axial load in the columns is 0.28 of the Euler load.

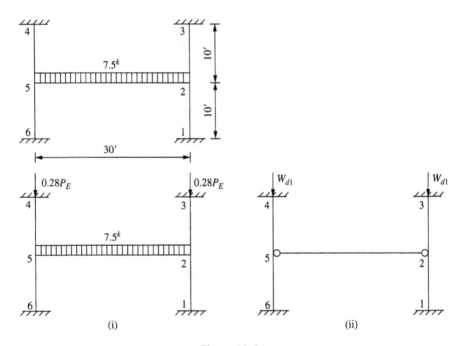

Figure 12.6

The stiffness of the members of the frame shown at (i) are:

$$s_{21} = s_{23}$$
$$= qE(0.6I)/10$$
$$= 3.62 \times 0.6EI/10$$
$$= 0.217EI$$
$$s_{25} = 2EI/30$$
$$= 0.067EI$$

The distribution factors at joint 2 are:

$$d_{21} = d_{23}$$
$$= 0.217/0.501$$
$$= 0.434$$
$$d_{25} = 0.067/0.501$$
$$= 0.132$$

The initial fixed-end moments in the frame are:

$$M_{25}^F = -M_{52}^F$$
$$= 7.5 \times 360/12$$
$$= 225 \text{ kip-in}$$

The final moments in the frame are:

$$M_{21} = M_{23}$$
$$= -225 \times 0.434$$
$$= -98 \text{ kip-in}$$
$$M_{25} = 196 \text{ kip-in}$$

Thus, as loading is increased, plastic hinges are produced simultaneously at both ends of the beam at a load factor of:

$$N = 180/196$$
$$= 0.92$$

The deteriorated structure, at this load factor, is shown at (ii), and the deteriorated critical load is obtained by equating the stiffness of joint 2 to zero. Thus:

$$2 \times 0.6n_{21}/10 + 0 = 0$$
$$n_{21} = 0$$

and:

$$W_{d1}/P_E = 0.25$$
$$N_{d1} = 0.25/0.28$$
$$= 0.89$$

This is less than the load factor required to produce the plastic hinges, and thus elastic-plastic failure occurs at a load factor of:

$$N_f = 0.92$$

12.4 Computer analysis

Methods, have been proposed for both the linear elastic-plastic analysis of frames and grids[9,10] and for the elastic-plastic analysis of frames[11,12]. An alternative method is available for the elastic-plastic analysis of frames[13,14,15,16], but this has the disadvantage of producing a larger stiffness matrix that is non-banded.

Neglecting instability effects, the procedure consists of setting up the stiffness matrix in the usual way and obtaining the linear elastic displacements by inverting and multiplying by the load vector. The member forces are obtained by back substitution in the stiffness sub-matrices for each member. The load vector is progressively increased until the first plastic hinge forms in the structure, when both the stiffness matrix and the load vector require modification.

(a) Linear elastic-plastic analysis

For the uniform frame shown in Figure 12.7, the linear elastic-plastic response curve has been obtained and is shown in Figure 12.18. The first plastic hinge forms at joint 7 in member 47. Prior to the formation of this hinge, the stiffness submatrix for member 47, for the displacements and forces shown in Figure 12.8, is given by:

$$
\begin{bmatrix} M_{47} \\ lP_{y47} \\ M_{74} \end{bmatrix} = EI/l \begin{bmatrix} 4 & 6 & 2 \\ 6 & 12 & 6 \\ 2 & 6 & 4 \end{bmatrix} \begin{bmatrix} \theta_4 \\ y_4/l \\ \theta_7 \end{bmatrix}
$$

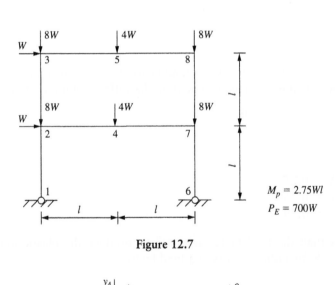

Figure 12.7

Figure 12.8

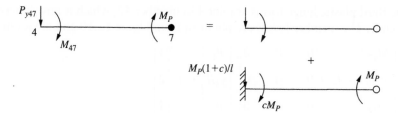

Figure 12.9

After the formation of the hinge at joint 7, the member may be replaced by a member hinged at 7 with an applied moment M_p. The member forces are obtained from Figure 12.9 and are given by:

$$\begin{bmatrix} M_{47} \\ lP_{y47} \\ M_{74} \end{bmatrix} = EI/l \begin{bmatrix} 3 & 3 & 0 \\ 3 & 3 & 0 \\ 0 & 0 & 0 \end{bmatrix} \begin{bmatrix} \theta_4 \\ y_4/l \\ \theta_7 \end{bmatrix} + M_p \begin{bmatrix} 1/2 \\ 3/2 \\ 1 \end{bmatrix}$$

or:

$$\{P_{47}\} = [S'_{47}]\{\Delta_{47}\} + \{C_{47}\}$$

where $\{P_{47}\}$ and $\{\Delta_{47}\}$ are the member forces and end displacements of member 47, $[S'_{47}]$ is the modified stiffness matrix, and $\{C_{47}\}$ is the plastic force vector for member 47. The stiffness sub-matrices of all the members are combined in the usual way to give the complete stiffness matrix of the deteriorated structure $[S]$, and the modified load vector is obtained by subtracting the plastic force vector from the corresponding terms of the vector of external loads $[W]$. In general:

$$\{W\} - \{C\} = [S]\{\Delta\}$$

and the analysis continues with a reduced loading system applied to the deteriorated structure. Frame displacements are obtained by inverting the stiffness matrix and multiplying by the modified load vector $\{W - C\}$. Member forces are obtained by back substitution in the stiffness submatrices for each member, modified where necessary.

As the modified load vector is progressively increased, the second plastic hinge forms at joint 8 in member 58. For subsequent loading the stiffness sub-matrix for member 5 is modified, the load vector is again modified, and the member forces are given by:

$$\{P_{58}\} = [S'_{58}]\{\Delta_{58}\} + \{C_{58}\}$$

The third plastic hinge forms at joint 4 in member 47, which is now hinged at both ends. The member forces are obtained from Figure 12.10 and are given by:

$$
\begin{bmatrix} M_{47} \\ lP_{y47} \\ M_{74} \end{bmatrix} = EI/l \begin{bmatrix} 0 & 0 & 0 \\ 0 & 0 & 0 \\ 0 & 0 & 0 \end{bmatrix} \begin{bmatrix} \theta_4 \\ y_4/l \\ \theta_7 \end{bmatrix} + M_p \begin{bmatrix} 1 \\ 2 \\ 1 \end{bmatrix}
$$

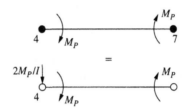

Figure 12.10

The complete stiffness matrix of the structure and the load vector are again modified, and loading proceeds until the last hinge forms at joint 5 in member 58. The collapse mechanism is shown in Figure 12.11, and the complete linear elastic-plastic response curve is shown in Figure 12.18.

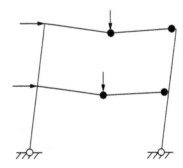

Figure 12.11

(b) Elastic-plastic analysis

For elastic-plastic analysis the procedure is similar and consists initially of assembling the stiffness matrix of the structure using the stiffness submatrices for each member, allowing for instability effects. For a member 12, considering the forces and displacements shown in Figure 12.12, this is given by:

$$
\begin{bmatrix} lP_{y12} \\ M_{12} \\ lP_{y21} \\ M_{21} \end{bmatrix} = EI/l \begin{bmatrix} j & (q+r) & -j & (q+r) \\ (q+r) & q & -(q+r) & r \\ -j & -(q+r) & j & -(q+r) \\ (q+r) & r & -(q+r) & q \end{bmatrix} \begin{bmatrix} y_1/l \\ \theta_1 \\ y_2/l \\ \theta_2 \end{bmatrix}
$$

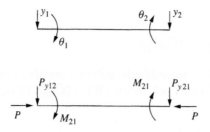

Figure 12.12

where the stability functions are defined in the notation. An initial estimate is made of the axial forces in the members, and the complete stiffness matrix is obtained. Solving for joint displacements and member forces provides revised values of the axial forces. The procedure continues until the revised values of the axial forces are in close agreement with the previous values.

As the applied loading is progressively increased, the first plastic hinge forms in a member; this corresponds to the insertion of a frictionless hinge in the member and the application of an external moment M_p.*The modified stiffness matrix for a member 12 with a plastic hinge at 2 is obtained from Figure 12.13 as:

$$\begin{bmatrix} lP_{y12} \\ M_{12} \\ lP_{y21} \\ M_{21} \end{bmatrix} = EI/l \begin{bmatrix} j' & q' & -j' & 0 \\ q' & q' & -q' & 0 \\ -j & -q' & j' & 0 \\ 0 & 0 & 0 & 0 \end{bmatrix} \begin{bmatrix} y_1/l \\ \theta_1 \\ y_2/l \\ \theta_2 \end{bmatrix} + M_p \begin{bmatrix} (1 + r/q) \\ r/q \\ -(1 + r/q) \\ 1 \end{bmatrix}$$

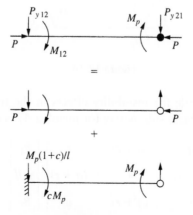

Figure 12.13

or:

$$\{P_{12}\} = [S'_{12}]\{\Delta_{12}\} + \{C_{12}\}$$

These values are used to assemble the stiffness matrix of the deteriorated structure $[S']$ and the reduced load vector $\{W\} - \{C_{12}\}$. Thus, for the deteriorated structure:

$$\{W\} - \{C_{12}\} = [S']\{\Delta\}$$

where $\{W\}$ is the vector of applied loading, which is progressively increased until the next plastic hinge forms, and so on.

The modified stiffness matrix for a member with a plastic hinge at both ends is obtained from Figure 12.14 as:

$$\begin{bmatrix} lP_{y12} \\ M_{12} \\ lP_{y21} \\ M_{21} \end{bmatrix} = EI/l \begin{bmatrix} -\rho\pi^2 & 0 & \rho\pi^2 & 0 \\ 0 & 0 & 0 & 0 \\ \rho\pi^2 & 0 & -\rho\pi^2 & 0 \\ 0 & 0 & 0 & 0 \end{bmatrix} \begin{bmatrix} y_1/l \\ \theta_1 \\ y_2/l \\ \theta_2 \end{bmatrix} + M_p \begin{bmatrix} 2 \\ 1 \\ -2 \\ 1 \end{bmatrix}$$

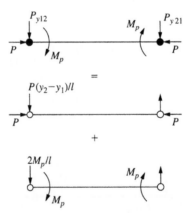

Figure 12.14

When allowance is made for instability effects in the frame shown in Figure 12.7, the modified stiffness sub-matrix for member 47 is given by:

$$\begin{bmatrix} M_{47} \\ lP_{y47} \\ M_{74} \end{bmatrix} = EI/l \begin{bmatrix} q & (q+r) & r \\ (q+r) & j & (q+r) \\ r & (q+r) & q \end{bmatrix} \begin{bmatrix} \theta_4 \\ y_4/l \\ \theta_7 \end{bmatrix}$$

An initial estimate is made of the axial forces in all members, and the complete stiffness matrix is obtained. Solving for displacements and member forces provides revised values of the axial forces. The procedure continues until the correct values are obtained.

As the loading is progressively increased, the elastic response curve is followed until the first plastic hinge forms at joint 7 in member 47. The member forces are given by:

$$\begin{bmatrix} M_{47} \\ lP_{y47} \\ M_{74} \end{bmatrix} = EI/l \begin{bmatrix} q' & q' & 0 \\ q' & j' & 0 \\ 0 & 0 & 0 \end{bmatrix} \begin{bmatrix} \theta_4 \\ y_4/l \\ \theta_7 \end{bmatrix} + M_p \begin{bmatrix} r/q \\ (1 + r/q) \\ 1 \end{bmatrix}$$

For subsequent loading, the stiffness matrix and the load vector are modified, and the response now follows a new elastic curve, which is asymptotic to the critical load factor of the first deteriorated structure with $N_c = 3.00$. Loading proceeds until the second plastic hinge forms at joint 8 in member 58.

The third plastic hinge forms at joint 2 in member 12 at the elastic-plastic load factor of $N_f = 1.56$. Prior to the formation of this hinge, the stiffness sub-matrix for member 12, for the displacements and forces shown in Figure 12.15, is given by:

$$\begin{bmatrix} M_{21} \\ lP_{x21} \end{bmatrix} = EI/l \begin{bmatrix} q' & -q' \\ -q' & j' \end{bmatrix} \begin{bmatrix} \theta_2 \\ x_2/l \end{bmatrix}$$

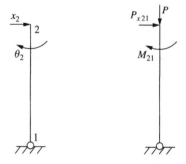

Figure 12.15

Subsequent to the formation of the plastic hinge at 2, the member forces are obtained from Figure 12.16 as:

$$M_{21} = -M_p$$
$$P_{x21} = (M_p - Px_2)/l$$
$$= M_p/l - \rho\pi^2 EIx_2/l^3$$

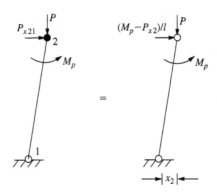

Figure 12.16

The modified stiffness sub-matrix and the plastic force vector are given by:

$$
\begin{bmatrix} M_{21} \\ lP_{x21} \end{bmatrix} = EI/l \begin{bmatrix} 0 & 0 \\ 0 & -\rho\pi^2 \end{bmatrix} \begin{bmatrix} \theta_2 \\ x_2/l \end{bmatrix} + M_p \begin{bmatrix} -1 \\ 1 \end{bmatrix}
$$

The structure becomes unstable on the formation of the plastic hinge at 2, and loading must be decreased to maintain equilibrium. The last plastic hinge forms at joint 7 in member 76, producing the collapse mechanism shown in Figure 12.17. The complete elastic-plastic response curve is shown in Figure 12.18.

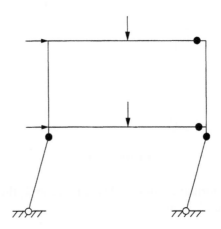

Figure 12.17

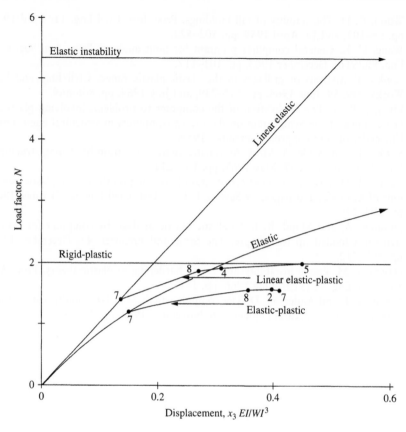

Figure 12.18

References

1. Merchant, W. A connection between Rayleigh's method and stiffness methods of determining critical loads. Proceedings of the International Congress of Applied Mechanics. Brussels. 1956. pp. 8–12.
2. Jennings, A. The elastic stability of rigidly jointed frames. Int. J. Mech. Sciences. 5. 1963. pp. 99–113.
3. Horn, M. R. Elastic plastic failure loads of plane frames. Proceedings, Royal Society of London, Series A. 274. July 1963. pp. 343–364.
4. Merchant, W. The failure load of rigidly jointed frame-works as influenced by stability. The Structural Engineer. 32. July 1954. pp. 185–190.
5. Merchant, W., Rashid, C. A., Bolton, A., Salem, A. The behavior of unclad frames. Proceedings of the Fiftieth Anniversary Conference, Institution of Structural Engineers. 1958. pp. 2–9.
6. Kamtekar, A. G., Little, G. H., Cunningham, A. The plastic design of steel frames. Proc. Inst. Civil Eng: Structures and Buildings. 146. August 2001. pp. 275–284.
7. Majid, K. I. An evaluation of the elastic critical load and the Rankine load of frames. Proc. Inst. Civil Eng. 36. March 1967. pp. 579–593.

8. Wood, R. H. The stability of tall buildings. Proc. Inst. Civil Eng. 11. April 1959. pp. 69–102, and 12. April 1959. pp. 502–522.

9. Wang, C. K. General computer program for limit analysis. Proc. Am. Soc. Civil Eng. 89 (ST6). December 1963. pp. 101–117.

10. Sawko, F. Analysis of grillages in the elastic-plastic range. Civil Eng. and Pub. Works. Rev. 59. June 1964. pp. 737–739, and July 1964. pp. 866-869.

11. Livesley, R. K. The application of the computer to problems involving plasticity. Proceedings of the Symposium on the use of computers in structural engineering. University of Southampton. September 1959.

12. Sawko, F. and Wilde, A. M. B. Automatic analysis of strain hardening structures. Proc. Inst. Civil Eng. 37. May 1967. pp. 195–211.

13. Davies, J. M. Discussion on effect of strain hardening on the elastic plastic behaviour of beams and grillages, by Sawko, F. Proc. Inst. Civil Eng 32. October 1965. pp. 294–299.

14. Jennings, A. and Majid, K. I. An elastic plastic analysis by computer for framed structures loaded up to collapse. The Structural Engineer. 43. December 1965. pp. 407–412.

15. Davies, J. M. Frame instability and frame hardening in plastic theory. Proc. Am. Soc. Civil Eng. 92 (ST3). June 1966. pp. 1–15.

16. Majid, K. I. and Anderson, D. The computer analysis of large multi-story framed structures. The Structural Engineer. 46. November 1968. pp. 357–365.

Answers to supplementary problems part 2

Chapter 1

The degree of indeterminacy of the frames are:

S1.1 $D = 1$

S1.2 $D = 1$

S1.3 $D = 12$

S1.4 $D = 22$

S1.5 (i) $D = 7$
(ii) $D = 7$

S1.6 (i) $D = 3$
(ii) $D = 0$
(iii) $D = 0$
(iv) $D = 0$

S1.7 (i) $D = 5$
(ii) $D = 9$
(iii) $D = 3$
(iv) $D = 1$
(v) $D = 1$

Chapter 2

S2.1 $\delta = wa^4(4l/3a - 1/3)/8EI$

S2.2 $\delta = 2.16$ in upward

S2.3 $\delta_2 = 0.114$ in
$\delta_3 = 0.068$ in
$W = 16.75$ kips

S2.4 $H/V = 0.65$

S2.5 $\delta_4 = 0.59$ in

S2.6 $M_{12} = -138\,\text{kip-in}$
$M_{21} = -102\,\text{kip-in}$
$M_{23} = -112\,\text{kip-in}$
$M_{32} = -128\,\text{kip-in}$
$Q_{12} = -Q_{21} = Q_{23} = -Q_{32} = 2\,\text{kips}$
$\delta_3 = 907,200/EI + 480/AG$

S2.7 $M_1 = 20\alpha t EI/al^2$
$V_1 = 20\alpha t EI/al^3$
$M_1 = 80\alpha t EI/a^2 l^4$

Chapter 3

S3.1 $P_{24} = 32.66\,\text{kips compression}$

S3.2 $P_{24} = 1.07\,\text{kips tension}$
$P_{26} = 10.33\,\text{kips compression}$

S3.3 $P_{13} = W/(5.407 + 0.472a^2 A/I)\,\text{compression}$

S3.4 $P_{34} = 11\,\text{kips compression}$

S3.5 $P_{12} = 132\,\text{kips tension}$

S3.6 $H = 62.5\,\text{kips}$

S3.7 $H = 8.28\,\text{kips}$

S3.8 $H = 0.166\,W$

S3.9 $T = 20.5\,\text{kips}$

Chapter 4

S4.1 $M_1 = Wab^2/l^2$

S4.2 $\theta_1 = 5Wl^2/128EI$
$\delta_3 = 3Wl^3/256EI$

S4.3 $\delta_2 = wa^3(l - a/4)/6EI$

S4.4 $M_{12} = 0.345M$
$M_{21} = -0.147M$

S4.5 $M_{12} = -47.5\,\text{kip-ft}$
$M_{21} = -31.3\,\text{kip-ft}$

S4.6 $M_{12} = -25.74\,\text{kip-ft}$
$M_{21} = -32.45\,\text{kip-ft}$

S4.7 $M_{12} = -101\,\text{kip-ft}$
$M_{21} = -113\,\text{kip-ft}$

S4.8 $M_{12} = -Pab(a + b)/l^2$
$M_{21} = -Pab(a + b)/l^2$

S4.9 $M_{12} = Pb(l^2 - b^2)/2l^2$

S4.10 $M_{12} = -21.45\,\text{kip-ft}$
$M_{21} = -7.10\,\text{kip-ft}$

Chapter 5

S5.1

x/l	0.2	0.4	0.6	0.8	1.0
V_2	0.36	0.64	0.84	0.96	1.00

S5.2   For member 12

x/l	0.2	0.4	0.6	0.8
M_2/l	0.048	0.084	0.096	0.072

For member 23

x/l	0.2	0.4	0.6	0.8
M_2/l	0.064	0.072	0.048	0.016

S5.3

x	0'	20'	40'	60'	80'
H_1	0.153	0.264	0.375	0.462	0.486

S5.4

Panel point	1	2	3	4	5	6	7
R_1	1.000	0.610	0.264	0.000	-0.069	-0.055	0.000

Chapter 6

S6.1 $M_{12} = 0.345M$
$M_{21} = -0.147M$

S6.2 $M_{12} = -47.6\,\text{kip-ft}$
        $M_{21} = -31.3\,\text{kip-ft}$

S6.3 $M_{12} = 2.54\,\text{kip-ft}$

S6.4 $M_{12} = -25.7\,\text{kip-ft}$
        $M_{21} = -32.5\,\text{kip-ft}$

S6.5 $M_{12} = -101\,\text{kip-ft}$
        $M_{21} = -133\,\text{kip-ft}$

S6.6 $M_{12} = -Pab^2/l^2$
        $M_{21} = -Pba^2/l^2$

S6.7 $M_{12} = Pb(l^2 - b^2)/2l^2$

S6.8 $s_{12} = -0.275EI$
        $c_{12} = 0.444$
        $M_{12} = -21.55\,\text{kip-ft}$
        $M_{21} = -7.05\,\text{kip-ft}$

S6.9 $H = 1.08\,\text{kips}$

S6.10 $M = 34.4\,\text{kip-in}$

S6.11 $M_{12} = -44\,\text{lb-ft}$
        $M_{21} = -36\,\text{lb-ft}$

Chapter 7

S7.1 $V_2 = 7.60\,\text{kips}$
        $V_3 = 8.32\,\text{kips}$
        $V_4 = 2.08\,\text{kips}$

S7.2 $M_1 = 19\,\text{kip-in}$
        $V_1 = 2.68\,\text{kips}$
        $H_1 = 0.48\,\text{kips} \ldots$ acting to the right
        $M_4 = 38\,\text{kip-in}$
        $V_4 = 3.32\,\text{kips}$
        $H_4 = 1.9\,\text{kips} \ldots$ acting to the left

S7.3 $M_2 = 1682\,\text{kip-in}$
        $M_3 = 421\,\text{kip-in}$

S7.4 $M_2 = 3.5\,\text{kip-in}$
$M_1 = 10.5\,\text{kip-in}$
$M_3 = 0.7\,\text{kip-in}$

S7.5 $M_2 = 75\,\text{kip-ft}$
$M_3 = 50\,\text{kip-ft}$

S7.6 $M_2 = 9.1\,\text{kip-ft}$
$M_3 = 13.5\,\text{kip-ft}$
$M_4 = 15.1\,\text{kip-ft}$

S7.7 $M_2 = 197\,\text{kip-in}$
$M_3 = 237\,\text{kip-in}$

S7.8 $M_{12} = -138\,\text{kip-in}$
$M_{21} = -102\,\text{kip-in}$
$M_{23} = -112\,\text{kip-in}$
$M_{32} = -128\,\text{kip-in}$
$M_{36} = 128\,\text{kip-in}$
$M_{25} = 214\,\text{kip-in}$
$P_{12} = 5.70\,\text{kips} \ldots \text{tension}$
$P_{23} = 2.13\,\text{kips} \ldots \text{tension}$
$P_{36} = 2.00\,\text{kips} \ldots \text{compression}$
$P_{25} = 0\,\text{kips}$
$Q_{12} = 2.00\,\text{kips}$
$Q_{23} = 2.00\,\text{kips}$
$Q_{36} = 2.13\,\text{kips}$
$Q_{25} = 3.57\,\text{kips}$

S7.9 $M_{12} = -687\,\text{kip-ft}$
$M_{21} = -811\,\text{kip-ft}$
$M_{23} = 347\,\text{kip-ft}$
$M_{32} = 154\,\text{kip-ft}$
$M_{34} = 257\,\text{kip-ft}$
$M_{43} = 242\,\text{kip-ft}$
$M_{45} = 239\,\text{kip-ft}$
$M_{54} = 259\,\text{kip-ft}$

S7.10 $M_{12} = -322\,\text{kip-ft}$
$M_{21} = -277\,\text{kip-ft}$
$M_{23} = -184\,\text{kip-ft}$
$M_{32} = -415\,\text{kip-ft}$
$M_{34} = 652\,\text{kip-ft}$
$M_{43} = 548\,\text{kip-ft}$

S7.11 $M_{21} = -225\,\text{kip-ft}$
         $M_{23} = -28\,\text{kip-ft}$
         $M_{32} = -80\,\text{kip-ft}$
         $M_{34} = -14\,\text{kip-ft}$
         $M_{43} = -22\,\text{kip-ft}$

Chapter 8

S8.1 $M_{30} = 1351.7 + 712.8 = 2064.5\,\text{kip-in}$
        $V_{30} = 94.7 + 40.5 = 135.2\,\text{kips}$

S8.2 $M_1 = 153\,\text{kip-in}$

S8.3 $k_W = 190$

S8.4 $P_{47} = 8.6\,\text{kips}$

S8.5   (i) $l_m = l_p/120 = 0.133(E_b I_b/A_t E_t)^{1/3}$
        (ii) $M_6 = 250 y_{12}/(5)^{1/2} + 50 y_{34}/(5)^{1/2} + 150 y_5$
        (iii) The influence line ordinates for M_6 are:

x,ft	0	50	100	150	200	250
I L ordinates for M_6	−40.5	−21.0	0	21.0	41.8	65.2

Chapter 9

S9.1 $l/a = 3.17$
        $w = 58.5 M_p/l^2$

S9.2 $S = 14.8\,\text{in}^3$

S9.3 $W/H \geq 8/3$ ... for the beam mode
        $W/H \leq 2/3$ ... for the sway mode
        $2/3 < W/H < 8/3$ ... for the combined mode

S9.4 $M_p = 9.42\,\text{kip-ft}$ ... for the columns
        $M_p = 18.84\,\text{kip-ft}$ ... for the beams

S9.5 $M_p = 25\,\text{kip-ft}$ ... for the posts
        $M_p = 50\,\text{kip-ft}$ ... for the chords

S9.6 $M_p = Wl$
        $x_2 = 5 M_p l^2 (5)^{0.5}/24EI$

S9.7   (i) $W = 2.0M_p/a$
        (ii) $W = 3M_p/a$
        (iii) $W = 2.80M_p/a$

S9.8 $M_p = Wa/2$

Chapter 10

S10.1   The stiffness matrix is:

$$\begin{bmatrix} P_{\theta 2} \\ P_{\theta 3} \end{bmatrix} = EI/l \begin{bmatrix} 20 & 6 \\ 6 & 28 \end{bmatrix} \begin{bmatrix} \theta_2 \\ \theta_3 \end{bmatrix}$$

The final moments are:

$M_{12} = 42.75$ kip-in
$M_{21} = 85.5$ kip-in
$M_{34} = 236.6$ kip-in
$M_{43} = 181.7$ kip-in

S10.2   The stiffness matrix is:

$$\begin{bmatrix} P_{\theta 2} \\ P_{\theta 1} \end{bmatrix} = EI/l \begin{bmatrix} 8 & 2 \\ 2 & 4 \end{bmatrix} \begin{bmatrix} \theta_2 \\ \theta_1 \end{bmatrix}$$

The final moments are:

$M_{21} = 12/7$ kip-ft
$M_{32} = 12/14$ kip-ft

S10.3   The stiffness matrix is:

$$\begin{bmatrix} P_{\theta 2} \\ P_{\theta 3} \\ P_{\theta 4} \\ lP_{x2} \end{bmatrix} = EI/l \begin{bmatrix} 8 & & \text{symmetric} & \\ 2 & 8 & & \\ 0 & 2 & 4 & \\ -6 & -6 & -6 & 24 \end{bmatrix} \begin{bmatrix} \theta_2 \\ \theta_3 \\ \theta_4 \\ x_2/l \end{bmatrix}$$

$x_2 = 0.136l^2/EI$
$\theta_4 = 0.477l/EI$
$M_4 = -0.285lH$

S10.4   The stiffness matrix is:

$$\begin{bmatrix} P_{x1} \\ P_{y1} \end{bmatrix} = EA/l \begin{bmatrix} 2.723 & 0.548 \\ 0.548 & 2.296 \end{bmatrix} \begin{bmatrix} x_1 \\ y_1 \end{bmatrix}$$

The member forces are:

$$P_{12} = -0.490W$$
$$P_{13} = 0.460W$$
$$P_{14} = 1.428W$$
$$P_{15} = -0.433W$$

S10.5   The stiffness matrix is:

$$\begin{bmatrix} P_{xo} \\ P_{yo} \\ P_{\theta o} \end{bmatrix} = EA/l \begin{bmatrix} 3/16 & \text{symmetric} & \\ 1/4 & 3 & \\ 9/8 & 1/2 & 41/4 \end{bmatrix} \begin{bmatrix} x_o \\ y_o \\ \theta_o \end{bmatrix}$$

The member forces are:

$$P_1 = 0.00 \text{ kips}$$
$$P_2 = 8.52 \text{ kips}$$
$$P_3 = 11.48 \text{ kips}$$

S10.6   The flexibility matrix is:

$$\begin{bmatrix} -\delta_1 \\ -\delta_2 \end{bmatrix} = \begin{bmatrix} 3002 & 0 \\ 0 & 3002 \end{bmatrix} \begin{bmatrix} R \\ R \end{bmatrix}$$

The cable force is:

$$R = 20.5 \text{ kips}$$

S10.7   The horizontal thrust H may be taken as the redundant. Then:

$$H = 11 \text{ kips}$$

Chapter 11

S11.1 $W_c = 2.55P_E$

S11.2 $W_c = 3.1P_E$

S11.3 The stiffness matrix is given by

$$\begin{bmatrix} P_{\theta 2} \\ P_{\theta 3} \end{bmatrix} = EI/l \begin{bmatrix} (4 + q'_{21}) & 2 \\ 2 & (6 + q'_{34}) \end{bmatrix} \begin{bmatrix} \theta_2 \\ \theta_3 \end{bmatrix}$$

$w_c = 1.55 P_E / l$

S11.4 The stiffness matrix is given by

$$\begin{bmatrix} P_{\theta 2} \\ P_{\theta 3} \\ P_{\theta 4} \end{bmatrix} = EI/l \begin{bmatrix} (2q'_{21} + q_{23} + q_{24} + q'_{25}) & r_{23} & r_{24} \\ r_{23} & (q_{34} + q_{23}) & r_{34} \\ r_{24} & r_{34} & (q_{24} + q_{34} + q'_{45}) \end{bmatrix} \begin{bmatrix} \theta_2 \\ \theta_3 \\ \theta_4 \end{bmatrix}$$

$\rho_{c21} = 1.055$
Load Factor = 2.81

S11.5 The stiffness matrix is given by

$$\begin{bmatrix} P_{\theta 2} \\ P_{\theta 3} \end{bmatrix} = EI/l \begin{bmatrix} (9 + 4n) & 2 \times o \\ 2 \times o & (9 \times 2n) \end{bmatrix} \begin{bmatrix} \theta_2 \\ \theta_3 \end{bmatrix}$$

$W_c = 0.455 P_E$

Author index

Subject index

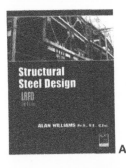
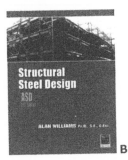
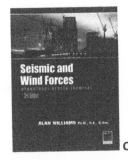

Printed and bound by CPI Group (UK) Ltd, Croydon, CR0 4YY

03/10/2024

01040437-0007